Batteries for Implantable Biomedical Devices

Batteries for Implantable Biomedical Devices

Edited by

Boone B. Owens

Boone B. Owens, Inc.
St. Paul, Minnesota

PLENUM PRESS • NEW YORK AND LONDON

Library of Congress Cataloging in Publication Data

Batteries for implantable biomedical devices.

Bibliography: p.
Includes index.
1. Implants, Artificial—Power supply. 2. Electric batteries. I. Owens, Boone B.
RD132.B37 1986 617′.95 86-582
ISBN 0-306-42148-8

A Division of Plenum Publishing Corporation
233 Spring Street, New York, N.Y. 10013

Printed in the United States of America

Contributors

BAROUH V. BERKOVITS • New England Research Center, Wellesley, Massachusetts 02181

MICHAEL BILITCH • USC Pacemaker Center, Los Angeles, California 90033

KENNETH R. BRENNEN • Medtronic, Inc., Minneapolis, Minnesota 55440

MICHAEL BROUSSELY • Departement Générateurs de Technologies Avancées, SAFT, Poitiers 86000, France

KEITH FESTER • Medtronic, Inc., Minneapolis, Minnesota 55440

JEAN-PAUL GABANO • Departement Générateurs de Technologies Avancées, SAFT, Poitiers 86000, France

MICHAEL GRIMM • Departement Générateurs de Technologies Avancées, SAFT, Poitiers 86000, France

GERHARD L. HOLLECK • Battery Division, EIC Laboratories, Inc., Norwood, Massachusetts 02062

CURTIS F. HOLMES • Research and Development, Wilson Greatbatch Limited, Clarence, New York 10431

JOHN S. KIM • Medtronic, Inc., Minneapolis, Minnesota 55440

SAMUEL C. LEVY • Sandia National Laboratories, Albuquerque, New Mexico 87185

BOONE B. OWENS • Department of Chemical Engineering and Materials Science, University of Minnesota, Minneapolis, Minnesota 55455

DAVID L. PURDY • Coratomic, Indiana, Pennsylvania 15701-0434

SAMUEL RUBEN • Reed College, Portland, Oregon 97202

ALVIN J. SALKIND • Department of Surgery, Bioengineering Section, UMDNJ–Rutgers Medical School, Piscataway, New Jersey 08854

PAUL M. SKARSTAD • Medtronic, Inc., Minneapolis, Minnesota 55440

ALAN J. SPOTNITZ • Department of Surgery, Thoracic Section, UMDNJ–Rutgers Medical School, Piscataway, New Jersey 08854

KENNETH B. STOKES • Medtronic, Inc., Minneapolis, Minnesota 55440

DARREL UNTEREKER • Medtronic, Inc., Minneapolis, Minnesota 55440

Foreword

Small sealed electrochemical power units have developed remarkably in the last two decades owing to improvements in technology and a greater understanding of the underlying basic sciences. These high-energy-density sealed battery systems have made possible the safe and rapid development of lightweight implantable electrical devices, some of which, such as heart pacers, have reached a large market. In most of these devices the battery constitutes the majority of the device volume and weight, and limits the useful life.

This book on *Batteries for Implantable Biomedical Devices* will be highly welcome to those interested in devices for heart pacing, pain suppression, bone repair, bone fusion, heart assist, and diabetes control, as well as numerous other biomedical devices that depend on sealed batteries. However, the material will also be extremely useful to a much broader audience, including those concerned with sealed batteries for such other difficult environments as space, the sea and remote locations.

The material included in this book is very comprehensive and contains both discussions of the excellent basic science and inside "know-how" on design and assembly operations. Dr. Boone Owens has selected scientists of worldwide reknown as chapter authors. Both he and the authors are active in research on and development and production of power sources and medical devices, and they have made significant contributions to the field. The first major implantable electrical devices (heart pacers) were based on mercury batteries, which were the invention of Dr. Samuel Ruben. It was a great pleasure for me to review this development with Dr. Ruben. The inventors and developers of the current advanced lithium batteries—as well as the co-inventor of the heart pacer, Dr. Greatbatch—have also contributed chapters of great educational and reference value.

I believe this book fills a unique niche in its coverage. It contains discussions of physiology, biological energy requirements, material science, electrochemistry, and battery technology treated in a cohesive manner. It will be helpful for power source researchers, medical students and physicians, and evaluators of present and future generations of implantable power sources and devices.

Alvin J. Salkind
Professor and Chief
Surgery/Bioengineering
UMDNJ-Rutgers Medical School
and Visiting Professor
Chemical and Biochemical Engineering
Rutgers University

Preface

One may speculate whether nearly 200 years ago as Volta constructed his famous Pile, he considered the future roles that would be played by portable stores of electricity. In our present society the battery serves consumers in many diverse ways—portable tools, radios, televisions, recorders, flashlights, wrist watches, emergency lights, smoke detectors, vehicle starting, portable telephones, pagers, hearing aids, telephone memory elements, utility metering, and the list continues on and on. The size of batteries varies over a surprising ten orders of magnitude, from ten milliwatt-hour memory cells to 100 Megawatt-hour batteries planned for utility load leveling. The electrochemical battery, always a component of some larger system and never the end item itself, best meets the user's needs when the user is not even aware of this collection of reactive chemicals that are contained in a manner ready to instantly convert chemical energy into electrical energy.

Military needs and space exploration prompted much of the first research and development to improve on the standard batteries which had been in use for decades. For example, the space suits of astronauts now contain highly advanced, lightweight lithium cells to power television lights and cameras as well as their communication units. Space exploration also has applications for rechargeable batteries to serve as ultra-reliable power sources, capable of operating over ten years and for many thousands of cycles under conditions that do not permit battery maintenance or replacement.

Another field driving recent battery development has been medicine. Many applications for implanted medical devices require batteries. These batteries power appliances to sustain life, alleviate pain, facilitate healing, dispense drugs, and complement other body functions. This area of application is unique in its combination of environmental, electrical, and reliability requirements.

The most well-known implantable medical device is the cardiac pacemaker. Mercury–zinc cells initially represented the best available combination of energy and reliability for the pacemaker of the 1960s. By the end of the sixties, it was clear pacemaker longevity was limited by the battery and further longevity improvement (necessary to reduce the need for device-replacement surgery) would depend upon power source improvements.

The decade of the 1970s saw the convergence of integrated circuit technology with the technology of small, hermetically-sealed high-energy-density lithium batteries as well as nuclear power sources. Reliability and longevity improvements paralleled reduction in device size and simplification of the medical procedures for using cardiac pacemaker therapy. The first commercial application of a lithium battery occurred in Veronna, Italy in 1972, when a lithium–iodine battery-powered pacemaker was implanted.

During the seventies, several new and diverse types of lithium battery systems were developed for this application. Both successes and failures were recorded, but as the 1980s have evolved, a maturing of lithium primary cell technology is in progress. Low-rate batteries for pacemakers have demonstrated highly reliable, real-time performance out to ten years. However, new therapies appear to require higher power levels in order to treat conditions such as tachyarrythmias, fibrillation, pain and scoliosis; and the artificial heart and implantable ventricular assist devices would benefit from a reliable hermetic, secondary battery that would be stable at 37°C and be of smaller volume and mass than presently available batteries of equivalent energy content. At this time, many laboratories are working on rechargeable high-energy-density lithium batteries, and medical batteries of this type should become available during the next decade. Thus the need for improved batteries continues and ongoing developmental efforts will satisfy these new requirements in the future.

The purpose of the present technical monograph is to summarize the technologies of batteries that have been developed and applied to implantable medical devices. This assessment is performed at a time corresponding to about ten years following the implementation of the various lithium primary batteries into pacemakers.

This book is the result of the efforts of many contributors. The editor wishes to thank all of the authors for their participation in this endeavor. The encouragement of the management at Medtronic is also gratefully acknowledged; Peter Mulier was especially instrumental in my early interest in implantable medical devices.

The able clerical assistance of Roxanne Olson, Linda Thomas, and Joanne Yates are all gratefully acknowledged. Special thanks must go to Diane Doyle who very reasonably and capably coordinated the frequent communications between the authors, the editor, and the publisher, and brought together the myriad of details that were required to bring order to the final manuscript.

Finally, the help of my wife, Tinie, who steadily encouraged my efforts, is acknowledged with sincere appreciation.

Boone B. Owens
St. Paul, Minnesota

Contents

1. Electrically Driven Implantable Prostheses

ALVIN J. SALKIND, ALAN J. SPOTNITZ, BAROUH V. BERKOVITS, BOONE B. OWENS, KENNETH B. STOKES, AND MICHAEL BILITCH

1. General Background 1
  1.1. Physiology, Medical Significance, and History 1
  1.2. Electronic Circuit Technology 5
2. Devices Background 7
  2.1. Heart Pacing Systems 7
  2.2. Cardiac Pacing Leads 14
  2.3. Automatic Implantable Defibrillator 21
  2.4. Bone Growth and Repair 22
  2.5. Other Devices 26
3. Business Aspects 31
4. Future Directions 32
References 33

2. Key Events in the Evolution of Implantable Pacemaker Batteries

BOONE B. OWENS AND ALVIN J. SALKIND

1. Introduction 37
2. An Interview with Samuel Ruben 38
3. An Interview with Wilson Greatbatch 44
References 49

3. Lithium Primary Cells for Power Sources

DARREL UNTEREKER

1. Introduction 51
2. The Elements of a Battery 53
   2.1. Anode 54
   2.2. Cathode 55
   2.3. Electrolyte/Separator 55
   2.4. Feedthrough 58
3. Battery Parameters 60
4. Battery Performance 62
5. Microcalorimetry 68
6. Implantable Battery Chemistries 72
   References 81

4. Evaluation Methods

KEITH FESTER AND SAMUEL C. LEVY

1. Evaluation Objectives 83
   1.1. Performance Data 83
   1.2. Reliability Data 84
   1.3. Quality Assurance 85
2. Accelerated Testing 85
   2.1. Empirical Approach 87
   2.2. Statistical Approach 87
   2.3. Physicochemical Approach 88
   2.4. Accelerated Testing without Failure 91
   2.5. Designing an Accelerated Life Test 93
   2.6. Other Acceleration Methods 93
3. Nonaccelerated Testing 96
   3.1. Real-Time Tests 96
   3.2. Materials Testing 97
   3.3. Microcalorimetry 98
4. Qualification Protocol 99
   4.1 Sample Qualification Plan 99
5. Data Analysis 101
   5.1. Longevity Projections 102
   5.2. Statistical Evaluation of Battery Longevity 103
   References 109

5. Battery Performance Modeling

KENNETH R. BRENNEN AND JOHN S. KIM

1. Description of the Problem 113
2. Importance of the Solution 114
3. Description of the Variables and Relationships 114
4. Classification of Models 117
5. Statistical Methods 117
 5.1. Self-Discharge 118
 5.2. Polarization 120
6. Modeling of the Lithium/Iodine Pacemaker Battery 120
7. Device Longevity 126
 7.1. Pulse Generator Hardware 126
8. Conclusion 130
 References 130

6. Lithium/Halogen Batteries

CURTIS F. HOLMES

1. Introduction 133
2. General Features of Lithium/Halogen Solid Electrolyte Batteries 134
 2.1. Thermodynamic Considerations 134
 2.2. Kinetic Considerations 135
3. The Lithium/Bromine System 136
 3.1. General Considerations 136
 3.2. The Li/Br_2–PVP Cell 138
 3.3. Other Cathode Formulations 138
 3.4. Summary 138
4. Chemistry of the Lithium/Iodine–Polyvinylpyridine System 139
 4.1. Cell Reaction 139
 4.2. The Lithium Anode 140
 4.3. The Cathode Material 141
 4.4. The Electrolyte/Separator 146
5. Construction of Lithium/Iodine–PVP Cells 151
 5.1. Principles of Cell Design 151
 5.2. The Central Anode/Case-Neutral Design 153
 5.3. The Central Cathode/Case-Neutral Design 154
 5.4. The Central Anode/Case-Grounded Design 155
 5.5. Central Anode/Case-Grounded Pelletized Cathode Cells .. 156

6. Discharge Characteristics of the Li/I_2–PVP Battery 157
6.1. General Considerations 157
6.2. Discharge Characteristics at Application Current Drain .. 158
6.3. The Effect of Current Drain on Cell Performance 164
6.4. Self-Discharge 165
6.5. Modeling and Accelerated Testing 167
7. Performance of the Li/I_2–PVP Cell 171
7.1. General Remarks 171
7.2. The Approach to Cell Reliability 171
7.3. Performance of Life Test Batteries 172
7.4. Performance of the Li/I_2–PVP Cell in Cardiac Pacemakers 173
8. Summary and Conclusion 175
References 177

7. Lithium Solid Cathode Batteries for Biomedical Implantable Applications

JEAN-PAUL GABANO, MICHAEL BROUSSELY, AND MICHAEL GRIMM

1. Introduction 181
2. General Features of Lithium Solid Cathode Systems 182
2.1. Thermodynamic Considerations 182
2.2. Some Properties of Electrodes and Electrolytes 182
2.3. Electrode and Cell Configurations 186
3. Specific Systems Used for Biomedical Applications 186
3.1. The Lithium–Silver Chromate Organic Electrolyte System 187
3.2. The Lithium–Cupric Sulfide Organic Electrolyte Battery ... 196
3.3. The Lithium–Vanadium Pentoxide Organic Electrolyte System 199
3.4. The Lithium–Manganese Dioxide Cell 203
3.5. Solid Electrolyte Lithium Cells 204
4. Use of Lithium Solid Cathode Systems in Implanted Medical Devices 208
4.1. Lithium–Silver Chromate 209
4.2. Lithium–Cupric Sulfide 210
4.3. Lithium–Vanadium Pentoxide 210
4.4. Lithium–Manganese Dioxide 211
4.5. Lithium–Lead Iodide, Lead Sulfide 211
5. Summary and Conclusions 211
References 212

8. Lithium–Liquid Oxidant Batteries

PAUL M. SKARSTAD

1. Introduction 215
2. Description of the System 217
   2.1. Liquid Oxidant Systems 217
   2.2. Cell Reaction 218
   2.3. Principles of Operation 219
3. Capacity and Energy Density 220
   3.1. Classification of Losses 220
   3.2. Stoichiometric Energy and Capacity Density 221
   3.3. Capacity Density of Practical Electrodes 224
   3.4. Packaging Efficiency 228
   3.5. Electrochemical Efficiency 230
4. State-of-Discharge Indication 248
5. Voltage Delay 250
   5.1. Anode Passivation 250
   5.2. Alleviation of Voltage Delay 251
6. Safety 252
   6.1. Short Circuit 252
   6.2. Overdischarge 253
   6.3. Charging 255
   6.4. Casual Storage 255
   6.5. Disposal 256
   6.6. Future 256
   References 257

9. Mercury Batteries for Pacemakers and Other Implantable Devices

ALVIN J. SALKIND AND SAMUEL RUBEN

1. Background 261
2. Chemistry 262
3. Cell Design and Performance Characteristics 266
   References 274

10. Rechargeable Electrochemical Cells as Implantable Power Sources

GERHARD L. HOLLECK

1. Introduction 275
2. Nickel Oxide/Cadmium Cells 276

2.1. Brief History 276
2.2. General Nickel Oxide/Cadmium Cell Characteristics 276
2.3. The Nickel Oxide/Cadmium Pacemaker Cell 278
3. Rechargeable Mercuric Oxide/Zinc Cells 279
3.1. Brief History 279
3.2. Cell Chemistry and Construction 280
3.3. Cell Performance 281
4. Prospects for Future Use of Rechargeable Cells 282
References 283

11. Nuclear Batteries for Implantable Applications

DAVID L. PURDY

1. General Description of Nuclear Batteries 285
1.1. Description of Isotopic Decay 285
1.2. Types of Nuclear Batteries 287
2. Isotope Selection 288
2.1. General Parameters 288
2.2. Isotope Longevity 288
2.3. Isotope Comparisons 290
3. Detailed Characteristics of the Plutonium-238 Isotope 292
3.1. Fuel Form 292
3.2. Types of Radiation 294
3.3. Helium Release 296
4. Thermoelectric Generator Systems 297
4.1. Nuclear Battery Subsystems 297
4.2. Biosphere Protection 299
4.3. Operating Environment Design Requirements 300
5. Thermopile Design 302
5.1. Seebeck Effect 302
5.2. Thermal and Electrical Performance 302
5.3. Material Characteristics 305
5.4. Design Optimization 307
6. Insulation Design and Selection 308
7. Fuel Capsule Design 311
7.1. General Description 311
7.2. Helium Pressure 311
7.3. Capsule Material 312
7.4. Capsule Geometry 312
7.5. Capsule Stress Analysis 313
7.6. Credible Accident Testing 322
8. Thermal Analysis 328
9. Electrical Characteristics 336

10. Radiation Effects 337
10.1. Somatic Effects 337
10.2. Genetic Effects 339
10.3. Public Exposure 342
11. Licensing Requirements 342
12. Applications of Nuclear Batteries 343
13. Nuclear Battery Reliability 349
References 350

Index 353

Batteries for Implantable Biomedical Devices

1

Electrically Driven Implantable Prostheses

ALVIN J. SALKIND, ALAN J. SPOTNITZ, BAROUH V. BERKOVITS, BOONE B. OWENS, KENNETH B. STOKES, and MICHAEL BILITCH

1. GENERAL BACKGROUND

1.1. Physiology, Medical Significance, and History

The electrochemical and electrical nature of muscular and biological reactions has been known for centuries. The work of Galvani in the eighteenth century in his famous frog leg experiment as a Professor of Anatomy at Padua University, led to Volta's experiments and epochal discovery of the production of electricity by electrochemical reactions. Galvani also observed what is now known as "injury potential," the voltage difference between an injured area and the surrounding tissue. The existence of dc or time-varying electrical activity with the majority of physical and chemical processes in living organisms has also been well established. More recently, the electrophysiological aspects of living tissue were investigated by Drs. Yasuda and Fukada.[1]

This early knowledge did not lead to a significant number of implantable

ALVIN J. SALKIND • Department of Surgery, Bioengineering Section, UMDNJ–Rutgers Medical School, Piscataway, New Jersey 08854. ALAN J. SPOTNITZ • Department of Surgery, Thoracic Section, UMDNJ–Rutgers Medical School, Piscataway, New Jersey 08854. BAROUH V. BERKOVITS • New England Research Center, Wellesley, Massachusetts 02181. BOONE B. OWENS • Department of Chemical Engineering and Materials Science, University of Minnesota, Minneapolis, Minnesota 55455. KENNETH B. STOKES • Medtronic, Inc. Minneapolis, Minnesota 55440. MICHAEL BILITCH • USC Pacemaker Center, Los Angeles, California 90033.

devices prior to 1960, because of the large size and power requirements of the pretransistor-age electronic components. With the advancement of electronic components in the 1960s and the development of high-energy-density batteries, instruments to measure and control human physiological activity could be made small, reliable, and implantable.

The early implant instruments were grouped in biomedical engineering texts into four categories[2]:

1. *Telemetry: in vivo* measurements of physiological signals and transmission to outside the body.
2. *Stimulation:* response of organs to *in vivo* electrical signals.
3. *Remote-controlled manipulation:* to control implanted machinery by remote control.
4. *Closed-loop control signals:* the use of one part of the body to control or alter the function of other parts (e.g., to control a paralyzed limb by implanted EMG transmitters).

The early implanted devices concentrated in the field of telemetry. For comparison, Table 1 shows the stages of medical treatment[3] (diagnostic, therapeutic, etc.) and the phase of medical use for selected specific technologies. The devices that are electrical, implantable, and available at the present time have a superscript *b* and those that are in the research phase have a superscript *a*. This shows that the nature of implanted devices has grown far beyond the telemetering stage and that therapeutic stage devices constitute the area of greatest commercial significance.

The variety of available implantable devices and materials is illustrated in Figure 1. Although some of the implantable devices are not electrically driven, there are electronic and electrochemical aspects[4] to the prevention of clotting and corrosion even with the unpowered devices, such as heart valves and hip joints.

In 1971, Brown, Jacobs, and Stark[2] listed the implant instruments desired, most of which were underdeveloped at the time. These are shown in Table 2. In the intervening years development of virtually all of these desired implantables has been achieved. The present list of devices available in clinical practice or research stages includes devices in the following areas:

- Heart pacing
- Bone growth and repair
- Hearing aids
- Chemical dosing
- Drug infusion and dispensing
- Defibrillation
- Heart assist
- Nerve stimulation
- Gut stimulation
- Artificial larynx
- Pain suppression
- Implanted sensors
- Scoliosis treatment
- Artificial vision
- Artificial heart

TABLE 1 Health Care Technology—Clinical Services

Function	Stage			
	Preventive	Diagnostic	Therapeutic	Rehabilitative
General	Multiphasic health testing Screening tests	Computer-aided diagnosis Diagnostic ultrasound Glucose monitor[a] Implanted sensors[a] Telemetering[b]	Computer-aided prognosis & therapy Drug dispenser[b] Insulin dispenser[b] Wound repair[a]	Artificial organs and parts[b] Iontophoresis[a] Tissue banks
Chest	Implanted defibrillator[b]	Cardiac catheterization and angiography Cardiac monitoring	Artificial heart[a] Coronary bypass Coronary care units External defibrillators Pacemakers[b] Respiratory therapy	
Gastrointestinal			Intestinal shunts Reflux device[a]	
Neurology		Implanted EEG[b]	Brain and spine surgery Implanted pain control[b] Spinal fusion	Neurostimulators[b] Paraplegic support[a]
Orthopedics			Bone repair[b] Hip and spine surgery	Limb prosthetics[b]
Otolaryngology	Glaucoma screening	Hearing tests Visual tests	Hearing aids[b] Internal ear surgery Laster beam therapy	Artificial larynx[b] Cochlear implants[a] Sensory aids[a]
Urology			Genitourinary surgery Penile implant[a] Testicular implant[a]	Kidney transplants

[a] In development, implantable device. [b] Commercially available, implantable device.

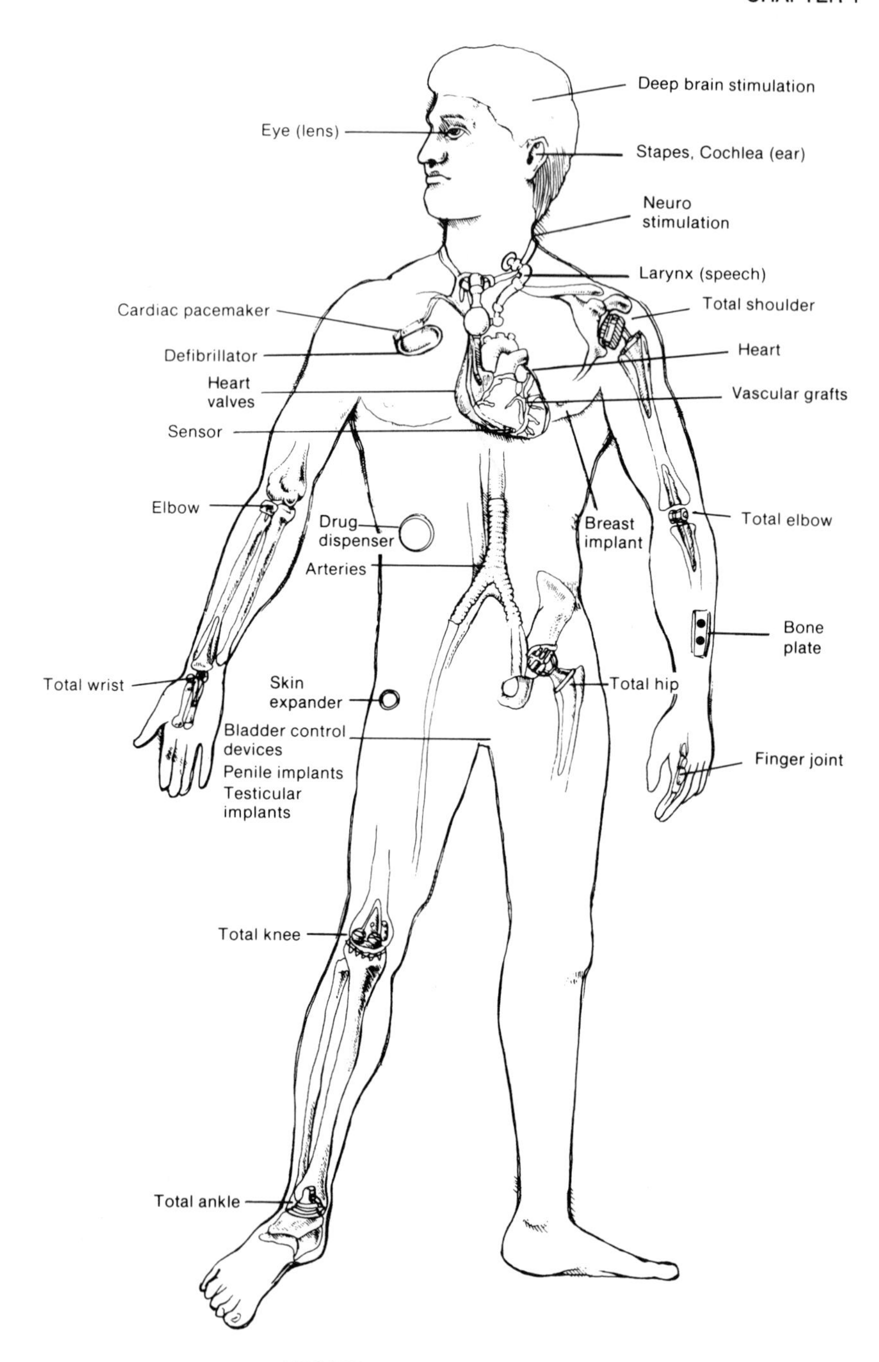

FIGURE 1. Implantable Devices.

TABLE 2 Desired Implant Instruments (1971)

Name	Function
Heart pacemaker	Adjusts rate on demand with redundancy for reliability, and telemeters alarm signals when difficulties develop
Bladder stimulator	Controlled by pressure inside, or can telemeter the critical pressure to outside for external control
Blood pressure regulator	Delivers stimulation automatically to control pressure, but can telemeter the condition to outside
Pain suppressor	At the spinal cord or peripheral nerve ending
Brain stimulation	For sleep and tranquilizing effect, controlled externally
Sensory stimulation	Hearing, vision, touch, etc., for prosthesis
Muscle stimulation	Analog control of force or displacement by external command
Stimulation of glands	Controls the secretion of hormones.
Telemetered measurements	Measure pressure, blood flow, pH, and chemical compositions inside the body or at the organs.
Telemetry of electrophysiological signals	Such measures as ECG, EMG, EGG, EOG
Implant manipulators	Release of chemicals or mechanical action upon external command and adjustment of the position of implant instruments such as brain electrodes or implant assistive devices

Source: From Brown, Jacobs, and Stark.[2]

1.2. Electronic Circuit Technology

The technology of electronic circuits has changed rapidly and dramatically in the last two decades. The earlier implantable prostheses, such as pacemakers in the 1960s, were fabricated from discrete components encapsulated in plastic (epoxy). This has evolved in stages to completely sealed integrated miniature microcircuits with very low power drains.

Implantable electrical prostheses became practical with the invention of the transistor. However, the discrete device prostheses were suitable for only a limited range of implanted pacemakers (and other implantables) because of power demands and size. The development of the integrated circuit in the early 1960s and the complementary-symmetry metal oxide semiconductor a decade later permitted a great reduction in size and power demands of electronic logic circuits. These were utilized in pacemakers in the early 1970s. Within a few years, virtually all implanted devices, particularly pacemakers, used these technologies, significantly reducing the size of the electronic assembly. This reduction in power and size permitted an increase in sophistication of the electronic circuit and simultaneously prolonged the life of battery-powered devices. Figure 2, from Gold,[5] is a comparison of an early electronic assembly using discrete transistor devices and components and a recent hybrid model with an integrated circuit chip.

The power requirement in a typical logic circuit used in a demand-type pacemaker was about 30 μW. Therefore, a circuit operating at 2–3 V draws only 10–15 μA (continuously). Considerably more energy was required for stimulating organs such as the heart. However, improvement in stimulating electrode technology has resulted in a great reduction of the power requirements for this aspect of the system. The earlier stimulating catheters delivered pulses of 1.5–3.0 ms at 5–6 V amplitude, which at nominal rates of 70 bpm consumed an average of 50–80 μW. Modern stimulating pulse widths are much shorter, in the range of 0.25–0.50 ms, and stimulating pulses are at lower voltages

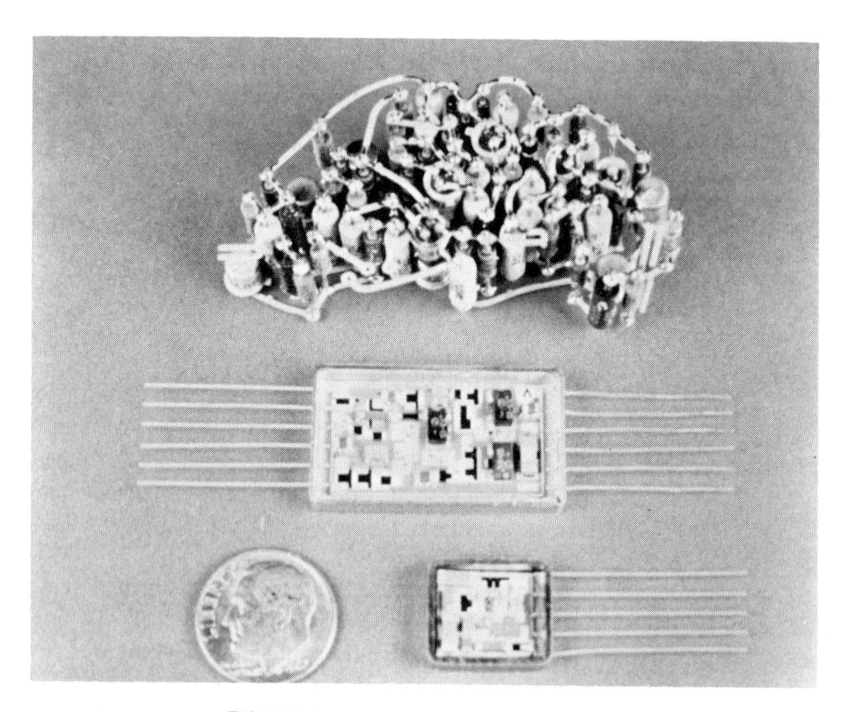

FIGURE 2. Electronic Circuit Comparison.

(3–5 V); consequently, the energy used in stimulation is reduced to the same order of magnitude as the energy for the logic circuit in a pacer.

With total power demands in the order of 50–60 μW or less, pacers have useful implanted lives of 7–10 years with power sources that can deliver 3–4 W-hours of energy. High-reliability components are obviously required to provide this long implant use. The earlier implanted stimulation devices, particularly pacemakers, exhibited premature failures because of the reaction of body fluids on transistor components. Corrosion, change of electronic values, and short-circuiting were observed in nonhermetically sealed units.

According to Escher,[6] one aspect of CMOS circuitry that may prove troublesome is "the destructive effect of gamma radiation on its integrity. Contrary to the situation with the old discrete circuit components, in which the major concern was the effect of electromagnetic force or magnetic force on the magnetic switch, in this case complete pacemaker malfunction may result."

2. DEVICES BACKGROUND

2.1. Heart Pacing Systems

The first successful use of the electrically driven implantable device came with the development of the permanent cardiac pacemaker in the early 1960s. Since its first successful clinical application, countless lives have been saved and many individuals have been relieved of otherwise incapacitating cardiac symptoms. Its widespread use attests to its efficacy. In 1981, in the United States alone, 118,000 individuals had permanent pacemakers inserted for the first time, while an additional 23,000 required generator changes in previously implanted systems. Similar numbers were implanted elsewhere in the world.

The basic pacemaker system consists of an electrode attached to the heart and connected by a flexible lead to the generator itself. This generator is a combination of a power source and the microelectronics required for the pacemaker system to perform its intended function.

To understand the requirements of the implantable cardiac pacemaker as well as the advances in the field since its earliest applications requires a basic understanding of cardiac physiology. The heart can be viewed as a pair of muscular pumps working in parallel to provide adequate amounts of oxygenated blood to the vital organs of the body. Each of these pumps (the right and left sides of the heart) consists of two separate chambers (an atrium and a ventricle) that function in series in a coordinated manner to produce the desired effect of adequate unidirectional blood flow. The "right heart" receives blood from the peripheral venous circulation via the right atrium. This chamber propels blood into the right ventricle, which pumps the blood to the lungs, where the red blood cells pick up the oxygen bound to the hemoglobin molecules. The oxygenated blood then drains via the pulmonary veins into the left atrium. Again, it is actively pumped into the left ventricle and then propelled to the peripheral organs. Unidirectionality of blood flow in the healthy heart is assured by four one-way valves placed between the major chambers of the heart as well as between the ventricles and their respective "great" vessels.

Optimal perfusion of the body is maintained by the coordinated pumping action of each of the separate chambers. In the absence of atrial contractions, the ventricles can only supply less than the normal output of the heart, which is sufficient to meet the normal demands of the body. However, significant increases in demand associated with exercise or stress may be met only partly. In the diseased heart the situation is more critical. At times, even the pumping requirements at rest or those associated with minimal activity may not be properly met by the heart in the absence of atrial contractions. The impetus in pacing technology has been toward duplicating normal cardiac physiology such that coordinated contractions of all four chambers result from the pacing stimuli.

All cardiac muscle fibers have the ability to depolarize with a resulting muscular contraction. This may occur in either of two ways. The fiber may depolarize spontaneously because of slow Phase 4 depolarization that characterizes cardiac muscle, or it may depolarize and contract as a result of a propagating wave of electrical activity originating elsewhere in the heart or from an external source. Once depolarization occurs, the fiber becomes refractory while it repolarizes and will not conduct electrical activity during this period. The normal spontaneous depolarization rate decreases as one goes down the pathways of normal electrical propagation through the heart. This spontaneous depolarization rate is greatest in the sinoatrial node and least in ventricular muscle.

In the normal-functioning heart, the sinoatrial node located at the junction of the right atrium and superior venacava serves as the origin of the impulse for distribution to the rest of the heart. Its rate, like that of other muscle tissue, may vary in response to many factors. These include temperature, blood volume, circulating catecholamines, vagal or sympathetic nerve impulses, and a number of other factors. Once initiated in the sinoatrial node, the impulse spreads throughout the heart over specialized conductive pathways. This results in a uniform depolarization and contraction of both atria and, in short order, arrival of the electrical impulse at the atrioventricular node. As its name implies, this is located between the atria and ventricles. During the delay of the impulse to the ventricle, the atria complete their contraction. From here a coordinated propagation of the electrical impulse follows specialized rapidly conducting fibers that pass through the septum and separate to supply the muscle fibers of the left and right ventricles. A uniform contraction of the chambers results. As long as a proper depolarization wave originates high in the conduction system at an appropriate rate, normal synchronous pumping will occur and spontaneous depolarizations elsewhere in the system will not occur. However, should the sinoatrial node, the atrioventricular node, or the various conduction bundles of the heart fail to function properly, slowing or cessation of the heartbeat will result. In the absence of an artificial pacemaker or the development of an adequate spontaneous rate elsewhere in the system, symptoms will result and death may well ensue.

The clinical application of an implantable cardiac pacemaker system as we know it today occurred first in the 1960s and was reported by Ekestrom, Johansson, and Lagegren[7] in 1962. The previous 30–40 years of investigation had been marked by multiple attempts at developing techniques to maintain an adequate heart rate in the face of inherent bradycardia (slowing of the heart) or asystole (cessation of the heartbeat). As understood at the time, these conditions were most commonly caused by a lack of proper propagation of the electrical signal through the atrioventricular node. The devices under development at this time were characterized by large, cumbersome external power sources attached to electrodes that delivered an electrical impulse to the heart. Originally, these electrodes were placed on the body surface and resulted in extreme discomfort to the patient when the generator delivered its impulse. Subsequently, electrodes were developed that could be placed directly on the surface of the heart and brought out to the body's surface. This resulted in lower energy requirements and no patient discomfort but required a major operation. Infection frequently ensued because of the transcutaneous leads. Successful clinical applications of these techniques were not practical. The limitations lay in a number of areas. The external power sources required were cumbersome and not appropriate for prolonged use. (Today's problems associated with designing a practical power source for an artificial heart are similar to those associated with the first implantable pacemakers.) The surgical risk associated with a major operation for placing wires on the surface of the heart rendered the procedure impractical. The pacing leads themselves were poor in durability and easily fractured during prolonged use.

Dramatic progress ensued, however, with the development of a durable, flexible transvenous endocardial lead by Furman, in 1958. This solved many of the problems inherent in previous lead designs and eliminated the need for major surgery to position the lead. Chardack and Greatbatch followed shortly with the development of an implantable pacing unit powered by an insulated mercury/zinc battery system. A combination of these developments resulted in the first successful implantable pacemaker insertions. Except for some technical modifications, the techniques used at that time are still in clinical use today.

Pacemaker technology has expanded dramatically in the 20-odd years since those first clinical applications, as shown in Figure 3. The progress in microcircuitry and microelectronics, along with the development of more reliable and longer-duration power sources, has had a profound impact. Many of the changes in pacemaker design have been in response to the rapidly expanding understanding of pacing physiology. Electrical stimulation of the heart continues to be the therapeutic choice for disorders of electric impulse formation and transmission

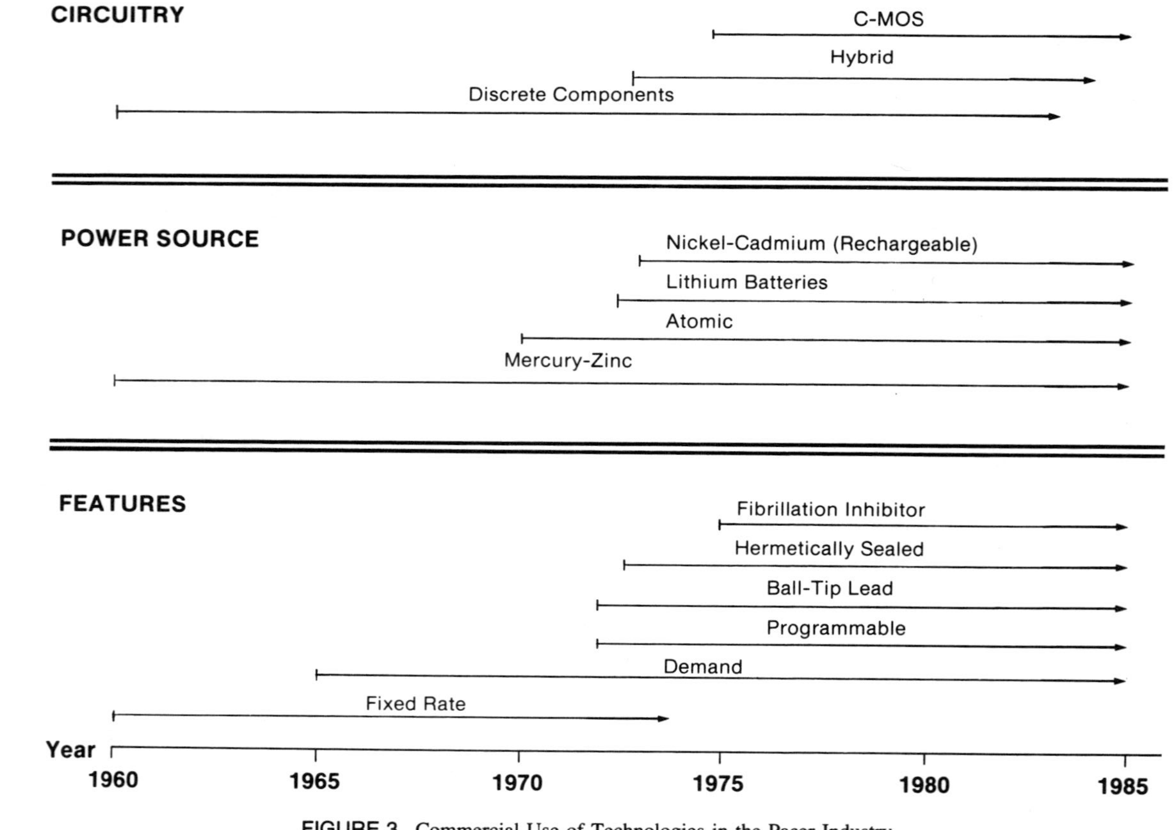

FIGURE 3. Commercial Use of Technologies in the Pacer Industry.

in the cardiac conduction system. The earliest pacemakers filled this objective to preserve life by preventing asystole through stimulation of the ventricle alone. These earliest pacemakers were "asynchronous" and fired at a predetermined fixed rate. Early modifications led to "demand" pacemakers that would not fire when a properly generated signal occurred in the myocardium. The increased understanding of pacing physiology has resulted in multiprogrammable dual-chamber pacemakers that can prevent the problems associated with stimulation of the cardiac ventricle alone. As noted earlier, this can result in compromise of the synchrony between the atria and the ventricles of the heart. It has been shown that proper AV synchrony provides regularized ventricular filling, increases cardiac output, and reduces the oxygen need of the heart.

Proper AV synchrony also has been found to afford protection against some complex arrhythmias and to prevent conduction in abnormal pathways. The "pacemaker syndrome," once considered rare, is receiving greater attention and is thought to be due to the loss of AV synchrony. This syndrome often resulted in a low-output state developing in patients with chronically implanted ventricular pacemakers. The hemodynamic consequences of the atrioventricular dissociation and other deleterious effects of ventricular pacing alone are now widely recognized, particularly in patients with "sick sinus syndrome," an alteration between bradycardia and tachycardia, as well as in patients of poorly contractile ventricles secondary to inherent cardiac disease.

In response to all of these developments, progress has been so dramatic that there are any number of clinical devices available to the cardiologist and cardiac surgeon today. Simple demand ventricular pacemakers similar to the units developed in the 1960s are available. The addition of programming capability and longer-duration, lighter-weight power sources has resulted in more reliable and universal pulse generators. These units are marked by simplicity, long life, and easy care for the patient. At the other end of the spectrum are the more dramatic dual-chamber multifunctional atrioventricular pacemakers that can sense as well as fire in both atria and ventricles of the heart. These can be externally programmed to perform any number of functions. They can distinguish the origin and routes of transmission of different impulses and distinguish between their relationship independent of each other.

The indications for pacing have expanded from the simple treatment of asystole and bradycardia to the more elaborate therapy of multiple rhythm disturbances. The most complex AV sequential pacemakers required for this type of treatment are such that up to 42 million programming combinations may be available in some of the more complex units. Though this would seem to offer many advantages and afford greater flexibility in techniques of treatment, new problems have been created. The most simple demand pacemakers can be programmed easily and last up to 10 years between generator changes. The more complex units require detailed record keeping, perhaps frequent programming, verification of proper generator settings, and a more complex system of followup.

Each of the many manufacturers of pacemakers has its own programming unit without interchangeability. In addition, the advantage of the longer-duration batteries may be lost because of the potentially high power requirements of these more complex units.

Compared to other electronic applications, the amount of energy required to operate a cardiac pacemaker is very small. In a typical stimulus, a pulse of 5–19 mA is delivered at 3–5 V for 0.25–0.5 ms. (In early units, pulse times ranged between 1 and 1.5 ms.) Each pulse, therefore, consumes power at approximately the 10–50 mW rate. Assuming a pacing rate of 75 pulses per minute, the average continuous energy drain is 15 μW. This compares to 60 mW, an energy drain 4000 times as great, for a six-transistor radio at reasonable volume.

Most older heart pacers contained a mercury/zinc battery of four to six cells with a total capacity of 5–8 Wh. This cell system, commonly used in hearing aids and cameras, has the virtue of relatively high volumetric energy density, 0.5 Wh/cm^3 at rates of one month to one year. Though the mercury/zinc battery has good properties in most regards, it does lose about 20% of its capacity from self-discharge during a four-year stand in the body. Additionally, zinc electrode cells in potassium hydroxide electrolyte vent small quantities of hydrogen gas on stand; consequently, the mercury/zinc cell cannot readily be hermetically sealed. As pacer life is limited by battery life, the desire to minimize surgical generator replacement, reinforced by the opportunity to make smaller, lighter pacers, led to research on other battery sources for pacers.

In the last five years, lithium anode batteries have almost universally become the power source for cardiac pacers. These systems operate in a nonaqueous electrolyte, such as propylene carbonate containing a mobile ionic charge carrying species such as lithium perchlorate. In some cases, such as iodine cathode cells, the electrolyte is a solid ionic conductor (LiI). The cathode, separator, and materials used determine many of the system properties. The lithium battery chemistries that have been implanted in humans are shown in Table 3.

In general, the lithium systems offer the main advantages of high-energy density in the range of 180–250 Wh/kg and 0.6 to 1.0 Wh/cm^3, high voltage per cell (2–3 V), hermetic sealability, very low loss of capacity on stand (1–2% per year), and the ability to be configured in thin shapes. Clinically, their use has had the benefit of reducing the size of a pacemaker from 200 g down to 40–50 g, mainly by providing a smaller battery and making possible hermetic sealing of the pacer, therefore eliminating heavy epoxy encapsulation. In addition, because of this greater energy storage, the time interval between battery changes of a ventricular demand pacer has been extended from an average of two years with the mercury/zinc batteries to as much as 8–10 years with some of the newer batteries. However, the greater energy demands of the multiprogrammable AV sequential pacemakers has shortened this time interval considerably. The lithium systems in common use[8] are lithium/iodine, lithium/copper

TABLE 3 Properties of Implantable Lithium Anode Cells

Cathode	Theoretical Limits[a]				Practical Limits[b]		Electrolyte Types
	Wh/cm³	Wh/g	Ah/cm³	E_{oc} V	E_L, V	Ah/cm³	
PbI_2, PbS	1.4	0.31	0.73	1.9	1.8	0.28	Solid
CuS	1.8	0.58	0.86	2.1	2.1	0.29	Liquid
I_2	1.9	0.57	0.61	2.8	2.7	0.28	Solid
MnO_2	2.5	0.86	0.70	3.0	3.0	0.26	Liquid
Ag_2CrO_4	2.1	0.5	0.64	3.3	3.1	0.19	Liquid
Br_2	2.5	1.1	0.71	3.5	3.0	<0.2	Liquid
$SOCl_2$	2.3	1.7	0.47	3.6	3.6	0.25	Liquid
V_2O_5	1.36	0.46	0.40	3.4	2.8	0.08	Liquid

[a] Volumetric energy density, weight energy density, volumetric specific capacity, and open-circuit voltage, respectively.
[b] Load voltage and practical volumetric specific capacity.
[c] Based on a high-rate design.
Source: From Brennen and Owens.[8]

sulfide, and lithium/silver chromate. Among these, the lithium/iodine system is the most widely used.

Many other approaches to energy storage have been experimentally used, including piezoelectric, biologic, nuclear, and biogalvanic, as well as other electrochemical cells containing solid electrolytes and nonaqueous electrolytes. Also other chemical cells with aqueous electrolytes have been tested. Besides the mercury/zinc and lithium anode battery systems, only three other battery types have been used in pacers implanted in humans. These are nuclear isotope batteries, biogalvanic batteries, and nickel/cadmium batteries. The nickel/cadmium batteries were radiofrequency recharged by transmission of energy to an implanted receiver. This was accomplished with both line-powered electric circuits and ordinary primary cells, replaced frequently (weekly) outside the body. Although these systems were rendered obsolete by the development of lithium cells for pacemakers, they are useful for other implantable devices.

Cardiac pacing has traveled a long road in the two decades since pacing was first employed for advanced symptomatic heartblock. As we look to the future of cardiac pacing, it is possible to predict that innovative pacing circuits and techniques will emerge as more specific electrophysiologic information becomes available. It is not too optimistic to think that many, if not all, electrophysiologic disorders of impulse formation and conduction could in some manner be treated by electrical control of the heart.

Patients with heartblock or slow heartbeat (asystole and bradycardia) were the first to be treated with pacemaker therapy and considerable statistical information is now available on the results of this treatment. For these patients, pacer therapy has proven to be very effective, and their statistical life expectancy is within 5% of the curve for their age population.[9] Expectations for the future are that treatments for tachycardia and other arrythmia dysfunctions as well as other electrical therapies (defibrillation) will continue to increase and with similar statistical results.

2.2. Cardiac Pacing Leads

Cardiac pacing leads are generally divided into two categories: unipolar and bipolar. Of course, to complete a circuit, a pacing system must have two electrodes. In the bipolar pacemaker both electrodes are located on the lead (in the heart). The so-called unipolar system has only one electrode on the lead. The other electrode is on the pulse generator in subcutaneous or intramuscular tissue.

In the past, a major problem was dislodgement of transvenous electrodes to affect pacing adversely. This problem has largely been solved through the use of several different fixation mechanisms, which can be classified as either passive (no damage to tissues at the stimulation site) or active (anchored by tissue-damaging mechanisms).[10] Passive fixation has been achieved by the use of pliable tines (Table 4) which are designed to catch on cardiac trabeculae.[11,12] This provides acute anchoring. Chronic fibrotic tissue formation around the tines provides further security. Another mechanism is the open helicoidal electrode wire loop (Helifix by Vitatron Medical, Dieren, Holland), which is designed to wrap atraumatically around trabeculae or papillary muscle.[13,14] Porous electrodes also can provide passive chronic anchoring because of the ingrowth of fibrous tissue.[12,15] Active fixation mechanisms include hook, barb, screw, or whisker-shaped electrodes that penetrate tissue.[10] Thus, there exists a bewildering array of electrode shapes and fixation mechanisms for cardiac electrodes.

Regardless of the complexity of electrode configurations, they all have a common purpose—to stimulate the heart reliably. During stimulation, however, it is most desirable to use as little battery charge as practical to prolong the pacemaker's useful life. Three parameters—threshold (amplitude), pulse width, and impedance—must be balanced properly to minimize charge drain without compromising patient safety.

Cardiac cells are caused to contract when their resting membrane potentials are raised to a critical (threshold) value.[16] The cell then reacts completely according to the "all-or-none principle" to generate an action potential. The action potential of the stimulated cell then stimulates neighboring cells and the response is propagated throughout the muscle mass. With cardiac pacing electrodes, membrane resting potentials are raised to threshold by an electric field generated around an implanted electrode.[17] The minimum field intensity

TABLE 4 Fixation Mechanisms for Transvenous Pacer Leads

Manufacturer and Model No.	Distal Tip
Medtronic 6904A[a]	
Medtronic 6901R and 6907R	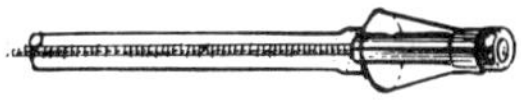
Medtronic 6961 and 6962	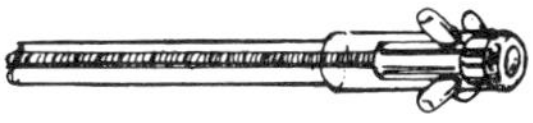
Vitatron Helifix	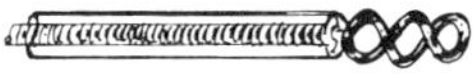
CPI 4130 and 4230 (porous electrode)	
Cordis Fin-Tip	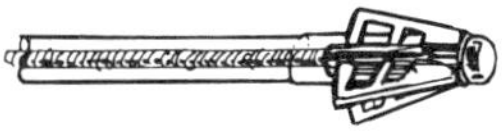
Medtronic 6957	
Biotronik IE651	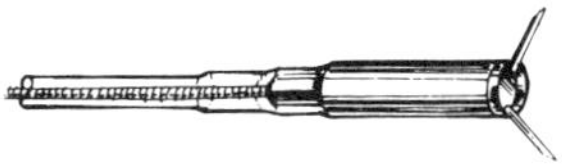
Coratomic Endo Loc L-40	

[a] Obsolete.

required to stimulate by a hemispherical electrode is given by

$$F = \frac{V}{r} \quad (1)$$

where V is the applied voltage (for a constant voltage generator) and r equals the electrode radius. With time, however, fibrotic tissue forms about the electrode, separating it from stimulatable tissues. Thus, the effective radius of the electrode increases according to:

$$F = \frac{V}{r}\left(\frac{r}{r + d}\right)^2 \quad (2)$$

where d is the thickness of the fibrotic capsule. To compensate for the increase in "virtual electrode" size, more voltage or current must be applied.[18] Thresholds, therefore, typically rise with implant time to higher peak values within a few weeks or months, then decrease to chronically stable values (Figure 4).

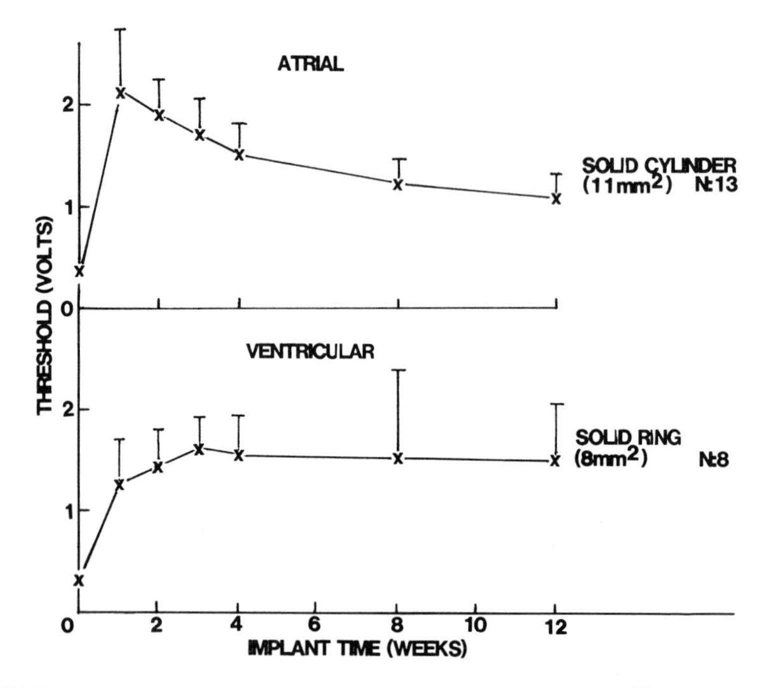

FIGURE 4. Threshold Voltage Variation with Implant Time for Heart Pacing.[10] Acute to Chronic Canine Constant Voltage Threshold Changes at 0.5 ms as a Function of Implant Time Are Shown for Tined Electrodes with Typical Solid Polished Platinum Electrodes. Electrode Shapes and Surface Areas Are Presented to the Right of Each Curve.

It was determined by Lapicque and Weiss in the early 1900s that the stimulus necessary to pace varies as a function of the time stimulus applied.[19,20] This defined the strength (stimulus amplitude)–duration (pulse width) curve, which has different shapes, depending on how threshold is measured.[10] Voltage and current thresholds follow hyperbolic curves described by Lapicque (for constant current generators) as:

$$I = \frac{a}{t} + b \tag{3}$$

where I equals the threshold current intensity at pulse width t and a and b are empirically determined constants. At wide pulse widths, thresholds remain virtually constant regardless of pulse width. This is defined as rheobase. A pulse width at twice rheobase was defined by Lapicque as chronaxy. Using these terms, the Lapicque equation becomes:

$$I = I_\infty \left(1 + \frac{t_c}{t}\right) \tag{4}$$

where I_∞ is the rheobasic current amplitude at pulse width t, and t_c is the chronaxy pulse width. Voltage strength–duration curves are similar in shape, typically with a higher rheobase and lower chronaxy pulse width.

As a stimulating electric field is generated around the electrode, charge is drawn from the power source. Charge is defined as:

$$Q = It = \frac{Vt}{Z} \tag{5}$$

where Z equals the lead/electrode/tissue impedance. The charge strength–duration curve is essentially a straight line, as is shown in Figure 5. Energy is given by the relationship:

$$E = VIt = \frac{V^2 t}{Z} = VQ \tag{6}$$

By inspection, therefore, one can see in Figure 5 that the energy strength–duration curve is parabolic in shape (voltage × charge).

It is clear that to minimize power source charge drain, the best place to stimulate on the strength–duration curve is at as low a pulse width as possible. But at too low a pulse width the stimulus falls on the ascending limb of the voltage or current strength–duration curve. Thresholds are known to fluctuate daily by as much as 50% in response to exercise, sleeping, eating, drugs, and

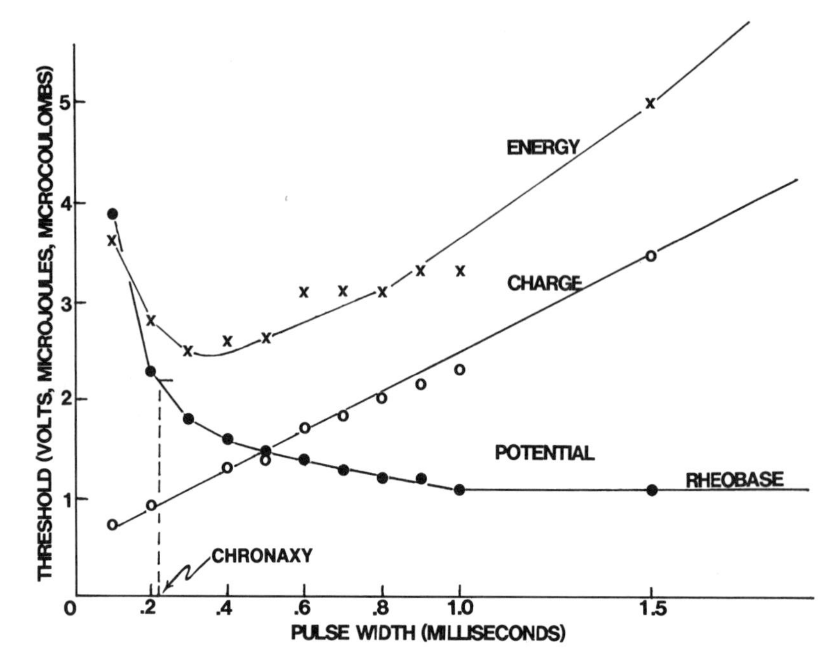

FIGURE 5. Threshold Energy Variation with Pulse Width for Heart Pacing.[10] Typical Relationships between Chronic Ventricular (Canine) Constant-Voltage Threshold Strength–Duration Curves Expressed in Terms of Potential (V), Charge (μC), and Energy (μJ) for a Tined Unipolar Lead with an 8-mm^2 Ring-Tip.

so on.[21,22] At very low pulse widths, therefore, slight changes in the patient's threshold can cause large variations to the extent that capture can be lost. There must be a place on the strength–duration curve where an optimum stimulation/longevity compromise is reached. This is the point where minimum energy is consumed, which is also about the chronaxy pulse width.[23,24]

Thresholds normally vary throughout the day. Chronic thresholds are higher than those measured at implant. Thus, the pulse generator must deliver a greater stimulus than the threshold measured at implant or at rest in the clinic. For chronic leads, according to Barold and Preston, a 1.75:1 safety factor is adequate for constant-voltage devices (or 3:1 energy safety factor for constant current).[25] Using constant-voltage devices, a 2:1 safety factor is easier to compute and to work with. Thus, this is becoming generally accepted as an acceptable bench mark for constant-voltage generators. Low thresholds are therefore important to minimize charge drain from the power source. They must be low enough that generator output can be raised to provide an adequate safety factor without overly draining the power source.

Pacing impedance is also important to provide effective stimulation without depleting the power supply too soon. Impedance is a complex function defined as the sum of all factors that retard current flow.[10] The lead/tissue/electrode impedance can be modeled as:

$$Z = R_1 + R_e + Z_p \tag{7}$$

where R_1 is the conductor resistance, R_e is the ohmic resistance of the electrode in tissue, and Z_p equals the polarization impedance of the electrode/tissue interface. Two of these components, R_1 and Z_p, while reducing charge drain, also waste that charge without stimulating the heart. R_1 represents energy wasted as heat in the conductor wire. When current is passed from an electrode into an electrolyte solution, a potential develops across the interface because of a separation of ions of one charge from ions with the opposite charge. This polarization increases as a function of pulse width (another reason to use low pulse widths in pacing). The power source must supply an additional voltage to overcome the polarization potential, thus wasting energy. The optimum electrode, therefore, has minimal R_1 and Z_p with reasonably maximized R_e.

The magnitude of R_e and Z_p are dependent upon several factors, but one of the most important is electrode surface area. The pacing impedance in tissue is a function of the apparent or geometric electrode surface area. Polarization, on the other hand, is limited effectively to a region within Angstroms or microns from the electrode surface. Thus, Z_p is dependent on the true microscopic surface area. Maximum R_e and minimum Z_p are achieved by making the electrode either microscopically porous or textured.

The optimum trade-off between stimulation safety and charge drain occurs at chronaxy, where:

$$Q_c = \frac{Vt_c}{Z} \tag{8}$$

The optimum electrode must have minimum threshold (V) at a low chronaxy (t_c) and a high impedance (high R_e with minimum R_1 and z_p). Until recently, a pulse generator output of 5 V and 0.5 ms at 5000 Ω was considered "standard." These values have been shown to be safe and reliable for state-of-the art electrodes shaped like rings, cylinders, hemispheres, screws, hooks, and so on with geometric surface areas between 5 and 20 mm^2. According to Eq. (5), at these settings 5 μC charge per pulse is drawn from the power source. The introduction of porous electrodes has allowed greater efficiency by reducing polarization losses, although claims of lower threshold appear to be controversial.[15,26–28] Next, "activated" carbon electrodes gave both reduced polarization losses and somewhat improved chronic thresholds.[29–31] As of May 1983, an even more effective electrode, the Target Tip™, was introduced.[32] This is a grooved hemispherical, platinum electrode that has its surface electroplated with submicron-sized particles of pure platinum. Human clinical studies following thresholds as a function of implant time demonstrate that this lead safely paces ≥98% of the patients with a 2 to 1 safety factor at pulse generator settings of 2.7 V and 0.5 ms, from the day of implant.[33,34] That is, threshold rise or "peaking" was not statistically significant as a function of time. Pacing impedance in the human has been about 530 ± 110 Ω at implant ($N = 31$). Thus, as shown in Table 5, a pulse generator setting of 2.7 V, 0.5 ms draws on the average only 2.5 μC charge per pulse.

The improvements possible with even newer leads still in clinical study are shown in Table 4. Truly low threshold electrodes are being developed rapidly. Thus, as electrode design improves, pacing becomes more efficient, requiring less and less power source output. Battery drains as low as 0.8 μC per pulse will become typical by 1986, an 84% improvement over the state of the art in electrode technology in 1983. The benefits of these low-threshold, high-impedance electrodes will be smaller (since less battery capacity is necessary), longer-lived pulse generators with more features (powered by the energy no longer required to stimulate the heart).

The threshold for stimulation of cardiac muscle is also affected by drugs.[35] Since stimulation is to a large extent current-density dependent, the reduction in size of pacemaker tips has permitted an overall reduction of electric drain. At the same time,[36,37] the margin of safety is increased, since the gap between stimulation and fibrillation threshold energies became greater.

TABLE 5 The Effect of Electrode Design on Pulse Generator Output Settings and Power Source Charge Consumption[a]

Electrode Type	Electrode Surface Area (mm²)	Pulse Generator Output[b]	Acute Human Pacing Impedance at Threshold (Ω)	Charge Drain per pulse (μ)
Polished[c] ring-tip	8	5.4 V, 0.5 ms	500	5.0
Platinized[c] target tip[d]	8–10	2.7 V, 0.5 ms	530	2.5
Porous steroid[e] tip[f]	8	2.7 V, 0.3 ms	470	1.7
Future design[g]	5	2.7 V, 0.2 ms	700	0.8

[a] Estimates based on multicenter clinical studies, following human thresholds as a function of implant time.
[b] Generator settings at implant giving a 2:1 voltage safety factor for ≥98% of patients. Note that chronic thresholds may be such that for many patients, pulse generator output can be reduced for even greater energy savings after 3–6 months.
[c] "Market released."
[d] Ref 24–26.
[e] Combined data from multicenter clinical study (per FDA's premarket approval regulations in United States).
[f] Refs. 27–29.
[g] Projected values based on canine data and early human clinical results.

The introduction of steroid-eluting, low-threshold, low-polarizing electrodes[38–40] promises to reduce the necessary energy of the pacer pulse and prolong pacer life or permit the use of smaller batteries.

2.3. Automatic Implantable Defibrillator

The prevention of sudden cardiac death due to the cessation of the heartbeat was the original impetus for developing the cardiac pacemaker. The prevention of sudden death due to very rapid, uncoordinated heartbeats (ventricular tachycardia and ventricular fibrillation) has prompted the development of the automatic implantable defibrillator. Currently, it is an experimental device that is undergoing clinical trials and has been implanted successfully in more than 300 individuals.

The automatic implantable defibrillator is similar in design to a pacemaker incorporating cardiac electrodes into a combination battery and defibrillator. While the implantable cardiac pacemaker fires in the absence of a cardiac beat, the defibrillator functions by sensing the configuration of the electrocardiographic impulse. The defibrillator then supplies a single electrical impulse of 25 J to convert the heartbeat to a regular rhythm. At times, multiple impulses may be required. This power of delivery is approximately a million times greater than that normally delivered by the cardiac pacemaker. As a result, the useful life of this unit is uncertain and will probably be directly related to the frequency with which defibrillation is necessary. It is estimated that in the average patient the unit would remain in place for approximately three years or 100 defibrillatory episodes before a generator change would be required.

As the techniques have evolved over the last few years, the implantation of this device has become similar to that of the cardiac pacemaker. It has evolved from requiring a major surgical thoracotomy to the current situation where a small skin incision is required to position an electrode plate on the heart via a subxiphoid approach. A second transvenous electrode is required. Both of these are attached to the implantable defibrillator.

The problems with this approach derive primarily from engineering constraints. This approach is really akin to the way the pacemaker industry began. A known and proved treatment modality, dc shock (which today can be delivered only externally, and only if the patient happens to be in the proximity of equipment and trained personnel), is engineered into a package small enough to be implanted and is able to stand by and be automatically activated when the patient requires treatment. The technological problem is, therefore, essentially a series of engineering challenges, the principal of which are (1) miniaturization to implantable size of a device that must deliver a fairly high amount of energy (a pulse that is almost 1 million times greater in energy than that of a standard pacemaker) and (2) incorporation of a fail-safe system that eliminates the chance of false positives (i.e., having the device fire when *not* needed). While these

design constraints may seem enormous, some progress has been made by certain investigators.

Future progress with this device is eagerly awaited by the cardiology community. The incidence of sudden cardiac death in this country is estimated at approximately 450,000 cases per year. At least part of this number is attributable to episodes of ventricular fibrillation. A number of patients are known to have recurrent episodes of ventricular fibrillation. It is hoped that the implantable ventricular defibrillator may prevent the incidence of sudden death in those individuals who can be identified as "at risk" for this type of episode.

2.4. Bone Growth and Repair

The pioneering research in the bioelectric properties of bone was performed by Yasuda and Fukada in Japan during the early 1950s.[1,41] Subsequently, bone-healing studies were performed by many investigators,[42–57] as shown in Table 6. In the United States, Bassett and Becker in the late 1950s and Shamos and Lavine in the early 1960s began similar experiments, as did Friedenberg and Brighton[42,49] in 1961, all working independently.

These studies demonstrated that the potentials that arise when bone is mechanically stressed are not dependent on cellular viability, that they arise from the organic component of bone, and that they are electronegative in areas of compression and electropositive in areas of tension. Nonstress potentials (known as bioelectric or standing potentials and originally detected by Friedenberg and Brighton) require cellular viability. These potentials are electronegative in areas of active growth and repair and electropositive in less active areas.

During the 1960s many investigators demonstrated that the application of a small amount of electric current stimulated osteogenesis at the negative electrode or cathode. Typical of these studies were three animal models studied by Brighton and colleagues.

The first osteogenic model demonstrated that maximum bone formation occurred around a stainless steel cathode inserted through a drill hole into the medullary canal of the adult rabbit at a current level of 5–20 μA.[42] The second model demonstrated that a stainless steel cathode delivering 10 μA directly into a fracture site in the adult rabbit fibula produced a statistically significant increase in mechanical resistance to bending when compared with controls.[43]

The third osteogenic model studied electrically induced osteogenesis in a region far removed through a drill hole in the proximal rabbit tibia, then advanced through the medullary canal into the distal tibia, which is normally filled with fatty tissue. Current magnitudes ranging from 1 to 100 μA were evaluated. Results under these experimental conditions demonstrated osteogenic activity in a region where bone was not previously being formed and indicated that 20 μA produced the optimal level of bone regeneration.

On the basis of these and other studies, clinical trials were launched in

TABLE 6 Early Bone Healing Studies

Investigator	Year	Current	Duration	Species	Clinical Trials	Fracture
Friendenberg[42,43]	1970	5–20 μA, dc	12 days	Dog		Femur
	1971	10 μA, dc	75 days	Rabbit		Fibula
	1971	10 μA, dc	11 weeks	Human	1 nonunion	Medial malleolus
Lavine[44,45]	1972	4 μA, dc	125 days	Human	1 nonunion	Tibia
	1977	2–4 μ, dc	6 weeks	Rabbit		Tibia
Jorgensen [46,47]	1972	20 μA, dc	12 weeks	Human	1 nonunion	Tibia
	1977	20 μA, dc	12 weeks	Human	28 nonunions	Tibia
Srivastava[48]	1977	15 μA, dc	8 weeks	Human	20 nonunions	Tibia
						Radius
					90% success	Ulna
Brighton [49,50]	1975	10 μA, dc	12 weeks	Human	24 nonunions	Tibia
					62% success	Clavicle
						Femur
	1977	20 μA, dc	12 weeks	Human	46 nonunions	Malleolus
					84% success	Ulna
						Humerus
Paterson[51]	1977	20 μA, dc	4 weeks	Dog		
Connolly[52]	1977	20 μA, dc	6–12 weeks	Human	2 nonunions	
Hassler[53]	1977	10 μA, ac	3 weeks	Rabbit		
Harris[54]	1977	20 μA, dc	17 weeks	Rabbit Dog		
McKnight[55]	1976	4 μA, dc	—	Dog		
Bassett[56,57]	1974	10–25 μA/cm^2, ac	28 days	Dog	20 nonunions	fibula
	1977	10 μA/cm^2, ac	3–6 mos	Human	70% success	

1970. Friedenberg *et al.*[58] at first reported the successful treatment of a nonunion of the medial malleolus after nine weeks of constant direct current therapy to the nonunion site. Reports of cases using various techniques and types of current followed shortly thereafter.

Substantial published documentation exists that various forms of electrical current, whether it be constant direct current, pulsed direct current, or electromagnetically induced current, can stimulate osteogenesis. To date, there is more published data available on the effects of osteogenesis by direct current stimulation[50] than for other forms of electrically induced osteogenesis.

2.4.1. *Physiology of Action*

The mechanisms by which electricity induces osteogenesis are not completely known. Studies of the local tissue microenvironment have revealed the following clues:

1. When cathodes are implanted into the nonunion site and an electrical potential of less than 1 V is applied, oxygen is consumed at the cathode and hydroxyl ions are produced, decreasing local tissue oxygen tension and increasing alkalinity.[49]
2. Low tissue oxygen tension has been shown to be favorable to bone formation.[59]
3. Bone formation follows a predominantly anaerobic metabolic pathway.[60]
4. Studies of bone-forming junctions have demonstrated that an alkaline pH is present in the zone of the hypertrophic cells of the growth plate when calcification begins.[61]

Obviously, other mechanisms must exist in electrically induced osteogenesis. While they are yet to be defined, clinical experiences to date have proven the efficacy of the direct current stimulation technique. Research is continuing in an attempt to isolate the physiological chain of events leading to osteogenesis in bone and cartilage cells.

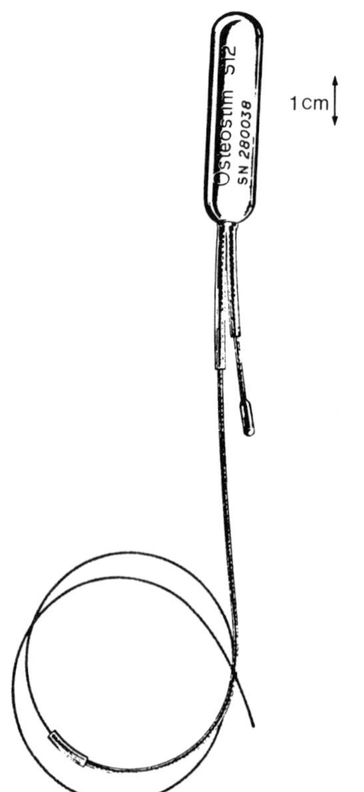

FIGURE 6. Implantable Bone Growth Stimulator.

2.4.2. *Bone Growth and Repair: Devices*

Three techniques have been developed for applying electrical current to bone: implantable, percutaneous, and transcutaneous.

Among these the invasive approach more closely related to pacer technology will be summarized briefly. One of the prominent pacer companies has developed a clinically applicable, implantable dc bone growth stimulator.[62] A bipolar lead electrode system is employed and a constant current of 20 μA is delivered to the fracture site. The preliminary results from the clinical trials indicate a level of success in excess of 70%.[63,64]

A typical implantable bone growth stimulator (Telectronics-Osteostim Model S12) intended for use with long bone fusion is shown in Figure 6, and the design characteristics of the unit are listed in Table 7.

Other units designed for posterior spinal fusion have slightly different characteristics and use multiple cathodes. They too use silver oxide/zinc alkaline cells of the button type with 60-mAh capacity.

As electrical stimulation of bone advances, there undoubtedly will be a

TABLE 7 Characteristics of Typical Implanted Bone Growth Stimulator

Power source	Two silver oxide alkaline zinc cells of 60 mAh capacity.
Electronics	Solid-state "constant-current/constant-field" generator with "failsafe" failure-mode protection. Circuitry is designed to maintain current between anode and cathode despite wide variations of bone/tissue resistance and electrode polarization.
Characteristics	The S12 is designed to produce a cathode current of 20 μA.
Electrodes	Tissue anode short (5 cm) length of three-strand titanium wire insulated with polyethylene and fitted with a pure platinum tip having a surface area of approximately 50 mm^2. Cathode triple-stranded titanium wire helix, insulated for first 15 cm with polyethylene sheath with a further 25 cm of titanium wire uninsulated to form actual cathode area.
Case	Medical-grade pure titanium of approximately 1 mm thickness with electronics encapsulated in clean silicone resin.
Identification	Each unit has model and serial number engraved on case. No filling.
Weight	14 g.
Size	11 × 45 mm.
Standard accessories	Each unit is supplied with a mandrel to form the helix and a 14-cm length of 1-cm-diameter silicone rubber tube to assist in "tunneling" electrode leads.
Operating period	Provided the model S12 is implanted within 12 months from date of manufacture it should deliver its full rated current for a period of 12–16 weeks.
Sterilization	The units are supplied ethylene oxide sterilized. If resterilization is needed, it can be done in ethylene-oxide gas but the temperature *must not* exceed 45–50°C.
Shelf life	A maximum of 12 months from date of manufacture and is determined by self-discharge of batteries. A "use before" date is displayed on the outer package.

marked increase in the complexity of the hardware. In fact, the electronic functions associated with telemetry and programmability that have been demonstrated to be efficacious in the pacing field will find additional applications in the stimulation of bone. Additionally, the technology associated with high-reliability implantable power sources used in cardiac pacers will find applications in bone growth stimulators.

2.5. Other Devices

This section discusses developments in the areas of pain control, CNS (central nervous system) stimulation, implanted drug dispensers, telemetering devices, hearing aids, health assist sensors, and so on.

2.5.1. CNS Stimulation

Implantable electrical stimulation devices have been used to treat cases of epilepsy, phantom limb pain, chronic nerve injury, and multiple sclerosis.[65–67] There are two basic approaches to CNS stimulation: self-contained implantable stimulators and transcutaneous RF-linked systems. The fully implantable approach utilizes current pacer technology. A cardiac pacer without sensing capability may be utilized as a CNS stimulator by changing the operating regimes of the device. Typical CNS applications require stimulation parameters of pulse duration over the range of 100 μs–1 ms; output voltage of 1–5 V, depending on electrode position; and pulse repetition frequencies of 10–100 Hz. These output requirements pose additional questions as to the longevity of the device as well as the power source. Typically, the energy requirements for CNS stimulation are an order of magnitude greater than pacemaker applications. In summary, cardiac pacer technology has been applied to many CNS applications. Further development of implantable devices will be centered in the areas of miniaturization, improved electrodes, and a high-rate power source.

2.5.2. Drug Delivery Systems (as in Treatment of Diabetes or Pain Control)

Numerous effects on body function have been associated with the rate of secretion and blood concentration of insulin. In the diabetic state, the pancreas is unable to regulate the levels of blood glucose and insulin. Two basic approaches have been taken in efforts to solve the problem of diabetes: pancreatic tissue transplantation and electromechanical insulin delivery systems.[68–70] The artificial devices consist of closed-loop and open-loop systems. The closed-loop insulin delivery system consists of a blood glucose transducer, a pump, and an electronic controller. The open-loop system delivers insulin independently of blood glucose levels. The self-contained electrically actuated systems appear promising. These systems are based on piezoelectric and electrodynamic transduction techniques and could be powered by pacer-type batteries.

In summary, a variety of systems have been developed for insulin delivery. The major problems lie in the areas of glucose sensor development and pump miniaturization. The packaging, power source, and electronic controller, on the other hand, have progressed more rapidly, based primarily on existing pacer technology.

Typical of the drug administration system is the one shown in Figure 7, manufactured by Medtronic. The tiny drug administration device is self-contained and consists of a 20-cm^3 reservoir, a pump and motor, microelectronic circuitry, and a lithium battery, all encased in lightweight titanium. The pump, which releases the medication, is driven by a motor that responds to signals from a microcomputer.

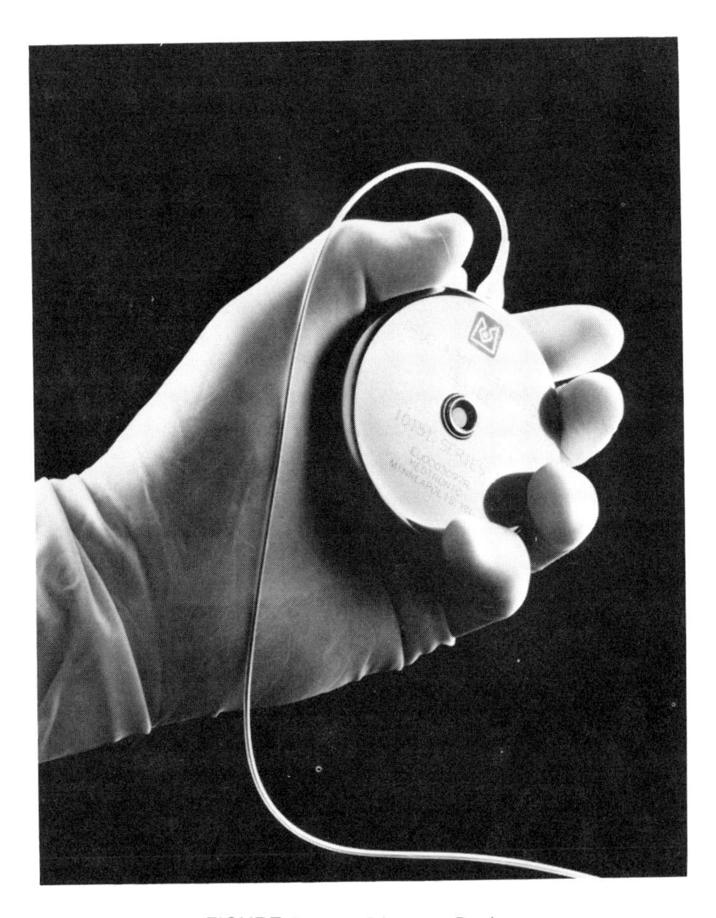

FIGURE 7. Drug Dispenser Device.

Once implanted, the unit can be programmed by the physician using a hand-held, computer-based controller to release medication at almost any rate. Through radiofrequency telemetery, the device can communicate with the physician's monitor to verify that the correct program has been received and to report the amount and rate of drug being dispensed.

"The failure mode is designed so that the device simply stops functioning, making it impossible for it to run away." The device also automatically records the amount of time that passes after a shutdown. When the patient reaches the doctor's office, all relevant information may be conveyed to the monitor by telemetry.

The device can be refilled through a self-sealing septum made of silicone. Clinical studies by investigators are focusing on the use of the device for control of intermittent or persistent chronic pain, chemotherapy, treatment of cardiovascular disease, and control of diabetes.

Preliminary evaluation suggests that the device promises to be advantageous in the following areas:

1. *Chronic pain:* Since the device permits smaller, more frequent dosages of morphine or other drugs, the patient receiving the treatment may be able to remain alert and continue normal functions.
2. *Chemotherapy:* Using such devices, drugs may be directed to the site of the tumor, thus reducing potential side effects. Time in the hospital may be reduced and the danger of infection, hemorrhage, arterial perforation, and other complications is lessened.
3. *Cardiovascular disease:* In cases where heparin is prescribed for patients with severe clotting problems, the device may be advantageous over previous methods of administering the drug through injection or infusion, since both require hospitalization.
4. *Diabetes:* The device may be helpful in reducing complications believed to be caused by wide fluctuations in blood glucose levels that occur with daily or twice-daily injections of insulin. Among those complications are blindness, damage to nerves, gangrene, amputation, and loss of kidney function. The device allows physicians to prescribe insulin in amounts that will maintain a steady level of glucose similar to that of nondiabetic individuals.

2.5.3. *Biotelemetry*

Implanted physiological signal telemetry devices have been designed for a variety of human and animal parameters. Typical examples of these devices are shown in Figure 8, from the research of S. Deutsch and colleagues.[71,72]

The availability of low-supply-voltage integrated-circuit amplifiers now makes

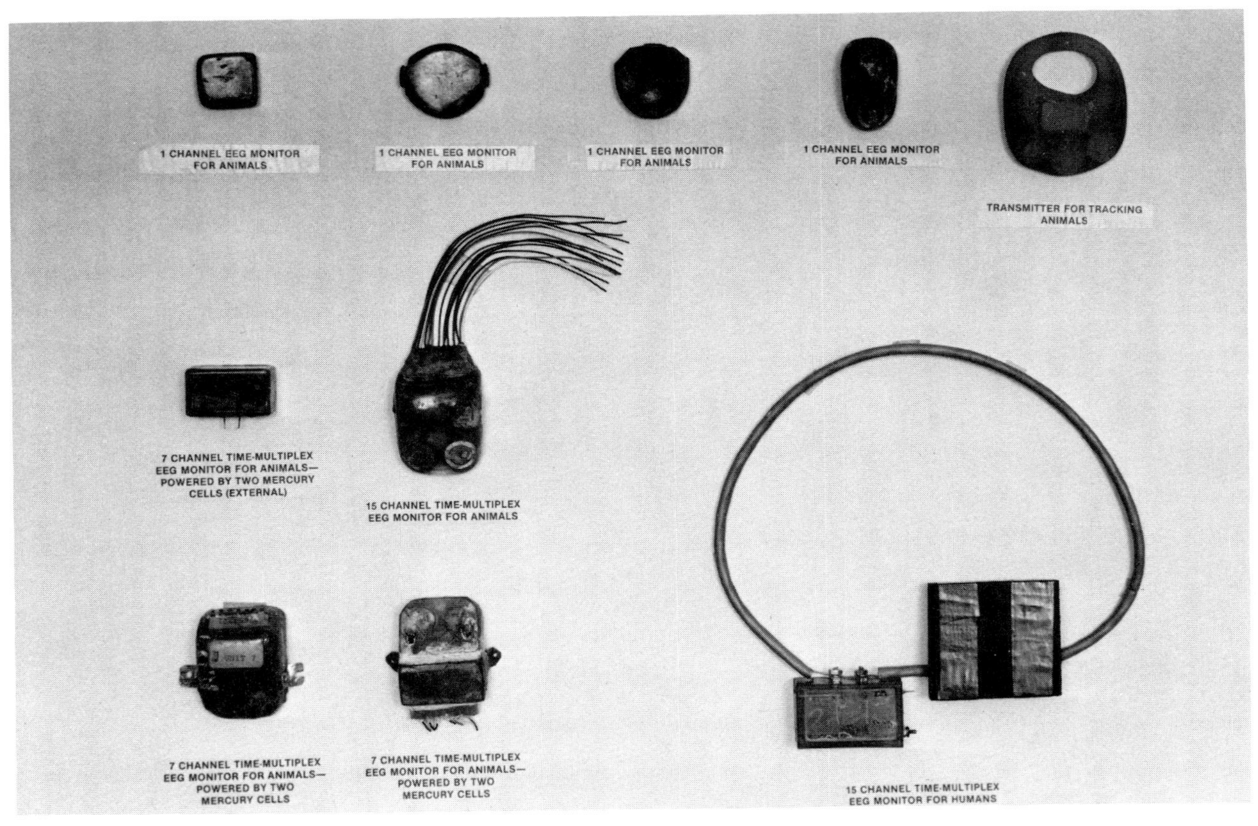

FIGURE 8. Telemetering Devices.

it possible to design low-noise implant units that operate on a single 1.35-V mercury cell (or any other cell of appropriate voltage). Designs feature transcutaneous radiofrequency turn-on and magnetic turn-off in order to conserve battery power. The signal-to-noise ratio of the system is high because amplification yields a relatively high degree of frequency modulation of the RF carrier. The advantage of a totally implanted unit, of course, is that the animal can be completely ambulatory, with only a small protuberance to reveal the presence of the transmitter.

A typical unit, including a battery that can operate 175 h before replacement, is 1 × 2.5 × 4 cm in size and weighs 15 g. Although it was designed for the electroencephalographic study of socially interacting epileptic monkeys, it can be used with any animal that can carry 15 g.

The implantable units shown are designed to handle the signals picked up by a relatively gross extracellular electrode that is embedded in brain or heart tissue, up to 2 mV peak. This is larger, by an order of magnitude, than the signals picked up by relatively remote surface electrodes. For intracellular work, where action potentials can have a peak value of some 100 mV, one can use a 50-to-1 input voltage divider; this also improves the high-frequency response.

Ideally, the physiological signal bandwidth should extend from dc to 100 Hz for EEG and ECG recordings. Experience shows, however, that the input may contain a relatively large spurious dc component because of (1) "battery" action if the electrode and ground strips are made of dissimilar materials and (2) amplifier rectification of the local RF field picked up by a long EEG or ECG lead. The spurious dc component is almost completely removed by an input capacitor that blocks rectifier current. A sufficiently large value is used so as to get a nominal low-frequency cutoff of 0.2 Hz.

2.5.4. *Pain Control and Other Stimulators*

It has been established that the sensation of pain can be eliminated or suppressed by scrambling with electrical message from the pain site or other pathway with an added electrical signal. Present work and commercial devices are in the field of transcutaneous electrical nerve stimulators (TENS), which supply the signal across intact skin as well as implantable devices similar to those discussed in the previous section on CNS stimulators.

In the research, development, or clinical trial period of other types of stimulators are implanted hearing aids, heart assist devices, artificial larynx, as well as a variety of sensors. The development of new generations of automatic heart stimulators and drug dispensers, however, depends on the perfection of small, reliable, implantable sensors that will directly measure physiological parameters as part of the control mechanism.

3. BUSINESS ASPECTS

The annual value of implantable battery-powered electronic medical devices was approximately $2 billion on a worldwide basis in 1984. This is especially remarkable considering that the key components of the devices, hermetically sealed batteries and sealed CMOS electronic packages, were available for only a decade. Yet this represents only a part of the total nonelectronic implantable business, such as sutures, valves, and knee and hip joints, although many of the electrochemical fundamentals are the same.

At this time more than half the total business of battery-powered implanted devices is represented by the pacemaker industry. The special nature of reliability, regulation (FDA), technology, and marketing aspects of the pacemakers has led to the devlopment of specialized companies for such implantable products. These in general are not the major electronic companies. The same logic is true for the battery components of implantable devices. Tables 8 and 9 list the major manufacturers of these devices. The implantable artificial heart pacemaker industry grew rapidly during the first two decades of its development. However, beginning in the early 1980s the rate of growth of units lessened, although the total patient population increased. This is accounted for by the dramatic increase in device life, from an average of 2–3 years in the 1960s to an average useful life of 5–10 years in the 1980s.

The other devices, such as the implantable defibrillator, which are outgrowths of the pacer technology, are still in the early stages of technical and commercial development and do not represent a substantial market at this time. However, by the 1990s they may represent a market of considerable size, particularly since they will probably have a much greater unit cost than the heart pacer.

TABLE 8 Major Manufacturers of Implanted Heart Pacemakers

Biotronik GmbH, West Germany
Cardiac Pacemakers, Inc., St. Paul, MN (Division of Eli Lilly)
Coratomic, Indiana, PA
Cordis Corp., Miami, FL
Ela Medical, France
Intermedics, Inc., Freeport, TX
Medtronic, Inc., Minneapolis, MN
Pacesetter Systems Inc., Sylmar, CA
Siemens-Elema, West Germany
Telectronics Ltd., Australia

Source: From Escher.[6]

TABLE 9 Manufacturers of Implantable Grade Batteries

Altus Corp., San Jose, CA
Battery Engineering Co., Inc., Hyde Park, MA
Cardiac Pacemaker Inc., St. Paul, MN
Catalyst Research Corp., Baltimore, MD
Cordis Corp., Miami, FL
Duracell Inc., Tarrytown, NY
Energy Technology Division, Medtronic, Inc., Brooklyn Center, MN
General Electric Co., Gainsville, FL
General Telephone & Electronics Inc., Waltham, MA
Wilson Greatbatch Ltd., Clarence, NY
Honeywell Inc., Ft. Washington, PA
Ray-O-Vac Corp., Madison, WI
SAFT, Paris, France and Valdosta, GA
Union Carbide Corp., Cleveland, OH

The power sources (batteries) represent costs equivalent to several percent of the total cost of typical pacemakers or other implantable devices. This substantial market for such special batteries is sufficient to justify the special designs to insure maximum energy density and reliability.

4. FUTURE DIRECTIONS

The future is probably unlimited for the development of additional electrically driven implantable devices. The implantable cardiac pacemaker and the implantable cardiac defibrillator have probably just touched the tip of the iceberg when it comes to the control and prevention of many cardiac arrhythimas. The artificial heart will continue to proceed in its development for a long-duration clinically application device. The holdup will be the development of a proper power source to drive the necessary hydraulics of the pump. Just as very crude developments preceded the successful insertion of an implantable device, more breakthroughs in the future should result in the development of a proper power source. Temporary devices to permit the hard-to-heal following a myocardial infarction will probably also be developed and function in much the way that a temporary cardiac pacemaker does today. Implantable devices, though somewhat crude, are available today for the continuous administration of medication to many individuals. This may range from the insulin pumps that are mounted externally on the body to pumps, such as the Infusaid pump, that can be implanted in the abdominal cavity for delivery of appropriate doses of medication at an appropriate time. Further progress will be in the areas of automatic feedback mechanisms to permit automatic infusion of these agents. Early experiments are in progress for the development of techniques of artificial sight for those individuals whose blindness is not of cortical origin. Blood electrodes implanted in the visual areas of the brain have already permitted sensing of light versus dark and may one day permit true artificial vision. Much is being performed to permit paraplegics to regain full function of their limbs. Again, this work is in its infancy and will require the development of techniques of graded stimulation of skeletal muscle as well as the nerves leading to it. The technology would appear to be available to solve this problem; all that should be required is further investigation in the area. Again, once the physiology has been understood and duplicated, the requirement will develop for miniaturization and implantability of the currently elaborate computers and the stimulating devices required to permit this today. Though the million-dollar man, with his artificial limbs and superhuman capacities, may represent the dreams of a TV scriptwriter, one would be hardpressed to believe that duplication and creation of artificial limbs having normal capabilities can occur sometime in the near future.

REFERENCES

1. E. Fukada and I. Yasuda, On the piezoelectricity of bone, *J. Phys. Soc. Jap. 12,* 1158–1162 (1957).
2. J. H. Brown, J. E. Jacobs, and L. Stark, Biomedical engineering in the United States, *Biomed. Eng. 6,* 405–408 (1971).
3. Committee on Technology and Health Care System, *Medical Technology and the Health Care System,* National Academy of Sciences, Washington, D.C. (1979).
4. P. N. Sawyer, Application of electrochemical techniques to the solution of problems in medicine, *J. Electrochem. Soc. 125,* 419C–436C (1978).
5. R. D. Gold, Cardiac pacing—from then to now, *Med. Instrum. 18,* 15–21 (1984).
6. D. J. W. Escher, Pacemakers of the 1980's, *Med. Instrum. 18,* 29–33 (1984).
7. S. Ekestrom, L. Johansson, and H. Lagergren, Behandling av Adams–Stokes syndrom med en intracardiell pacemaker elektro, *Opusc. Med. 7,* 1–3 (1962).
8. K. Brennen and B. Owens, in *Medical Batteries: Lithium Battery Technology* (H. V. Venkatasetty, ed.), pp. 139–158, Wiley, New York (1984).
9. S. Furman and R. Whitman, Cardiac pacing and pacemakers IX: Statistical analysis of pacemaker data, *Am. Heart J. 95,* 115–125 (1978).
10. K. Stokes and N. L. Stephenson, The implantable cardiac pacing lead—just a simple wire? in: *The Third Decade of Cardiac Pacing* (S. Barold and J. Mugica, eds.), pp. 365–416, Futura, Mount Kisco, New York (1982).
11. P. Citron and E. A. Dickhudt, Endocardial electrode, U.S. Patent No. 3902501 (1975).
12. H. Mond and G. Sloman, The small-tined pacemaker lead—absence of dislodgment, *PACE 3,* 171–177 (1980).
13. I. Babotai and W. E. Meier, Erste klinische ehrfahrungen mit der neuen intrakardialen electrode helifix, *Schweiz Med. Wochenschr. 107,* 1592–1593 (1977).
14. H. D. Dahl, H. Lubbing, D. W. Behrenbeck, B. Schorn, and H. Dalichau, Clinical experiences with electrodes for endocardial implantation with helically coiled tips, *PACE 4,* A-39, Abst. (1981).
15. D. C. MacGregor, G. J. Wilson, W. Lixfeld, R. M. Pilliar, J. D. Bobyn, M. D. Silver, S. Smardon, and S. L. Miller, The porous-surfaced electrode: A new concept in pacemaker lead design, *J. Thorac. Cardiovasc. Surg. 78,* 281–291 (1979).

16. A. G. Wallace, Electrical activity of the heart, in: *The Heart, Arteries and Veins,* 5th Ed., (J. W. Hurst, ed.), pp. 115–127, McGraw-Hill, New York (1983).
17. W. Irnich, Engineering concepts of pacemaker electrodes in: *Advances in Pacemaker Technology* (M. Schaldach and S. Furman, eds.), pp. 241–272, Springer-Verlag, New York (1975).
18. S. Furman, P. Hurzeler, and B. Parker, Clinical thresholds of endocardial cardiac stimulation: A long term study, *J. Surg. Res. 19,* 149–155 (1975).
19. L. Lapicque, Definition experimentale de l'excitabilite, *C. R. Soc. Biol. 67,* 280–283 (1909).
20. G. Weiss, Sur la possibilité de rendre comparable entre eux les appareils cervant à l'excitation électrique, *Arch. Ital. Biol. 35,* 413–446 (1901).
21. E. Sowton and J. Norman, Variation in cardiac stimulation threshold in patients with pacing electrodes, in *Digest, 7th International Conference on Medical and Biological Engineering, Stockholm, 1967* (B. Jacobson, ed.), pp. 74, Stockholm, Sweden (1967).
22. T. A. Preston, R. D. Judge, B. R. Lucchesi, and D. L. Bowers, Myocardial threshold in patients with artificial pacemakers, *Am. J. Cardiol. 18,* 83–89 (1966).
23. W. Irnich, The chronaxie time and its practical importance, *PACE 3,* 292–301 (1980).
24. A. Ripart and J. Mugica, Electrode–heart interface: Definition of the ideal electrode, *PACE 6,* 410–421 (1983).
25. S. S. Barold, L. S. Ong, and R. A. Heinle, Stimulation and sensing thresholds for cariac pacing: Electrophysiologic and technical aspects, *Prog. Cardiovasc. Dis. 24,* 1–24 (1981).
26. D. C. Amundson, W. McArthur, and M. Mostafa, The porous endocardial electrode, *PACE 2,* 40–50 (1979).
27. K. Breivik, O. J. Ohm, and H. Engedal, Acute and chronic pulse-width thresholds in solid versus porous tip electrodes, *PACE 5,* 650–655 (1982).
28. N. D. Berman, S. E. Dickson, and I. M. Lipton, Acute and chronic clinical performance comparison of a porous and solid electrode design, *PACE 5,* 67–71 (1982).
29. G. J. Richter, E. Weidlich, F. V. Sturm, E. David, G. Brandt, M. Elmqvist, and A. Thoren, Nonpolarizable vitreous carbon pacing electrodes in animal experiments, in: *Proceedings of the VI World Symposium on Cardiac Pacing, Montreal 1979* (C. Merre, ed.), Pacesymp. Chapter 29, pp. 13, Montreal (1979).
30. H. Elmqvist, H. Schueller, and G. Richter, The carbon tip electrode, *PACE 6,* 436–439 (1983).
31. M. P. Kleinert, H. R. Bartsch, and K. G. Muhlenpfordt, Comparative studies of ventricular and atrial stimulation thresholds of carbon-tip electrodes, *PACE 6,* A-64, Abst. (1983).
32. G. A. Bornzin, K. B. Stokes, and W. A. Wiebusch, A low threshold, low polarization platinized endocardial electrode, *PACE 6,* A-70, Abst. (1983).
33. F. Heinemann, M. Davis, and J. Helland, Clinical performance of a pacing lead with a platinized target tip electrode, *PACE 7,* 471, Abst. (1984).
34. K. Breivik, P. I. Hoff, A. Tronstad, H. Eugedal, and O. J. Ohm, Promising new pacemaker lead, *PACE 7,* 465, Abst. (1984).
35. P. Chen, A. Salkind, S. Fich, and V. Parsonnet, A basic study of pacemaker electrodes, in *Proceedings, 30th Annual Conference on Engineering in Medicine and Biology, Los Angeles, CA, 1977,* pp. 133, AEMB, Bethesda, Maryland (1977).
36. P. Chen, K. Chatterjee, P. Katz, G. Myers, and V. Parsonnet, Effects of electrode size and location on pacemaker induced fibrillation in acute myocardial infarction, in: *Proceedings, 28th Annual Conference on Engineering in Medicine and Biology, New Orleans, LA, 1975,* pp. 87, AEMB, Chevy Chase, Maryland (1975).
37. P. C. Chen, *A Study of Factors Influencing Pacemaker's Stimulation Threshold, Fibrillation Threshold and R-Wave Detection,* PhD. thesis, Rutgers University (1977).
38. K. B. Stokes, G. A. Bornzin, and W. A. Wiebusch, A steroid-eluting, low threshold, low polarizing electrode, in: *Cardiac Pacing, Proceedings of the VIIth World Symposium on Cardiac Pacing, Vienna, 1983* (K. Steinbach, D. Glogar, A. Laszkovics, W. Scheibelhofer, and H. Weber, eds.), pp. 369–376, Steinkopff Verlag, Darmstadt (1983).

39. G. C. Timmis, S. Gordon, D. C. Westveer, J. R. Stewart, K. B. Stokes, and J. R. Helland, A new steroid-eluting, low threshold pacemaker lead, in: *Cardiac Pacing, Proceedings of the VIIth World Symposium on Cardiac Pacing, Vienna, 1983* (K. Steinbach, D. Glogar, D. Laszkovics, W. Scheibelhofer, and H. Weber, eds.), pp. 361–367, Steinkopff Verlag, Darmstadt (1983).
40. G. C. Timmis, V. Parsonnet, D. C. Westveer, J. Stewart, and S. Gordon, Late effects of a steroid-eluting porous titanium pacemaker lead electrode in man, *PACE 7*, 479, Abst. (1984).
41. I. Yasuda, The classic fundamental aspects of fracture treatment, *J. Kyoto Med. Soc. 4*, 395–406 (1953). Translation: *Clin. Orthop. 124*, 5–8 (1977).
42. Z. B. Friedenberg, E. T. Andrews, B. I. Smolenski, B. W. Pearl, and C. T. Brighton, Bone reaction to varying amounts of direct current, *Surg. Gynecol. Obstet. 131*, 894–899 (1970).
43. Z. B. Friedenberg, P. G. Roberts, N. H. Didizian, and C. T. Brighton, Stimulation of fracture healing by direct current in the rabbit fibula, *J. Bone Joint Surg. 53A*, 1400–1408 (1971).
44. L. S. Lavine, I. Lustrin, M. H. Shamos, R. A. Rinaldi, and A. R. Liboff, Electrical enhancement of bone healing, *Science 175*, 1118–1121 (1972).
45. L. S. Lavine, I. Lustrin, M. H. Shamos, Treatment of congenital pseudoarthrosis of the tibia with direct current, *Clin. Orthop. 124*, 69–74 (1977).
46. T. E. Jorgensen, Effect of electric current on the healing time of crural fractures, *Acta Orthop. Scand. 43*, 421–437 (1972).
47. T. E. Jorgensen, Electrical stimulation of human fracture healing by means of a slow pulsing asymetrical direct current, *Clin. Orthop. 124*, 124–127 (1977).
48. K. P. Srivastava and A. K. Saxena, Electrical stimulation in delayed union of long bones, *Acta Orthop. Scand. 48*, 561–565 (1977).
49. C. T. Brighton, S. Adler, J. Black, N. Itada, and Z. B. Friedenberg, Cathodic oxygen consumption and electrically induced osteogenesis, *Clin. Orthop. 107*, 277–282 (1975).
50. C. T. Brighton, Z. B. Friedenberg, E. I. Mitchell, and R. E. Booth, Treatment of non-union with constant direct current, *Clin. Orthop. 124*, 106–123 (1977).
51. D. C. Paterson, T. M. Hillier, R. F. Carter, J. Ludbrook, G. M. Maxwell, and J. P. Savage, Electrical bone-growth stimulation in an experimental mode of delayed union, *Lancet, 1*, 1278–1281, June 18 (1977).
52. J. F. Connolly, H. Hahn, and O. M. Jardon, The electrical enhancement of periosteal proliferation in normal and delayed fracture healing, *Clin. Orthop. 124*, 97–105 (1977).
53. C. R. Hassler, E. F. Rybicki, R. B. Diegle, and L. C. Clark, Studies of enhanced bone healing via electrical stimuli, *Clin. Orthop. 124*, 9–19 (1977).
54. W. H. Harris, B. J. L. Moyen, E. L. Thrasher, L. A. Davis, R. H. Cobden, D. A. MacKenzie, and J. K. Cywinski, Differential response to electrical stimulation, *Clin. Orthop. 124*, 31–40 (1977).
55. J. McKnight, Private communication to A. J. Salkind (1976).
56. C. A. L. Bassett, R. J. Pawluk, and A. A. Pilla, Acceleration of fraction repair by electromagnetic fields: A surgically noninvasive method, *Ann NY Acad Sci. 238*, 242–262 (1974).
57. C. A. L. Bassett, A. A. Pilla, R. J. Pawluk, A nonoperative salvage of surgically resistant pseudoarthrosis and nonunions by pulsing electromagnetic fields, *Clin. Orthop. 124*, 128–143 (1977).
58. Z. B. Friedenberg, M. C. Harlow, and C. T. Brighton, Healing of nonunion of the medial malleolus by means of direct current: A case report, *J. Trauma 11*, 883 (1971).
59. C. T. Brighton, R. D. Ray, L. W. Soble, and K. E. Kuettner, In vitro epiphyseal plate growth in various oxygen tensions, *J. Bone Joint Surg. 51A*, 1383–1396 (1969).
60. A. B. Borle, N. Nichols, and G. Nichols, Metabolic studies of bone in vitro: I. Normal bone, *J. Biol. Chem. 235*, 1206–1210 (1960).
61. D. S. Howell, J. C. Pita, J. F. Marquez, and J. E. Mandruga, Partition of calcium, phosphate and protein in the fluid phase aspirated at calcifying sites in epiphyseal cartilage, *J. Clin. Invest. 47*, 1121–1132 (1968).

62. Osteostim Implantable Bone Growth Stimulator Model S12—For Use with Long Bone Fusion, Telectronics Pty. Ltd., Commercial Literature Z Sirius Road, Lane Cove, N.S.W. 2066 Australia.
63. Osteostim-implantable Bone Growth Stimulators—A Summary of Clinical Results, Product Literature, Osteostim Division, Telectronics Pty. Ltd., 8515 E. Orchard Road, Englewood, Colorado 80111, May (1981).
64. D. C. Paterson, G. N. Lewis, and C. A. Cass, Treatment of delayed union and nonunion with an implanted direct current stimulator, *Clin. Orthop. 148,* 117–128 (1980).
65. K. D. Nielson, J. E. Adams, and Y. Hosobuchi, Phantom limb pain, treatment with dorsal column stimulation, *J. Neurosurg. 42,* 301–307 (1975).
66. W. H. Sweet and J. G. Wepsic, Stimulation of the posterior columns of the spinal cord for pain control: Indications, technique and results, *Clin. Neurosurg. 21,* 278–310 (1974).
67. L. J. Seligman, P. P. Tarjan, and R. Davis, A totally implantable pulse generator for the CNS, in: *Proceedings, 31st Annual Conference on Engineering in Medicine and Biology, Atlanta, GA, 1978,* pp. 270, AEMB, Bethesda, Maryland (1978).
68. A. J. Matas, D. E. R. Sutherland, and J. S. Najarian, Current status of islet and pancreas transplantation in diabetes, *Diabetes 25,* 785–795 (1976).
69. J. V. Santiago, A. H. Clemens, W. L. Clarke, and D. M. Kipnis, Closed-loop and open-loop devices for blood glucose control in normal and diabetic subjects, *Diabetes 28,* 71–84 (1979).
70. J. S. Soeldner, K. W. Chang, S. Aisenberg, and J. M. Hiebert, Progress towards an implantable glucose sensor and an artificial beta cell, *Temporal Aspects of Therapeutics* (J. Urquhart and F. E. Yates, eds.), Plenum, New York, pp. 181–207 (1973).
71. S. Deutsch, An implanted telemetry unit for ambulatory animals, *IEEE Trans. Commun.* com-23, 983–987 (1975).
72. S. Deutsch and J. W. Mackenzie, Time-multiplex telemetry of seven intensive care parameters, in: *Proceedings, 31st Annual Conference on Engineering in Medicine and Biology, Atlanta, GA, 1978,* pp. 282, AEMB, Bethesda, Maryland (1978).

2

Key Events in the Evolution of Implantable Pacemaker Batteries

BOONE B. OWENS and ALVIN J. SALKIND

1. INTRODUCTION

The successful development of implantable cardiac pacemakers has been facilitated by the merging of the relevant technologies in electronics and power sources. The mercury/zinc oxide battery was a key component in the initial development of implantable pulse generators and that battery system is closely associated with the name of Dr. Samuel Ruben (Fig. 1). Subsequently, the lithium/iodine battery was developed for hermetically sealed, longer-lived pulse generators, and this development has been closely associated with the name of Dr. Wilson Greatbatch (Fig. 2). Both of these men are recognized as outstanding inventors and have been creative in a number of technologies.

With both Drs. Ruben and Greatbatch the innovations in specialized power sources resulted from identified needs for improved batteries. These inventors were able to assess the existing technology and make improvements and innovation as required to achieve a product that satisfied the need. The present chapter records interviews that occurred in January 1983; the purpose was to explore

BOONE B. OWENS • Department of Chemical Engineering and Materials Science, University of Minnesota, Minneapolis, Minnesota 55455. ALVIN J. SALKIND • Department of Surgery, Bioengineering Section, UMDNJ–Rutgers Medical School, Piscataway, New Jersey 08854.

FIGURE 1. Dr. Samuel Ruben.

these individuals' thoughts on the subject of creativity in general and their experiences in being so innovative in the area of power sources.

2. AN INTERVIEW WITH SAMUEL RUBEN[1]

A. SALKIND: Dr. Ruben, what was your background and training that led to your interest in high-energy-density and long-shelf-life batteries?

S. RUBEN: I obtained my first job with the Electrochemical Products Company in New York City. It was a small organization that was attempting to produce nitrous oxide from the air by high-frequency, high-density capacitor discharge. I was a licensed radio "ham" and also had a chemical lab at home. Because I had some background and knowledge with high-frequency techniques from my amateur radio hobby, I was selected for this job as a researcher with the Electrochemical Products Company over the other applicants.

A. SALKIND: You have mentioned to me in the past that you owe much of your formal training in electrochemistry to Professor Bergen Davis of Columbia University. How did this come about?

S. RUBEN: Professor Bergen Davis, who had worked with J. J. Thompson at the turn of the century, became a consultant to Electrochemical Products Company and his background was in electrical discharge through gasses. The work was moved from the original laboratory in Brooklyn to a laboratory on the Columbia campus. Professor Davis took a personal interest in me and arranged for special lectures at Columbia University, where I was able to sit in on colloquia and formal courses. He spent evenings with me and assigned readings. He would then interrogate me to make sure that I understood and carried out the assignments. Over the years I attended evening noncredit college-level lectures.

A. SALKIND: What happened after Electrochemical Products Company?

S. RUBEN: At the end of World War I, the Haber Process from Germany became known and the electrical discharge process could not compete with it. Bergen Davis got the president of Electrochemical Products Company, Mr. Malcolm W. Clephane, to sponsor me and to carry out research on my idea for electron bombardment relays. This was the beginning of Ruben Laboratories.

A. SALKIND: More than 100 independent patents have been issued to you personally. Of these, which do you believe is the most significant?

S. RUBEN: The dry electrolytic aluminum capacitor, the solid-state magnesium/cupric sulfide rectifier, the vacuum tube relay, the quick heater vacuum tube, and the concept of a balanced-cell mercury battery.

A. SALKIND: What led you to the research on mercury batteries and how did it proceed? (cf. Ref. 1,2)

S. RUBEN: The need for a high-temperature stable power source, particularly for electronic and military devices that could stand high-temperature storage and would deliver a constant potential under load. My previous work on the dry electrolytic capacitor, which utilized an immobile glycoborate electrolyte in a wound electrode and spacer structure and the solid-state rectifier (Mg–CuS), led to the work on the sealed mercury–zinc cell. In the early 1930s I was searching for a better C-bias cell for radio tubes. The small commercial LeClanche cells at the time had very poor shelf life. The first bias cell in commercial production had a vanadium pentoxide resin-impregnated pellet as the positive electrode, a glycoborate immobilized electrolyte, and a cadmium negative electrode. It would deliver one volt for many years. For somewhat higher voltages, zinc was substituted as the negative electrode. I wanted an immobilized alkaline electrolyte rather than a LeClanche type (NH_4Cl–$ZnCl_2$) for high rate performance. My early cells had a cathode of sand-blasted steel, strip-coated with a cupric oxide–graphite mixture with a porous plastic binder. This coated strip was rolled to provide minimum impedance and corrugated to allow for expansion space for end products and at the same time provided increased area for maximum current density. The wound structure was similar to my earlier magnesium cells and dry electrolytic capacitors. However, the capacity and voltage of the wound structure using copper oxide was not satisfactory for military applications and in the early 1940s I turned to the use of mercuric oxide positives. The coulombic capacity of the cell was increased by the use of consolidated pellets of the positive electrode and a zinc anode of large effective area having a barrier between the cathode and the cellulose absorbent containing potassium hydroxide saturated with potassium zincate.

A. SALKIND: What were the most significant events in your innovative work with mercury power sources?

S. RUBEN: The concept of the balanced cell between anode and cathode coulombic capacities, to eliminate generation of hydrogen when the cathode was completely reduced.

A number of other factors were also found necessary to produce a sealed zinc/mercuric oxide alkaline cell (grommet sealed). These included: (a) amalgamated anodes, (b) high surface area anodes to reduce polarization, and (c) alkali-resistant and absorptive barriers to eliminate bridging paths between the electrodes.

A. SALKIND: Do you have a guiding philosophy toward technical innovation?

S. RUBEN [paraphased from Refs. 3–5]: Regardless of the psychological analyses of the cause and motivation of invention, the basic fact remains that necessity is the mother of invention. It is the catalyzer for the primary concept and the drive for development of the imaginative concept into a useful end product. The process of translation of the imaginative concept in today's technology requires understanding it in relation to the existing science and technology. This is materially facilitated by the accumulation of as broad a mental tool storage of the facts of technology as possible, by continuous study and experience. The most important mental tool I have worked with was the result of a very early recognition of the value of the science of materials, or more specifically, the periodic classification of the elements and their relative properties, with particular reference to the electronic construction of matter so that the difference between physics and chemistry becomes an energy relation. That this is not a hindsight observation may be noted in my rectifier patents of 1925, in which I classify the suitable materials in relation to their valence position in the periodic table of elements.

Inventors today are recognized by society as a positive factor in the progress of our civilization. This was not always so, for the inventors who initiated the beginning of the industrial revolution were treated with suspicion and hostility, often by the very people who in the long run benefited by their efforts. Today the climate is different and sometimes the inventor is afforded recognition and honors, particularly if the product of his efforts results in widespread acceptance. The growth of industries is dependent upon invention and innovation, for one reads in company statements that some of their major revenue producers were not in existence 20 years ago.

We have come to take inventions for granted, and the beginnings of many have been forgotten in the dimmed vision of time. The progeny of the wheel, namely, the vehicles, are hardly thought of as inventions. Today the younger generation takes such a popular utility as the television set or other modern apparatus as something related to advertising programs and not as an application of imaginative thought in the use of modern technology.

The modern technological inventions of today are in a different class from those earlier days of our patent system on account of the tremendous expansion in science and technology, with accompanying increase in the diffusion of knowledge. They tend to become a more sophisticated project in order to properly integrate the imaginative concept to present technology.

A. SALKIND: What qualities do you feel separate innovative from noninnovative people?

S. RUBEN: Self-motivation is a necessary factor to catalyze the generation of imaginative concepts. The innovator is motivated by the intellectual excitement of the thought process and will resist adverse premature opinions of others who lack imaginative thinking.

In general, progress has always required creative thinking and a determined effort in the pursuit of bold dreams, however imaginative. Goals become reality only when ideas are complemented by working technology.

The most important factor that an imaginative individual is endowed with is an inner sense of direction. The effectiveness of his inner sense of direction requires the understanding of a concept in relation to the existing science and technology. This involves the planning and practical embodiment of the imagined concept and persistence in order to bring it to effective realization as a real-world invention.

For inventors, the nature of the time and state of technology are most important. If an inventor is ahead of the general state of the art, it may well happen that his innovation will not be developed or put to large-scale commercial use until after the patent has expired. I have experienced the advantage of development at a time when a recognized need existed, but I have also seen some of my inventions go into large-volume use after my patents had run out.

A. SALKIND: Looking to the future, do you foresee other significant changes or developments in power sources for implantable devices and other developments?

S. RUBEN: There are several organic and nonalkaline electrolyte battery systems that may develop in the future for special applications. These include the zinc/mercuric dioxysulfate cell[7] in zinc sulfate electrolyte. This system might make better pacer cells or be suitable for devices with higher current demands. It has an extremely long shelf life for very-low-power devices. Solid-state cells may also be commercially important. In organic electrolyte batteries, calcium and calcium silicides may be important anode materials.

For the past 55 years I have been engaged in research and invention in electric batteries and allied fields, and I believe I can convey the importance of chemical inventions in response to industrial needs.

The needs for new industry can be seen as motivating inventors toward conceiving and developing new technologies. Industrial need creates an inspiring and encouraging atmosphere, conducive to invention, and the same need provides a market when the conceptual work is translated into a working device or system. New battery systems have always evolved in response to the needs of electric devices.

Comments by Alvin J. Salkind

Samuel Ruben is an extremely motivated individual with a strong driving force to obtain knowledge. His search of knowledge and broad basic thinking, starting with atomic structure, has led to inventions that have affected all of our lives. Dr. Emanuel Piore[(5)], Vice President of IBM, said: "Dr. Ruben's inventions are of comparable importance to our civilization . . . , as the steam engine, the telephone, and the airplane, yet are not known to the public at large and Dr. Ruben's name is known only to his scientific peers, rather than to the general public. The reason for this is that his inventions, which always made certain things possible as practical devices, are buried in the black boxes. One can almost say that the radio as we know it today would not have been possible without his invention [the dry electrolytic capacitor]."

Samuel Ruben continues to pursue new ideas and fields. Among his latest inventions, U.S. Pat. 4,297,232 (1981) is a new method of complexing iodine for use as an antiseptic. This is now being sold in veterinary medicine and leaves no stain.

For his research and development work in electrochemistry and electronics, Samuel Ruben holds the honorary degrees of Doctor of Science from Butler University, Indianapolis; Doctor of Engineering, *honoris causa,* from Polytechnic Institute of Brooklyn; Doctor of Science, *honoris causa,* Columbia University, N.Y.; Medal and Certificate of Inventor of the Year, 1965, from George Washington University; Golden Plate Award from the American Academy of Achievement; the Acheson Gold Medal of the Electromechanical Society; the Longstreth Medal and Life Fellowship of the Franklin Institute; Fellow of the American Association for the Advancement of Science; Fellow of American Institute of Chemists; Fellow of the Polytechnic Institute of New York; member of Columbia University Engineering Council; Honorary Professor, Polytechnic Institute of New York; and Senior Staff Associate, Columbia University Department of Chemical Engineering and Applied Science. In 1982 Columbia University established the Samuel Ruben–Peter Viele Chair in Electrochemistry. In 1985 he became an Adjunct Professor at Reed College.

He has contributed papers to the Electrochemical Society, the New York Academy of Sciences, A.I.E.E., and C.I.T.C.E. He is the author of four books—*Handbook of the Elements, Electronics of Materials, The Founders of Electrochemistry* (Dorrance & Co., 1975), and *The Evolution of Electric Batteries in Response to Industrial Needs* (Dorrance & Co., 1978), and contributing author of "The Mercuric Oxide Cell" in *The Primary Battery,* pp. 207–227 (John Wiley & Sons, New York, 1971), and "Sealed Zinc/Mercury Cells" in the 10th Edition of *Standard Handbook for Electrical Engineers,* section 24, pp. 33–40 (McGraw-Hill, 1968); contributing author of "Primary Batteries-Sealed Mercurial Cathode Dry Cell," section 7, pp. 223–245, *Comprehensive Treatise of Electrochemistry,* Vol. 3 (Plenum Press, New York and London, 1981).

FIGURE 2. Dr. Wilson Greatbatch.

3. AN INTERVIEW WITH WILSON GREATBATCH[7]

The first time a lithium-battery-powered medical device was implanted into a human body was in Ferrara, Italy, in March of 1972.[8] The battery was a Wilson Greatbatch model 702-C Li/I_2 cell. The man whose name is associated with this first cell is an inventor who is internationally respected as one of the key innovators and leaders in cardiac pacemaker technology.

B. OWENS: Bill, your formal training was in electronics, and naturally enough the early phase of your inventiveness in biomedical technology appears to have been directed toward pacemaker circuits. How did you, with no specific training or background in power sources technology, come to identify key concepts in this field?

W. GREATBATCH: Fortunately, one thing the Lord doesn't require of you to do His work is that you know what you're doing! My initiation to power source technology was forced, informal, and hurried. Our first pacemakers in 1958–1960 immediately encountered battery problems. The mercury-zinc Ruben cell could not survive the years of exposure to a 37°C, 100% humidity environment at a low current drain. I was quickly forced to learn how to disassemble and analyze failed cells and to try to talk like a chemist to the Mallory battery people at Tarrytown to try to get them to modify the design. After five years of futile effort we began to seek out new systems. I then had to familiarize myself with nuclear decay systems such as plutonium and promethium and conversion devices such as photovoltaic targets and semiconductor thermopiles. Two years and some $200,000 later, we abandoned nuclear batteries because of regulatory requirements and made a total commitment to the lithium–iodine system.

Thus, our introduction to power source technology was neither planned nor desired. It was an absolute necessity if we were to achieve a satisfactory pacemaker. I credit very broad interdisciplinary training at Cornell University with my being able to subsequently swing with the problems and the times. Thus, we were able to make a few right decisions when the chips were really down.

B. OWENS: Then is your experience in this area of getting into pacemaker batteries in agreement with the old adage, "Necessity is the mother of invention?"

W. GREATBATCH: Yes, very definitely. It has always been our experience that if you can identify a real need and can satisfy that need in an eminently satisfactory way, you don't really have to worry about the cost. If the need was real, people will buy the device.

B. OWENS: For the record, how many different areas of inventions and numbers of patents have you come up with?

W. GREATBATCH: I have been granted over 100 U.S. and foreign patents, either solely or jointly with others. They are in areas of analog circuit design, digital circuit design, nuclear batteries, chemical batteries, electronic control of growth and infection, and more recently in the field of renewable biomass energy.

B. OWENS: What was the most significant event in your innovative work with pacemaker power sources?

W. GREATBATCH: I think there were two. One was the collaboration with Jim Moser and Al Schneider in 1968–1970 that resulted in our first 702C lithium–iodine battery. The other was our acquisition of the assets and personnel of the Wurlitzer Company's battery division in the early 1970s. This brought us Ralph Mead and Frank Rudolph, who got us into the battery business. Both are still with us today and still just as obstreperous!

B. OWENS: If I recall correctly, each of these individuals had some experience with thermal batteries. Was that technology particularly a key building block in the design of the lithium–iodine pacemaker cell?

W. GREATBATCH: Yes, I think it was. The lithium–iodine cell is based on lithium technology, whereas thermal batteries are based on the reactive materials technology. Many of the same considerations are necessary in building each. For example, you have to maintain a very-low-humidity dry room for the manufacture of the cells; you have to be very careful of fire and other constraints. The background that these people had with thermal batteries, I think, led directly into the technology that was needed in lithium–iodine batteries.

B. OWENS: I suppose one other similarity between the thermal and the pacemaker batteries is that they both must exhibit a very high reliability.

W. GREATBATCH: That's certainly true.

B. OWENS: Have you a guiding philosophy with regard to technical innovation?

W. GREATBATCH: First, I never look for a problem to solve. Rather, I take something I can do very well and find a place to use it. For example, two great needs today are an implantable sugar sensor and an automatic blood pressure measuring device. Many professional careers and many millions of dollars have been wasted away looking for solutions to them. That's not for me! In contrast, one thing I can do very well is to generate pulses, in sonar, in radar, in computers, and so on. It is really not a large step to use such a pulse to drive a heart. Yet this small step created a billion-dollar industry.

I do not accept government support for any of my research. We pay our own way and call our own shots. Of course this means that we have to scrimp and starve, but it gives us a truly beautiful freedom and a very fast reaction time when we come across something really exciting.

I always like to put a lot of safety factor into a new innovative design. In 1958 we had no idea what long-term cardiac threshold would be. Thus, we put ten cells into our pacemakers. They delivered a 13.5-V pulse! Later we dropped to as low as four cells. It's always easier to take something out later than to wish you had a little more.

B. OWENS: What qualities do you feel separate an "innovator" from a "noninnovator"?

W. GREATBATCH: Mostly curiosity, but also a rebellious spirit. I think the true innovator is bored and unproductive when things are going well. He really doesn't get interested until everyone agrees that the project is impossible.

B. OWENS: Regards the first implant, how did you decide when the cell was ready to go?

W. GREATBATCH: When our accumulated data proved that our implant life would be better than what we were seeing with zinc-mercury.

B. OWENS: In March of 1972 mercury–zinc-powered pacemakers were lasting about one to two years. At that time did you have real-time data for lithium–iodine batteries that was in excess of this or was some extrapolation of performance data required when the decision was made to go ahead and implant this very new lithium battery technology?

W. GREATBATCH: In 1972 the average life of a mercury–zinc-pacemaker was about 24 months. At that time we didn't have real-time data for lithium–iodine cells out that far, but we did have sufficient confidence in the system and a sufficient mathematical background in statistical analysis that we were able to extrapolate the data and be confident that we had a superior battery. Time has proved that was so.

B. OWENS: Why was this first unit implanted in Italy?

W. GREATBATCH: We did not know of the implant until later. We had not yet cleared the cell for clinical use, but this particular user decided from his own data that the cell was ready. The pacemakers were explanted within a year because of a circuit problem but there were no battery failures.

B. OWENS: The lithium–iodine battery has been used widely in pacemakers during the past 10 years. In 1972 did you expect such wide usage when nuclear and rechargeable power sources were also being utilized?

W. GREATBATCH: Yes. I knew in 1970 when I saw the Moser–Schneider prototypes that lithium–iodine was the answer. We soon decided that the medical profession would never accept the Nuclear Regulatory Commission requirements on the isotopic battery. Then, as in 1958 (and as now), a rechargeable cell, with recharging, had less life than a primary cell without recharging. Neither the rechargeable cell nor the isotopic battery ever achieved 1% of the pacemaker market.

B. OWENS: The deliverable energy density of a lithium–iodine pacemaker cell has been increased nearly five-fold during the past 10 years. Present cells can deliver 0.8 to 1.0 Wh/cm^3 as compared with about 0.2 for the first cells. What key factor do you identify in this significant size reduction of these cells?

W. GREATBATCH: Two factors; the increase in the iodine/polymer ratio and the gradual elimination of much nonactive material such as insulation, caps, and retainers. A good example of the latter is the transition from an insulated, floating steel case to a grounded case construction. No attempt was made to make the early cells efficient. All the effort went into making them reliable. Even so, the first lithium pacemakers had half the weight and twice the life of their zinc-mercury counterparts. Some 1972 cells are still working in Australia today, powering early pacemakers.

B. OWENS: As of the present time, January of 1983, is it accurate to state that some of those very early implanted lithium–iodine-powered pacemakers have been in continuous operation for over 10 years?

W. GREATBATCH: I'll have to check that out with the records. I know that very recently there were some 702C-powered pacemakers still in service in Australia that had been implanted in 1972 and I know of one—I believe, 702E-powered—pacemaker being carried by a technician at CPI in Minneapolis that was implanted in 1973. So those both would be over 10 years, but we'll have to check that out.

B. OWENS: That certainly does give confirmation to the decision to implant these batteries back in the very early days of lithium battery technology.

B. OWENS: Is there a practical device capable of converting the body's own inherent energy into electrical power for artificial pacemakers?

W. GREATBATCH: No. For example, the fuel cell and galvanic designs of Roy and of Cywinski and Hahn have made much progress but have not approached the reliability and energy efficiency of conventional batteries. I do not see any future in this approach.

B. OWENS: Looking to the future, do you foresee other significant changes or developments in power sources for implantable devices?

W. GREATBATCH: Yes. Implantable devices of the future will need more current for short periods of time to operate telemetry, antiarrythmia pulse trains and to operate the programming circuits. Pacemakers currently in design have up to 100,000 transistors with all the computing capability of a conventional desk-top computer. This will probably require new battery systems with soluble cathodes, perhaps in bromine or sulfuryl chloride, or solid cathode systems such as silver vanadium pentoxide. We see, too, the need for autoclavable batteries, particularly in implantable drug delivery devices. Conventional ethylene oxide gas sterilization may well prove inadequate for such devices, forcing a return to steam sterilization.

B. OWENS: Pacemakers presently are not amenable to autoclaving because of possible battery degradation; yet these devices have proved to be extremely useful. Are you inferring that in the future autoclavability will be of more concern than it has been in the past, or is it just a continuing desire for the option of resterilization?

W. GREATBATCH: Ethylene oxide sterilization has never been the optimum, and not too long ago there was at least one movement to reduce the amount of time that technicians would be allowed to operate these sterilizers each day because of exposure to the gas. Ethylene oxide sterilizers must be very carefully maintained and very carefully operated to be safe. However, ethylene oxide sterilization of present pacemakers is probably quite satisfactory, and I see no change in that in the future. On the other hand, some devices such as implantable drug delivery devices are a little different. These devices have cavities inside them that must be sterilized. Ethylene oxide doesn't get into such cavities very easily, and if it does get in, it is very hard to get out. Ethylene oxide must be thoroughly flushed from all surfaces and all cavities before a device is implanted or it will result in necrosis of any tissue to which it comes in contact. So I feel that there is a real problem with ethylene oxide sterilization of implantable infusion pumps. I am chairing a session at AAMI in Dallas at which this will be one of the key questions to be discussed. It may be that methods can be worked out whereby these devices can be safely gas sterilized, but at the moment I think our position is that we need to think about using autoclavable batteries and autoclavable integrated circuits in any implantable drug infusion device so that the whole assembly can be autoclaved.

REFERENCES

1. Interview with Dr. Samuel Ruben by A. J. Salkind, Palm Springs, California, January 23, 1983.
2. S. U. Falk and A. J. Salkind, *Alkaline Storage Batteries,* pp. 32–34, Wiley, New York (1969).
3. Samuel Ruben, "Inventors and Society—Response to Industrial Needs," Columbia University, Armstrong Memorial Lecture, (1981).
4. Samuel Ruben, *Inventors and Society, Columbia Eng. Alumni Times, Vol. 23,* No. 1 (1982).
5. Henry B. Linford, *Samuel Ruben—Acheson Medalist J. Electrochem. Soc., 118,* 11c–13c (1971).
6. Samuel Ruben, U.S. Patent Capacitor (1925).
7. Interview with Dr. Wilson Greatbatch by Boone B. Owens, Clarence, New York. January 13, 1983.
8. G. Antonioli, F. Baggioni, F. Consiglio, G. Grassi, R. Lerrun, and F. Zanardi, Implantable cardiac pacemakers with a new solid state lithium battery, *Minerva Med. 64,* 2298–2304 (1973).

3

Lithium Primary Cells for Power Sources

DARREL UNTEREKER

1. INTRODUCTION

A battery is a chemical device that serves as a source of electrical power. Potential energy contained in the chemicals inside the battery is released by carrying out a chemical reaction in a controlled manner so electrical energy is produced rather than some other form of energy such as heat. Batteries are very important to all of our lives and are one of mankind's more important developments. They are used for thousands of applications where a portable source of power is advantageous or energy must be stored for future use. Batteries are economically important and millions are sold each month in a wide variety of types and configurations, ranging from very small flashlight cells to large automotive and load-leveling batteries. There are two major classifications of batteries. The first is *primary* cells or batteries and the second is *secondary* or *rechargeable* cells or batteries. The distinction between batteries and cells is that the cell is the fundamental building block of a battery. A battery may have one or more cells. Most often cells are combined in series to yield a battery with a higher voltage than is produced by an individual cell. Sometimes, however, cells are put in parallel to give higher capacity or current capability.

Rechargeable batteries can be discharged; then their chemical energy can be rejuvenated from an energy source external to the battery. They are rejuvenated or recharged by forcing a current to flow through the battery opposite to its normal direction. This causes the chemical reaction to occur in the reverse direction and eventually results in the battery reverting to its initial state, where-

---

DARREL UNTEREKER • Medtronic, Inc., Minneapolis, Minnesota 55440.

upon it can again be discharged. Certainly the most familiar application for a rechargeable battery is in the automobile. There the battery is used to start the car and maintain functions while the motor is off. The energy used to perform these tasks is supplied via the engine, which turns a generator or alternator to replenish the battery.

Primary batteries can be discharged only once. They are manufactured in their most effective state and supplied to the user who discharges them until they no longer meet the application's needs. They are then replaced with new batteries. The flashlight battery is probably the most familiar form of the primary battery.

However, while there are only three or four chemistries currently available in rechargeable batteries, there are at least a dozen chemical systems available in primary batteries. The main reason for this is the need for secondary batteries to recharge very efficiently. For example, even a battery that can recharge with 95% efficiency will have lost half its capacity after only 13 cycles if it started with stoichiometric amounts of chemicals in the battery. Changing the relative amounts of chemicals in the battery can compensate for inefficiencies of recharge, but this also lowers the amount of energy available in a given size of battery.

In general, primary batteries have much higher energy densities than secondary batteries because they can be optimized for a single discharge. Secondary batteries, on the other hand, are usually optimized to achieve a large number of cycles.

Since batteries provide a portable source of power, it was inevitable they should be used to power many medical devices, especially implantable ones, where there is a major benefit for the device plus its power supply to be packaged in a sealed container inside the body. The major use for implantable batteries over the last 20 years has been for cardiac pacemakers. During this time pacemakers have evolved from rather unsophisticated devices that simply put out an electrical pulse to state-of-art devices that are extremely sophisticated and provide physicians and patients with a multitude of features. During the early years, the mean longevity of pacemakers was on the order of a year. By contrast, the modern pacemaker may last as long as 10 years.

Implantable medical devices have unusual power requirements. These devices usually require low levels of power for long periods of time. Because of this, an implantable battery has an interesting set of constraints and latitudes. Operation over an extended temperature range is a major challenge for most batteries; however, because of the constant temperature inside the human body, implantable batteries need only operate at 37°C. In fact, the major constraints on medical batteries are in the areas of safety, reliability, and energy density. Since access is a major problem, these batteries must operate reliably and predictably for long periods of time. This means implantable battery systems must be chosen on the basis of consistent, predictable behavior and engineered to the highest quality standards. This, in turn, means such batteries must be manufactured to very exacting tolerances and undergo the scrutiny of very stringent quality-control procedures.

This chapter is intended to provide general knowledge about battery systems and particular information about the battery systems that have been used for implantable medical products. This chapter begins with a general discussion of how batteries operate. It then discusses battery chemistries that have been used for implantable applications. Microcalorimetry will also be discussed as a valuable tool for evaluating and understanding battery systems.

The discussions attempt to focus on implantable medical batteries, although examples of other batteries systems will be discussed when useful.

2. THE ELEMENTS OF A BATTERY

All batteries, whether they be flashlight cells, car batteries, or the implantable battery powering a heart pacemaker, operate because of the same fundamental physical chemical laws and processes. Thus, a set of terminology has evolved to describe the essential operations. To begin, a battery operates because of the chemical energy contained in the reacting entities in the battery. This energy becomes available because of a chemical reaction forming a reaction product. In common situations familiar to our senses we observe the energy from such chemical reactions in the form of heat. Probably the most familiar chemical reaction of this sort is combustion, where oxygen combines with a material to form a product and give off heat, often very rapidly. The chemistry occurring in a battery is very similar to the oxidation occurring during combustion and the amount of energy available is of a comparable magnitude. Thus, if the chemicals in a battery were simply reacting in the most expedient manner, a lot of heat would be produced. The secret of a battery, and its cleverness, lies in the way the chemical energy is transformed into electricity rather than heat.

To understand how this happens one must understand the nature of the chemical reactions that occur. In combustion the chemical process is called oxidation. During the course of the reaction the valence state of the reactant chemicals changes. As an example, let's consider a basic chemical reaction of this type, the reaction of elemental hydrogen and elemental oxygen to form their reaction product, water. The reactants, hydrogen and oxygen gas, are both elemental in nature and have valences of zero. In water, however, hydrogen has been changed to a $+1$ valence while oxygen has been changed to a -2 state. The chemical terminology is that hydrogen has been oxidized and oxygen reduced. To form the neutral water molecule two atoms of hydrogen are combined with one atom of oxygen. When this occurs, electrons originally associated with the hydrogen gas molecules are transferred to the oxygen and the resulting hydrogen and oxygen ions bond together to form water. Thus, there is a direct transfer of electrons from one species to the other. During this process a much more stable molecule is formed and the difference in energy between the reactants and the product is released in the form of kinetic energy. Since temperature is proportional to kinetic energy, the observation is that hot steam is produced from the chemical reaction.

Our experience tells us that when a battery is normally operated there is no dramatic rise in temperature as the chemical reaction occurs inside. This is consistent with conservation of energy, since most of the energy produced by the chemical reaction in the battery is simply being made available in the useful form of electricity.

During a redox (oxidation–reduction) chemical reaction electrons are transferred directly from one reactant to the other and the resulting species combine to form the reaction product. The battery operates by intervening in this electron transfer process and rerouting the electrons outside the battery, through some external device, and then back into the battery. The energy imparted to the electrons via the chemical reaction can be used by the external device to perform some work. Examples of this would be to run a motor, power a light bulb, or convert the energy back to heat by passing it through a resistance.

The chemical reaction in a battery can be thought of as occurring in a segmented fashion called half-reactions. One half-reaction is associated with the material that loses electrons and the other half-reaction is associated with the material that gains electrons. The material that loses electrons is said to undergo oxidation or be oxidized. The material that gains electrons is said to undergo reduction or be reduced. These half-reactions occur at specific physical locations in the battery and these locations are tied to the terminology used to describe the inside of a battery.

2.1. Anode

An anode is a site where oxidation takes place, that is, where one reactant loses electrons. However, in battery terminology *anode* is also commonly used to refer to the entire reservoir of material that will eventually be oxidized during discharge (*chemical anode*) and in fact may be used to refer to any or all parts of the negative side of a battery or cell. The actual *electrochemical anode* is usually the surface of the chemical anode. This is because most anode materials are metallic and thus electronic conductors and their oxidation will occur at their surface. In addition, there is usually an *anode current collector* in physical contact with the chemical anode. The purpose of this current collector is to transmit to the outside of the battery those electrons produced by the oxidation of the chemical anode. The anode current collector forms a battery's negative terminal.

2.2. Cathode

The terminology for the *cathode* is analogous to the anode. There is an *electrochemical cathode* that is the site of reduction and a *chemical cathode* that is a reservoir of material to be reduced. Similarly, there must be a *cathode current collector* to return electrons to the battery after their energy has been dissipated in an external device. However, unlike most anode materials, which are metallic (and electronic conductors), most cathodes are insulators rather than conductors. Because of this, the *cathode current collector* plays an extremely important role in the electrochemical activity of a battery. The cathode current collector may be a screen, wire, or even the battery container itself. Since its function is to permit electrons to complete their journey to the cathode material, and many cathode material are insulators, the current collector must reach effectively to all areas in the cathode. This is done by adding materials such as graphite or carbon to the cathode material in a high enough proportion to allow electronic conductivity to all parts of the chemical cathode and allow the cell to discharge efficiently. Thus, these materials are really extensions of the cathode current collector. The mixture of chemical cathode plus material to render it conductive is often termed *cathode depolarizer* or *cathode mix*. Like *anode, cathode* is a term often loosely used to describe the chemical cathode, the cathode depolarizer, or the positive side of a battery in general.

2.3. Electrolyte/Separator

Previously, the electrochemical reaction occurring in the battery was distinguished from a chemical reaction that might occur between the same materials by the fact the battery somehow keeps the reactants from directly combining, and temporarily reroutes the electrons being transferred between species outside of the battery. The material that prevents this direct combination is called the *electrolyte*. The electrolyte performs two functions. First, it physically separates the reactants and minimizes the possibility of them reacting directly. Second, it allows the ionic charges that form at the anode and cathode to combine and neutralize each other, forming the battery discharge product in the process. If something did not permit these ions to move and combine, the anode and cathode would quickly build up opposite charges that would cause the battery to cease operation.

Thus, the electrolyte must have two properties. First, it must not be an electronic conductor. If the electrolyte were an electronic conductor, the battery would be shorted (i.e., the direct chemical reaction would appear to occur with the appropriate quantity of heat being produced). Second, the electrolyte must be conductive to either or both of the ions formed at the anode and cathode so there can be a physical migration of charge to keep the cell functioning. The conductivity of the ions in the electrolyte determines where the discharge product accumulates in the cell.

Many materials can serve as electrolytes. The most common is a solvent or solution capable of dissolving the ions of interest. In many implantable medical batteries, the electrolyte is a nonaqueous solution (because water reacts with lithium) containing a dissolved lithium salt. Such solutions are lithium ion conductors and consequently cause the discharge product to accumulate on the cathode side of the battery.

Some solid materials can also conduct specific ions. The lithium/iodine medical battery uses such a solid electrolyte, as will be discussed below.

A term often confused with electrolyte is *separator*. The separator is a mechanical element that physically helps keep the chemical anode and chemical cathode apart when the electrolyte is a solution with no mechanical strength. Often the separator is a porous paper or polymeric material such as polypropylene. A separator is always used in conjunction with a solution electrolyte. When the electrolyte is a solid, the electrolyte itself can serve as the separator. Examples of both situations follow. In either case the electrolyte must (1) be an electronic insulator, (2) be an ion conductor, and (3) not be a solvent for the chemical anode or cathode.

In summary, a battery is a device to liberate electrical energy from a controlled chemical reaction and to divert it to the external world to do useful work. The discharge product from a battery is the same as that formed by the direct chemical reaction of the chemical anode and cathode, and the energy liberated is nearly the same as that from their chemical reaction. (Actually, this is only an approximation, but usually a good one, since the enthalpy and free energy of most battery reactions are nearly the same. In fact, they differ by a term $T\Delta S$, where ΔS is the entropy of the reaction. In most battery reactions both the reactants and products are reduced phases and ΔS is then very small.) However, in an ordinary chemical reaction the energy is usually given off as heat, while in a battery most of the energy is tapped and brought to the external world by electrons. It is very important to realize that a battery is simply a clever configuration to carry out a chemical reaction in a controlled and useful manner. Figure 1 schematically illustrates the elements of a battery as they have been discussed here. The battery illustrated in Figure 1 contains the salt M^+X^- of the anode material M. If the reduced form of the cathode is not soluble in the electrolyte, the discharge product (M^+N^-) accumulates on the cathode (+) side of the battery as shown.

One further comment about the chemical anode and the chemical cathode is in order. It is obvious that if a battery is a device for controlling a chemical reaction, there is a specific ratio of the chemical anode and chemical cathode that will react completely. This ratio is determined by the stoichiometry of the reaction. When batteries are manufactured, the amounts of chemical anode and chemical cathode are controlled to optimize performance. For example, it would

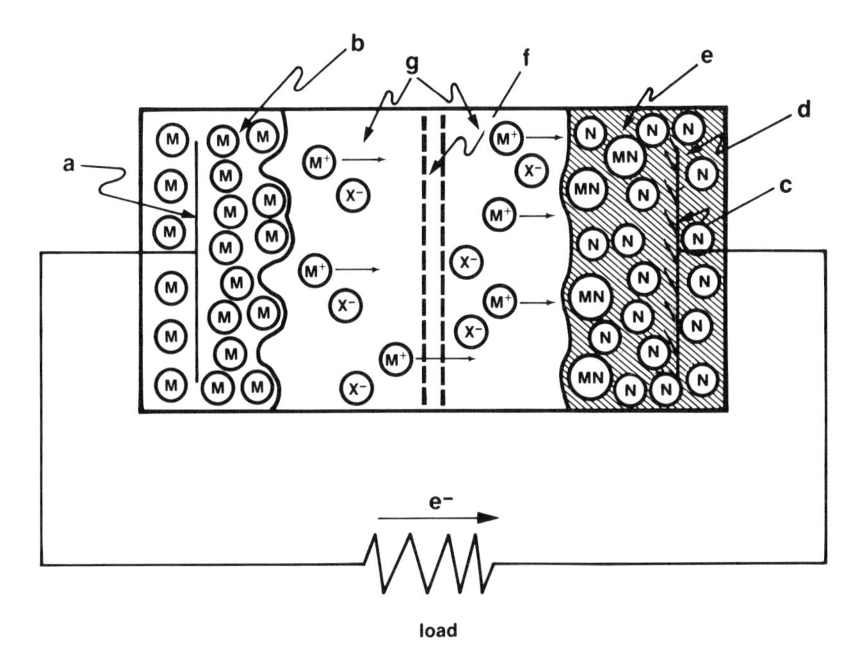

FIGURE 1. Schematic View of a Battery. *a*: Anode Current Collector; *b*: Chemical Anode (M); *c*: Cathode Current Collector; *d*: Cathode Depolarizer Mix (N + Carbon); *e*: Discharge Product (MN); *f*: Separator; *g*: Electrolyte Containing M^+, X^- Ions.

not make sense to have either a great abundance of cathodic or anodic material. Therefore, batteries are engineered with amounts of material near the stoichiometric value for the reaction. However, the amounts of materials are very seldom matched exactly. This is done for a purpose. For most applications it is desirable to control the *end-of-life (EOL)* of the battery in some specific way. For example, running out of anode material, which is usually a metal, may lead to a sudden loss of voltage (see Figure 2a). But cathode materials, which are often solutions or mixtures, may lose activity much more slowly. Thus, batteries with limited amounts of cathode material may exhibit a more gentle voltage decline as the battery wears out (see Figure 2c). Therefore, an engineer who is interested in having an indication that a battery is near depletion might design the battery so that the voltage begins to decline gradually near its end-of-life, or even to go through a voltage step (Figure 2b). Figure 2 illustrates several points. First, all batteries will wear out sometime. When this happens their voltage will fall toward zero. But as mentioned, this can occur in various ways, and the way it occurs is very important for an implantable medical device. Two specific voltages are important to designers of battery-powered implantable medical devices. The first is the *end-of-life indicator voltage*. The second is the *cut-off voltage*. The first

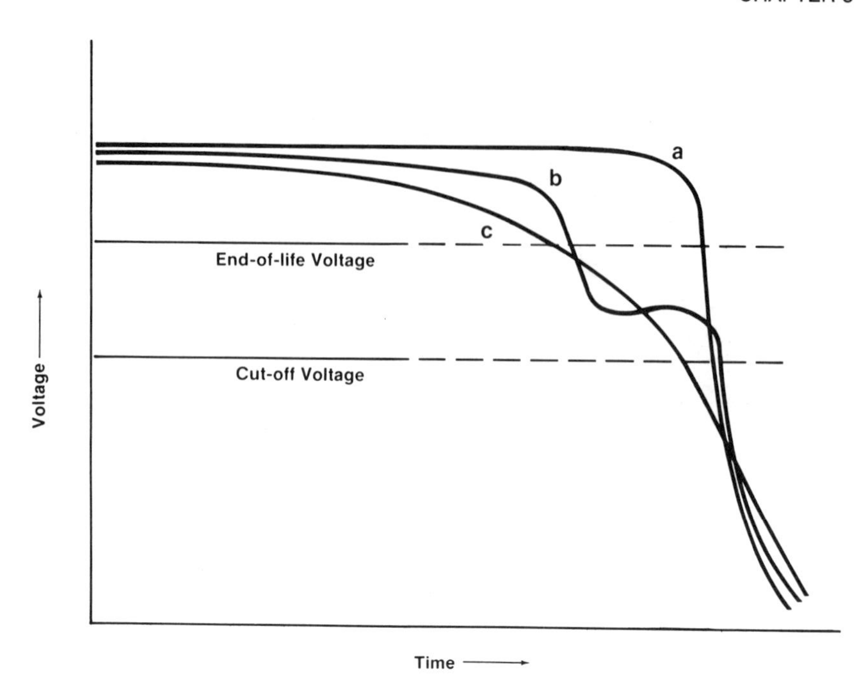

FIGURE 2. Battery Discharge Curve Showing End-of-Life Characteristics and Their Relationship To End-of-Life Indication and Cut-Off Voltages.

voltage is a "trigger" set below the normal operating voltage so it will only be encountered toward the end of useful life. The second is the voltage where the device no longer functions. The time between the occurrence of these two voltages is the *replacement interval*. In Figure 2 the replacement interval is very short for curve *a* but is reasonably long for curves *b* and *c*. For a cardiac pacemaker the replacement interval should be on the order of six months.

2.4. Feedthrough

One other battery term that is useful to discuss is *feedthrough*. A feedthrough has nothing to do with battery chemistry, but it is a very important part of a battery. We have already mentioned that electrons are routed outside the battery at the anode side and returned at the cathode side. This obviously means the two electrodes cannot be in electrical contact. The feedthrough keeps them apart. In its simplest form it is formed by a piece of insulating material (e.g., a piece of plastic) sandwiched between the anode and cathode sides of a battery. Many low-cost batteries are assembled in such a manner using a crimped seal.

Implantable medical batteries almost always use feedthroughs consisting of a metal pin surrounded by a ferrule, the two parts being isolated by a glass or ceramic insulator. Such seals are strong and can be made hermetic.

There are three ways batteries can be built. First, the anode may be in direct contact with the battery case and the cathode connected to a feedthrough pin. This style is called a *can-negative* design. The opposite may also be made. This is called a *can-positive* design. In this situation the can is positive with respect to the feedthrough pin, and a central anode is employed. This is the most commonly used design for modern implantable medical batteries. The third configuration is called *case-neutral* and both the anode and cathode are isolated from the battery case. This requires two feedthroughs plus some insulating material to keep the anode and/or cathode away from the case. Concerns about corrosion by the very reactive materials in lithium/iodine batteries led to the use of this design in the mid-1970s. However, it proved feasible to control the corrosion reactions in other ways, and this more complicated, expensive, and lower energy density configuration was replaced by more efficient designs.

Some examples may help illustrate the battery terms we have been discussing. The Leclanché cell is the most common small primary battery in commercial use. It is used to power flashlights, toys, and numerous consumer devices. In this battery the anode is zinc metal, which also serves as the anode current collector and the battery case.

The chemical cathode is manganese dioxide. Manganese dioxide is an electrical insulator, so carbon is blended with it to allow the electrochemical reduction to occur. The electrolyte is a 10% aqueous solution of ammonium chloride and a piece of porous paper is the physical separator between the zinc anode and the mixture of carbon and manganese dioxide. The Leclanché battery is electrochemically very similar to the currently popular alkaline batteries, which differ mainly in having a potassium hydroxide solution as the electrolyte and the zinc anode in the form of a granular amalgam.

A very different physical configuration exists for the lithium/iodine battery system that is currently used to power many cardiac pacemakers. In this battery system lithium metal, which is very reactive, is the chemical anode. The lithium is also the electrochemical anode as the oxidation of lithium takes place at its own surface. The lithium metal is usually pressed around a stainless steel or other nonreactive metal grid or mesh for support. This grid or mesh serves as the anode current collector and leads to the outside of this cell via a feedthrough. The chemical cathode in the lithium/iodine cell is iodine that is compounded with poly-2-vinylpyridine to make it conductive. This is analogous to the mixture of carbon and maganese dioxide in the Leclanché cell except the materials react chemically rather than forming a simple physical mixture as with the carbon and manganese dioxide. In both cases it was necessary to make the nonconductive chemical cathode (manganese dioxide or iodine) conductive so that the discharge could proceed.

Whereas in the Leclanché cell the electrolyte was an aqueous liquid, the in the lithium/iodine cell is a solid material, lithium iodide. Lithium so the discharge product of this battery. Thus, as the lithium/iodine battery discharges, the lithium iodide discharge product accumulates between the anode and cathode sides of the cell. This has the advantage that it makes the cell very reliable and immune from internal damage, but it also has the disadvantage that the internal resistance of this battery increases as discharge proceeds and more lithium iodide builds up. The cathode current collector for the lithium/iodine cell may be a metal, such as stainless steel, embedded in the cathode, or it may simply be the battery case made of stainless steel or another nonreactive metal in contact with the cathode depolarizer.

There is one other major difference between the two batteries chosen as examples. The Leclanché cell is built in a zinc can and has its anode and cathode separated by an insulating material such as paper or plastic. This battery is held together by what is called a crimp seal, which is tight but does allow some exchange with the atmosphere. On the other hand, the lithium/iodine battery, which is typical of most implantable medical batteries, is hermetically sealed using a ceramic or glass-to-metal feedthrough to keep the exchange of the atmosphere and battery contents to an absolute minimum. This is very important in high-energy batteries where battery components are very reactive and would be severely degraded by the ingression of atmosphere and moisture.

3. BATTERY PARAMETERS

There are several properties of batteries that are important to discuss and understand. Some of these are characteristic of the system itself. Others depend on the size of the battery and how it is constructed. Most important of the intrinsic parameters is the *open-circuit voltage,* $E°$. The open-circuit voltage is the maximum voltage that can be obtained from a battery and is characteristic of the electrochemical reaction occurring during the discharge of the battery. Open-circuit voltage is directly proportional to the partial molar free energy of the battery's electrochemical reaction at any time via the equation $\Delta G = -nFE$. The free energy of the reaction, ΔG, the maximum electrochemical energy content, is related to the energy available from the direct combination of the chemical reactants by the equation $\Delta G = \Delta H - T\Delta S$. The enthalpy ΔH is the energy available in the form of heat from the chemical reaction. The $T\Delta S$ term, which corrects for entropy effects, is often small and thus ΔG and ΔH are often nearly the same value. This simply means that most of the chemical energy can be converted into electricity under ideal conditions. The open-circuit voltage of the implantable battery systems presently in use range from approximately 2 to 4 V.

Other intrinsic properties of a battery system include the *stoichiometric capacity density* and the *theoretical energy density*. The stoichiometric capacity of a battery can be described on a weight or volume basis. In either case the amount of chemical anode and chemical cathode material required to react stoichiometrically is calculated and reported as the amount of charge theoretically available per unit weight or per unit volume of the reactants. The most widely used units for these specific capacity numbers are Ah/cm^3, Ah/kg, and Ah/lb. The stoichiometric capacity values do not take into account the size of the case or other interior components of the battery, nor do they account for the inefficiency of the electrochemical reactions. Thus, this value gives the most optimistic view of possible performance. Measured values are as much as 40–90% lower than the stoichiometric value. The energy content of the battery is the product of voltage and capacity. Therefore, the maximum energy density can be calculated by multiplying the theoretical capacity by the open-circuit voltage. This is also a very optimistic number. The usual units are Wh/cm^3, Wh/kg, and Wh/lb.

The large difference between the theoretical stoichiometric capacity or energy density and the actual deliverable values for a given system is caused by a variety of factors such as nonreactive materials within the cell, which include the battery container, current collectors and the like, various additives necessary to make the cell function, and the presence of an electrolyte. In addition, inefficiencies in the electrochemistry lower this number too. These will be discussed in more detail elsewhere.

Extrinsic properties depend on how a battery is constructed. These properties depend on how much material is used and how the materials are put together. They are probably the most important parameters a battery user needs to know in order to determine if a certain battery is appropriate for an intended application. Examples are *load voltage, end-of-life characteristics, rate or current capability,* and *deliverable capacity and energy*.

Properties such as load voltage and *internal resistance* of a battery have their inherent ranges determined by the battery system itself, but the exact values usually depend on construction. For example, the Leclanché cell discussed earlier has a well-defined discharge product that does not interfere with the performance of the cell, as well as a conductive electrolyte. Therefore, this battery does not suffer a large amount of voltage drop because of internal resistance. On the other hand, the lithium/iodine system discharges in a manner such that the lithium iodide discharge product is precipitated between the anode and the cathode. While this gives the battery some desirable reliability characteristics, the increasing amount of discharge product accumulating between the anode and the cathode ensures the internal resistance of the battery will increase with discharge. Since lithium iodide is only moderately conductive, the load voltage exhibited by this battery is very dependent on the state of discharge and the current that is being drawn from it. Therefore, the load voltage depends to a great extent on how the battery is constructed, particularly its specific geometric characteristics.

Several chemical properties are also factors in determining the internal resistance of a battery. For example, systems with solid-state electrolytes usually have much higher internal resistances than those with liquid electrolytes. Hence, solid-state batteries very seldom have high current capabilities. On the other hand, many batteries with conductive liquid electrolytes have excellent rate capabilities. This is one of the differences among the various implantable lithium anode systems that are commercially available today.

Generally, batteries are designed with specific usage rates in mind. The designer will consider the application requirements and the inherent rate capability of various chemistries and then construct a specific battery based on this information. For example, if an application requires a high rate compared to the inherent capabilities of the chemistry, the surface area of the electrodes might be increased to provide more current. Similarly, other important parameters such as the deliverable capacity and energy density are also a function of the cell design and engineering as well as inherent properties of the chemistry. The thickness of the anode and cathode per unit area, called the *cell loading,* is an important design parameter for both rate and deliverable energy considerations. The cell loading is actually the capacity per unit of electrode surface area, expressed as mAh/cm^2 or Ah/cm^2.

4. BATTERY PERFORMANCE

There are several ways to indicate battery performance. One of the most common ways is simply to plot the load voltage versus time or versus charge drawn from the battery at a particular discharge rate. A family of these curves characterizes a battery's performance at a given temperature. Figures 3 and 4 illustrate two typical sets of discharge plots often observed. The curves in Figure 3 are for a Li/I_2 battery where the internal resistance increases with time. The delivered capacity decreases with, and is strongly dependent on, rate. The data in Figure 4 are for a $Li/SOCl_2$ battery. This type of battery has a low internal resistance. It delivers about the same capacity at all the rates shown. The slightly higher deliverable capacity with increasing discharge rate indicates the importance of self-discharge reactions in this battery system.

For most applications the device under power will either cease to function or will fall below specifications at some voltage above zero. Hence, the shape of the battery discharge curve may also be very important in determining longevity of the device. Furthermore, since the shape of the discharge curves may change with the load on the battery (see Figure 3), it is necessary to characterize a battery's performance at a specified load and cut-off voltage. It is apparent that an estimate of device longevity for a rate other than that shown would depend a lot on which discharge curve and what cut-off voltage was chosen to make the estimation. On the other hand, the data in Figure 4 much more readily lend themselves to this estimation, since the deliverable capacity varies considerably less with both rate and voltage. As a practical example, consider the Leclanché

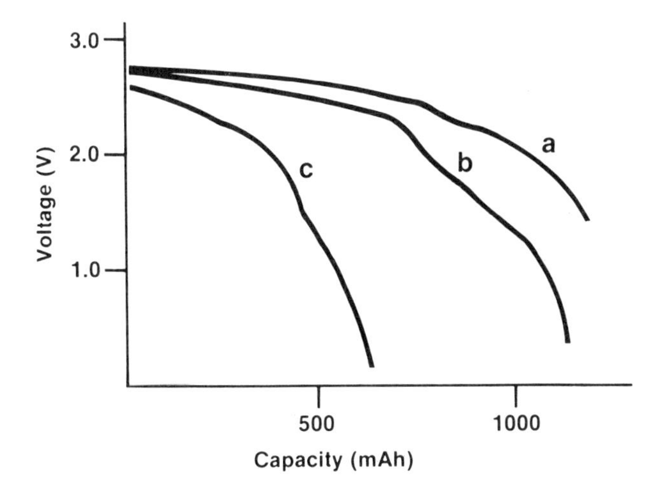

FIGURE 3. Voltage Capacity Curves for Li/I_2 Battery at $a = 4$, $b = 8$, $c = 20$ μA/cm^2 Discharge Current Density.

cell. The typical flashlight uses two batteries in series (3 V) and draws about half of an ampere current. When that voltage falls much below 2 V, the light becomes very dim and the flashlight is not very useful. Thus, a meaningful rating for a flashlight battery should reflect the time or capacity for the battery to reach 1 V at half an ampere current.

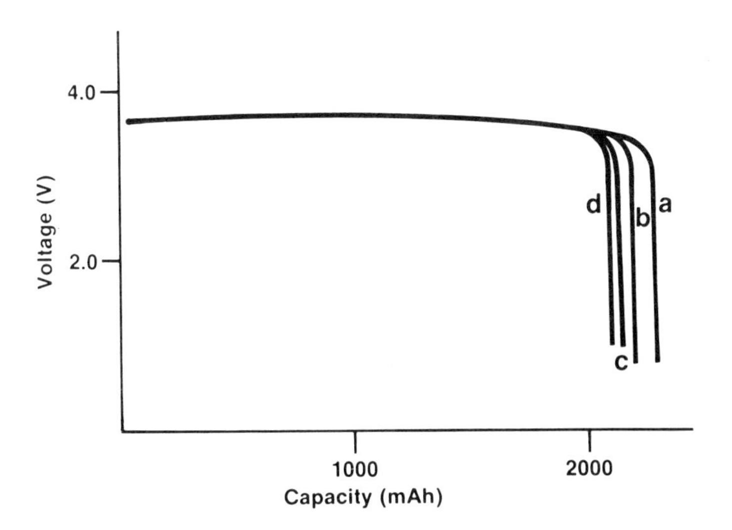

FIGURE 4. Voltage Capacity Curves For $Li/SOCl_2$ Battery at $a = 160$, $b = 80$, $c = 40$, $d = 20$ μA/cm^2 Discharge Current Density.

It would appear that rating a battery for a particular application would be reasonably straightforward. However, several additional complications occur when a battery is intended for a very long-term application.

First consider inherent problems with long-term applications. A battery for powering an implantable medical device such as a cardiac pacemaker is expected to operate for a long time. With today's technology these operating times are often 5–10 years. Directly determing how long a battery will operate would take too long. It is not usually practical for manufacturers to wait up to 10 years to put a new product on the market. As a first approximation the battery could be discharged at an accelerated rate and the test time multiplied by the acceleration factor to determine the real-time rating for the battery. But if the shape of the discharge curves is dependent on current as shown in Figure 3, this will not give an accurate indication for the extrapolation. Also, the time-dependent loss mechanisms may not be appropriately considered. Battery evaluators have addressed these problems by coming up with rating systems that characterize battery performance in various standard ways. These methods will be described in detail in subsequent chapters on accelerated testing and battery performance modeling.

A discussion of the influence of chemical parameters on discharge curves is in order. An example is shown for the lithium/iodine system. A family of discharge curves for the Li/I_2 system is shown in Figure 5 for discharge current

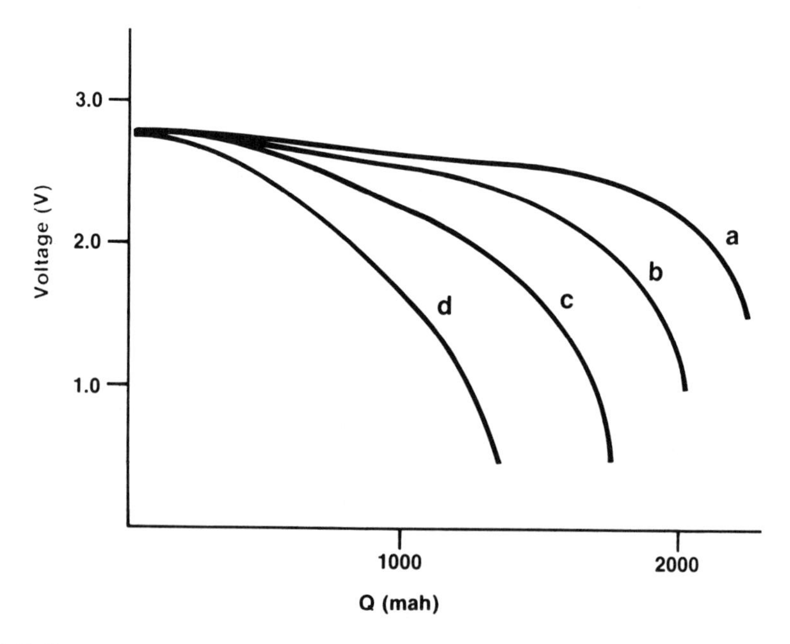

FIGURE 5. Family of Li/I_2 Discharge Curves at Current Densities of 5–50 μA/cm² (a = 5 μA/cm², b = 10 μA/cm², c = 25 μA/cm², d = 50 μA/cm²).

densities ranging from 5 to 50 μA/cm^2. Notice the utilizable capacity depends a lot on current, particularly to a 2-V limit, which is often used as a lower bound to determine the useful capacity for this battery system. The capacity varies by more than a factor of 3 over this range of rates. There are three reasons these curves vary from one another. They apply to all battery systems to some degree.

The first two are polarization phenomena. Polarization is caused by a deviation from equilibrium conditions. In a battery, polarization is observed via a lowering of the battery voltage from its thermodynamic or open-circuit value. Polarization always acts against a system and thus will always be observed to lower the battery voltage. Polarization can be arbitrarily divided into two types.

The first type of polarization may be termed *ohmic*. An ideal battery system would have no internal resistance and thus there would be no energy loss within the battery other than that predicted by thermodynamics. However, the conductivity of any battery is not infinite, and in some practical batteries the conductivity may be quite low. Because of this, the current flowing through the battery will cause some voltage drop that is proportional to the current according to Ohm's law. This form of polarization may be readily determined and corrected for.

The second type of polarization is more complicated. It is called nonohmic polarization and is not linearly proportional to the amount of the current flowing. In fact, voltage drop from nonohmic polarization usually goes up exponentially with current. This type of polarization may be caused by any of several processes occurring within the battery. However, all of them are similar in that this polarization occurs because one or more processes in the battery cannot happen fast enough to keep up with the demand placed on the battery by the external load. Two common causes are slow diffusion of a reactant species to or from an electrode surface and the slow transfer of an electron to or from a reactant at the electrode surface. There are other similar types of nonohmic polarization that are traceable to processes such as the initiation of crystal growth. However, in all situations the polarization is observed because at least one process cannot keep up with the external demand being placed on the battery. The only response the battery can have to such a situation is to lower its output voltage. Polarization phenomena can be treated individually and then summed up to get the net effect on the battery or cell.

A third important parameter is called *self-discharge*. Self-discharge is another nonideal process in the battery. Self-discharge arises because of direct chemical reactions inside the battery. Often these reactions are between the chemical anode and chemical cathode, although other elements may be involved, such as the electrolyte, the solvent, or the can material. In any case they are detrimental to battery performance because they utilize some of the battery components in a nonproductive manner.

Consider the example of self-discharge in the lithium/iodine system discussed earlier. Self-discharge in this battery occurs because of direct reaction of iodine and lithium. This happens because the lithium iodide discharge product is not a perfect barrier to iodine diffusion. It is formed with small cracks and grain boundaries, and the iodine can diffuse along them and react directly with the lithium. Since the lithium and iodine are the chemical reactants of the battery, the self-discharge reaction decreases the amount of useful capacity in the battery. But because the self-discharge reaction occurs at a slow rate, its effect is not noticeable during high discharge rates. However, its effect at low discharge rates can be appreciable. This will be discussed more thoroughly later.

As stated earlier, the rating desired by the user is the deliverable capacity or the operating time that can be obtained from a battery used in a specific application. Since these parameters may be functions of rate, temperature, and storage time, as well as battery chemistry and construction, battery manufacturers cannot easily calculate a deliverable capacity or operating time. The manufacturer often does not know the exact conditions of use for the battery. Thus, in their attempt to give the user some information about a battery, manufacturers have developed a number of arbitrary rating systems. In the following discussions each subsequent capacity value is smaller than the preceding ones and reflects a higher level of sophistication or understanding. In general, each value is also more difficult to obtain than the preceding ones.

The simplest way to rate a battery is by *stoichiometric capacity*. The stoichiometric capacity of a battery can be calculated directly from the amount of reactants put into the battery. Usually the amounts of anodic and cathodic material are not stoichiometrically equivalent. One or the other is in slight excess, so the performance of the battery near its end-of-life (EOL) will be well characterized. Thus, the stoichiometric capacity is calculated from the lesser amount of the anodic or cathodic material present. Very little of the detail of the battery's chemistry has to be known to obtain a stoichiometric capacity value.

The stoichiometric capacity can be calculated for any battery if the discharge reaction is known. The problem with using this value is that it does not take into account any of the inefficiencies that always occur during the discharge of a battery. Thus, stoichiometric capacity is usually very different from the eventual realizable capacity. Energy density may also be calculated from the stoichiometric capacity. Often this is done using the open-circuit voltage to derive the energy value. Since the discharge voltage is always less than the open-circuit value, this gives an even more optimistic view of possible performance. In general, capacity and energy density values derived from stoichiometric values should be used only for comparison with similar types of systems, or even better, within a single battery chemistry, to estimate relative performance.

Sometimes manufacturers know that a certain amount of the anode or cathode material will not be available for discharge. This is usually because it enters into a chemical reaction that is either needed or unavoidable. In an attempt to give a better value for the capacity of the battery, a manufacturer may calculate how much of the reacting material will be unavailable for discharge and correct the stoichiometric capacity accordingly. This value is related to the stoichiometric capacity but is smaller. It is a better approximation to realizable performance but can still be too optimistic. For the implantable lithium/iodine system this capacity value has been termed the *maximum available capacity*[1] and reflects the fact that some iodine is irreversibly bound to the organic material used to make the iodine cathode conductive.

For some range of rates and conditions the actual capacity can be measured. Capacity values obtained from direct tests are called *rated capacities*. Batteries for consumer items are typically specified this way, since continuous discharge is completed over a period of a few days at most. Often terminology such as C/5 or C/7 is used to indicate the rate of discharging a battery to deliver its capacity over a 5- or 7-h period, respectively. This notation is only commonly applied for rapid discharges.

A special value of the rated capacity is the *deliverable capacity*. The deliverable capacity is the rated capacity at a particular current the user is interested in. For implantable medical batteries the application rate is often very small (i.e., very small currents for periods of several years). Figure 6 illustrates the relation of these capacities to rate for a Li/I_2 cell.[2] This figure is a plot of capacity versus (log) current. The upper line is the stoichiometric capacity and is calculated irrespective of current. The next lower line is the maximum available capacity,[1] which is obtained by correcting the stoichiometric capacity for known inefficiencies. Each data point on the lower curve is a measured capacity. The

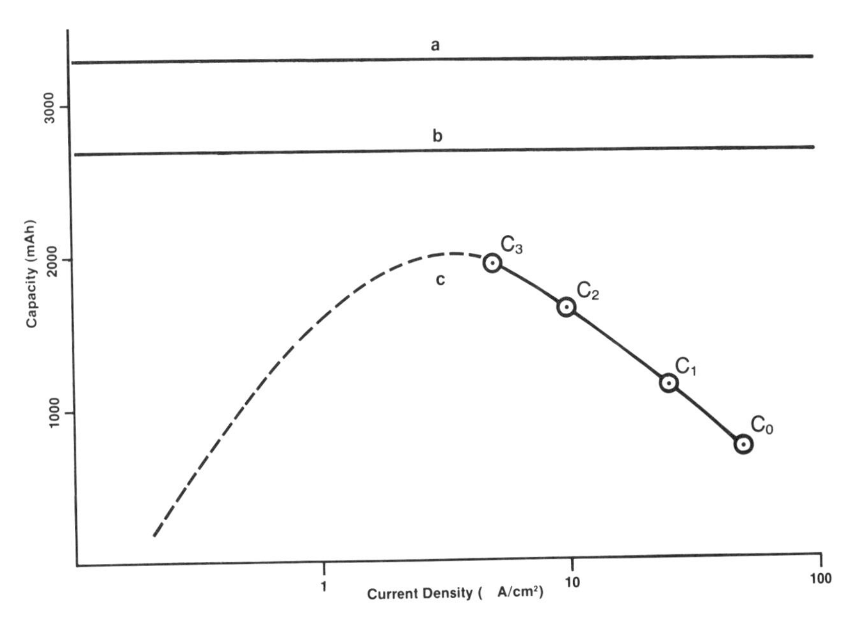

FIGURE 6. Relationship of Various Capacity Ratings and Current for Li/I_2 Battery in Figure 5. *a*: Stoichiometric Capacity (3300 mAh); *b*: Maximum Available Capacity (2700 mAh); *c*: Deliverable Capacity (C_0 = 50 μA/cm², C_1 = 25 μA/cm², C_2 = 10 μA/cm², C_3 = 5 μA/cm²), Dashed Portion of *c* is Projected Curve.

solid extension of the curve to the left is an estimation based on the rated capacities, the maximum available capacity, and modeling techniques, which will be discussed in a subsequent chapter. The general shape of this lower curve, falling off at both low and high currents, is inherent to all batteries. The lowering of capacity at small currents is due to the increasing importance of self-discharge. The lowering at high current is due to increasing polarization.

Energy density values may be calculated analogously to capacity values. The most useful values are those either measured or projected from data that take into account the likely deliverable capacity of the battery and the load voltage obtained during the discharge. Other methods of estimating energy density will be overly optimistic.

Sometimes special performance characteristics are desired for batteries. This is especially true of medical device batteries, where the characteristics near the battery's end-of-life (EOL) are important. For medical batteries the shape of the discharge curve near the EOL should be such that there is an advance warning that wearout is eminent, allowing ample time for replacement (see Figure 2). In some battery systems, such as lithium/iodine, the slope and nature of the discharge curve and the gently decreasing voltage near its EOL give adequate warning of the impending wearout. However, for other systems, in particular the higher rate systems with flatter discharge curves, the impending EOL may be very abrupt. For this reason, it is often desirable to alter the battery chemistry so that a voltage plateau will occur when the system is nearly depleted. This second plateau, which may be 0.2–0.5 V lower than the principal discharge voltage, gives warning in time to replace the battery or device. The type of EOL indicator used is generally specific for each type of chemistry. In many cases EOL indicators are currently under development. If the chemistry is not amenable to any chemical EOL, mechanical devices have been proposed, since depletion of the electrode material occurs.[3] In the future the nature of the EOL will become an even more important specification for an implantable battery.

In summary, the battery specifications that are usually most useful are the load voltage or operating voltage range, which includes information about the discharge curves such as whether it is flat, stepped, or a ramp; the rate capability, which gives the user some idea of the useful range of rates for a particular battery or system; the pulse current capability, which is the battery's capability to take higher than normal current loads for short periods of time; the operating temperature range, which includes a maximum and minimum acceptable temperature (including storage and processing); internal resistance; self-discharge; and storage characteristics.

5. MICROCALORIMETRY

Self-discharge was discussed as one of the important factors determining the deliverable capacity of a battery intended for long-term use. Historically, determining self-discharge has been difficult. Zinc/mercury oxide batteries were used to power early implantable devices. Self-discharge in these batteries involved several reactions such as the formation of hydrogen gas and the solubilization of some of the active materials. These reactions could be studied by determining the amount of active material present after a specified storage time or by collecting the volume of hydrogen gas produced. In fact, these chemical methods were used to estimate self-discharge losses in this battery system. However, with the advent of the lithium/iodine battery chemistry, it became much more difficult to study self-discharge processes. Corrosion of internal components and the direct combination of lithium with iodine were the principal suspected reactions. It soon became evident that corrosion, which involved reaction of iodine with metals present inside the battery, was self-limiting because the corrosion product is a passive film and a barrier to further reaction. Experiments showed this type of parasitic self-discharge reaction stopped after a few months time and involved no significant amount of material.[4] The other reaction, a direct combination of iodine with lithium, was much more difficult to study. This is because the product of this reaction, lithium iodide, is the same as the product for the electrochemical discharge reaction. Thus, normal analytical procedures could not be used to determine self-discharge losses. This situation was further complicated by the fact that self-discharge reactions in this battery system appeared to occur quite slowly.

In 1978, self-discharge was the largest unanswered question hindering attempts to predict the long-term behavior of this battery system. New techniques were sought to address the problem. One approach was to measure the heat evolved by the chemical reactions of interest and relate this back to the rate of the reactions. The problem, of course, was that very little heat is given off by these reactions and it was difficult to measure. Prosen[5] was one of the first to demonstrate that measurable quantities of heat could be detected in a calorimeter. Later, Hart[6] designed the first commercial battery calorimeter for use with small batteries. Other early workers in this field included Hanson,[7] Holmes,[8] and Untereker.[9] The stability and sensitivity of the calorimeter developed by Hart soon made it the most widely used instrument of this kind.

The battery calorimeter works because a battery undergoing self-discharge will always be at a slightly different temperature than its surroundings if the enthalpy of the parasitic reaction(s) is not zero. The battery calorimeter provides a very stable reference against which to read this temperature difference. A schematic diagram of a modern battery calorimeter is shown in Figure 7. The heart of the calorimeter is a central reference block of material, usually aluminum (see Figure 7). This block is held at a very constant temperature through a series of air and water baths (b, c, and d). The air and water baths provide progressively more stable temperature environments as one proceeds toward the center of the calorimeter. Often the outer bath (b) is an air bath held to $\pm 1°C$. The next bath is usually a high-capacity water bath (c). The temperature stability of this bath has to be very good, about $\pm 0.0001°C$ on commercial units. Inside the water

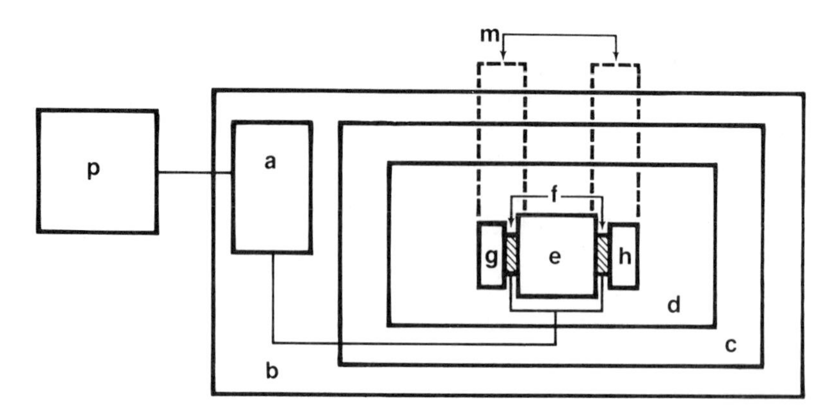

FIGURE 7. Schematic Representation of a Modern Battery Calorimeter. *a*: Electronic Module; *b*: Outer Air Bath; *c*: Water Bath; *d*: Inner Static Air Bath; *e*: Calorimeter Block; *f*: Thermopile (2); *g*: Test Battery Holder; *h*: Reference Battery Holder; *m*: Calorimeter Access Port (2); *p*: Recorder.

bath there are one or two dead-air chambers (d), which provide a further temperature buffer. These precautions for temperature stability are still not adequate because the heat flux from the battery is often just too small. Therefore, the measurements are made differentially. The battery being measured (g) is compared with a blank (h) of equal heat capacity in which no chemical reactions are occurring. This blank is often a piece of aluminum about the same mass as the battery. The actual measurements are made using thermopiles (f) that are like a series of thermocouples. They are arranged in series with a physical configuration that puts all of one type of junction on one side of the pile and all of the second type of junction on the other side. One side of each thermopile is placed against a reference block. The other side is put in intimate contact with the battery being measured or a reference blank. The outputs of the thermopiles are then measured differentially by a microvoltmeter (a), which is also kept in a temperature-controlled environment. The thermopiles generate about 0.1 μV output per microwatt of heat flux. The sensitivity of these instruments is on the order of 1–2 μW. Calorimeters can be designed to accommodate many sizes and shapes of batteries but are most sensitive for smaller batteries. Usually a typical measurement requires 3–5 h to complete.

This technique, commonly termed *microcalorimetry,* was applied very early to lithium/iodine pacemaker cells. It was suspected that self-discharge would consume no more than 10% of the active material over the entire life of the battery. Using microcalorimetry, it was observed that the steady-state heat flux from the battery started out very high and decreased exponentially as discharge progressed. These data fit very nicely with the concept of the chemical diffusion of iodine through lithium iodide. The rate is rather high through the thin layer of lithium iodide in a fresh battery. As discharge proceeds, the layer of discharge product becomes thicker, and the rate of self-discharge decreases dramatically and approaches zero when about a third of the discharge is completed. It soon became apparent that other interesting experiments could be done. Not only could this technique be used to evaluate the static reactions occurring in a battery system, but measurements could be made while the battery inside the calorimeter was being discharged via an external resistor. Using this technique only the difference between the enthaply and free energy would ideally appear as heat inside the calorimeter. Since this difference is quite small for many battery systems, it is a straightforward process to determine the effect of load on rates of parasitic processes.

Such experiments with the lithium-iodine system showed that self-discharge acutally decreased when current was being drawn from the battery.[9] The interpretation of these data is that iodine in the vicinity of the anode can be either electrochemically discharged or chemically discharged. When the battery discharge rate is increased, the competition for this iodine is moved in favor of the electrochemical path; thus, less iodine is available for self-discharge. This, of course, was very important because it showed the static experiments that had been done first actually led to pessimistic estimates of self-discharge losses for this battery system.

Microcalorimetry does have its limitations. It can be applied successfully only when the battery chemistry is understood. This appears to be the case with the lithium/iodine system, where there is little evidence that chemical reactions other than the direct combination of lithium and iodine are occurring. However, even in this system, which is certainly one of the better understood systems, there is a suspicion that some of the heat observed very early in battery life is due to chemical reactions other than self-discharge. These reactions are thought to be between the iodine and the poly-2-vinylpyridine additive used to render the iodine conductive. Even though the iodine and polymer are reacted during a heating process before they are put in the battery, there is some evidence these reactions proceed for some time after the battery is made.[10] For the lithium/iodine system this reaction leads to no appreciable error in estimating self-discharge. However, this may not be true for all battery chemistries. Perhaps the best example of this is the lithium/thionyl chloride battery system. Static calorimetry measurements made at various times during discharge yield a curve that exhibits at least one maximum.[11] This curve does not fit any simple self-discharge model and might represent a situation where the discharge products of the battery react further to form a more stable species and involve heat in the process. However, some workers in this field believe this heat is due to the reaction of thionyl chloride and lithium, and the increases and decreases in heat flux are due to variations in the thickness of the protective film coating the lithium anode. The answer to this question remains unknown, but further calorimetric experiments, such as studying the discharge current–heat output relationship, may shed some light on it. Other straightforward experiments, such as monitoring heat output with various electrolytes and additives that modify the lithium chloride protective film, may also help answer the question.

Regardless of the answer to this particular question, calorimetry has proved to be a very valuable tool for studying parasitic processes in many battery systems. Virtually every important battery system has probably now been studied in some detail by calorimetric methods. In most cases enough is known about the parasitic reactions to make a useful estimate of the effect of these processes on battery performance.

Microcalorimetry is a relatively slow technique and lends itself most readily to small batteries or cells. However, changes in technique are allowing the experimentalist to obtain more and higher-quality data. Calorimeters that will handle larger cells are now becoming available. This means a greater variety of batteries can be studied. In addition, calorimeters with larger cavities can be used to study several smaller batteries at one time. Thus, data for routine and screening applications can be obtained more quickly by averaging the output from a number of cells simultaneously. Calorimetry is also finding a place studying secondary battery systems and for general engineering studies involving battery heat management. These problems are of great practical importance and can be studied in a very fundamental way using this technique. Certainly, microcalorimetry has proved to be a very valuable tool for the battery researcher, designer, and engineer.

6. IMPLANTABLE BATTERY CHEMISTRIES

The majority of battery chemistries currently used for implantable medical devices use lithium metal for the chemical anode material. Lithium has become the preferred anode material for these batteries since the mid-1970s. The reason for this is that batteries made with lithium anodes offer significant advantages over the aqueous alkaline battery systems, primarily zinc/mercury oxide, which were used before, and nothing better has yet been found. Lithium is active and light and forms many lithium-ion-conducting compounds; moreover, its ions are soluble in many nonaqueous solvents. Its low weight and high activity make it very desirable for battery applications. In fact, it is so well suited for use in medical devices that the pacemaker industry was one of the early important forces in developing lithium battery technology to its present state.

Advantages of battery systems based on lithium include high voltages, high capacities, and high energy densities. But these systems exhibit one even greater advantage for the medical device industry: reliability. Experience with these batteries since their introduction has proved them to be extremely reliable. Overall they have extended the typical operating time of implantable medical devices severalfold.

Discussions in this chapter have dealt with various electrochemical aspects of batteries. The purpose has been to acquaint the reader with the general principles and considerations necessary to understand the electrochemical behavior of batteries. It seems beneficial to briefly review each of the chemistries that have been used to power implantable medical devices. Most of these chemistries will be discussed in more detail in subsequent chapters of this book; therefore, the scope of this discussion will be limited to general information about each chemical system. The first two battery systems are not based on a lithium anode and are included for historical completeness.

The first battery chemistry to be implanted to any great extent was the *Zn/HgO* system. This is the so-called mercury battery.[(12)] Although the chemical system has been known for a long time, the modern form of the battery was developed in the 1940s by S. Ruben.[(12)] It met the need for a reliable, high-energy-density battery for military applications.

From the early 1960s until the mid-1970s, during the time the pacemaker industry was growing from infancy to adolescence, the mercury battery was the principal power source for pacemaker products. This is because it was the most reliable design available and it had many desirable properties.

It is generally accepted that the discharge reaction of this battery is:

$$Zn + HgO \rightarrow ZnO + Hg \tag{1}$$

This discharge reaction has a thermodynamic voltage of 1.35 V, which is in good agreement with the observed value. The battery is usually made using a zinc amalgam anode and mercury oxide mixed with graphite as a cathode depolarizer. A concentrated solution of potassium or sodium hydroxide is used as the electrolyte. The most common cell configurations are button cells. In such cells the chemical anode and chemical cathode are separated from each other by a barrier of paper or polymeric material. Since the discharge reaction involves only the formation and consumption of solid phases of fixed composition, the voltage is constant throughout discharge. This cell system is capable of quite high discharge rates and has an excellent energy density, about 0.6 Wh/cm^3, even in small cells, when discharged at moderate to fast rates.

While the Zn/HgO cell system has high energy and very stable discharge characteristics, other properties made it less desirable as an implantable power source. One such problem was its propensity to evolve gas during service life, giving off small amounts of hydrogen. This made it difficult to seal these batteries hermetically. For those reasons most cardiac pulse generators powered by Zn/HgO batteries were epoxy encapsulated, which permitted the hydrogen produced by the battery to escape through the casing material. Another problem was their tendency to form small dendrites of liquid mercury, causing short circuits between the anode and cathode. Engineering features such as multiple wraps of separator material and additives to the cathode helped minimize the shorting problem. In their most advanced low-rate configurations these batteries would typically operate for one to three years, although examples of operation as long as five to seven years are known. After the mid-1970s very few Zn/HgO batteries were used in implantable medical products. Still this battery deserves a place of prominence in the history of the development of the cardiac pacemaker and was instrumental in the development of the pacemaker to its present advanced state.

Only one secondary or rechargeable battery chemistry has been implanted. This is the nickel/cadmium battery. A nickel/cadminum battery was used to power the first implanted cardiac pacemaker in Sweden in 1958.[13] But more important, it was used in the mid-1970s as part of a commercial pacemaker system. In this system the batteries were inductively charged by an exterior RF coil placed over the unit. The large number of possible recharges of a nickel/cadmium battery made this a potential lifetime pacemaker battery. However, patient compliance and acceptance and the inconvenience of frequent charging made it difficult to compete with the newer long-lived lithium primary systems that were becoming available.

Chemistry for the nickel/cadminum battery is based on the oxidation of cadminum metal to cadmium hydroxide and the reduction of Ni(III) to Ni(II).

$$2NiOOH + 2H_2O + Cd \underset{\text{charge}}{\overset{\text{discharge}}{\rightleftharpoons}} 2Ni(OH)_2 + Cd(OH)_2 \qquad (2)$$

The voltage of this battery is about 1.2 V. It is capable of quite high currents. The electrolyte is potassium hydroxide, which keeps the pH high enough to solubilize the metal ions of interest. This system is capable of a great number of cycles, sometimes approaching 1000 or even more if the charge and discharge cycles are controlled properly. Problems with the system include small amounts of oxygen formed during overcharge. For this reason it is necessary to limit the charge and discharge voltages in the battery very carefully. Today nickel/cadmium batteries are being used for a wide variety of applications where high currents are needed and rechargeable batteries are economically effective.

The cell system that largely replaced the Zn/HgO battery for implantable pacemakers was, and still is, the Li/I_2 battery. This battery, based on an iodine cathode and an active metal anode, can be traced back to the late 1960s.[14] Herman's anode, magnesium, was soon extended to lithium and a commercialized form of this battery was made available, first by Catalyst Research Corporation and soon after by Wilson Greatbatch Ltd. This chemistry was ideally suited for powering cardiac pacemakers. It has become the dominant battery chemistry for this application. During the 10 years since its introduction, the fundamental chemistry has remained relatively unchanged, although many changes have been incorporated that have boosted the energy density about three to four times.

The cell discharge reaction is:

$$Li + \tfrac{1}{2}I_2 \rightarrow LiI \tag{3}$$

The thermodynamic voltage for this discharge reaction is 2.80 V, which is the value experimentally observed. The discharge reaction predicts a constant voltage during cell life, as was the case for the Zn/HgO battery. The open-circuit voltage is observed to be relatively constant during most of the cell's discharge. The iodine is made conductive by the addition of poly-2-vinylpyridine. The iodine and polymer react in various ways to form a mixture of polymer–iodine compounds plus excess iodine. During the time there is pure iodine in the cell, the open-circuit voltage remains near 2.80 V. However, near the end of discharge the activity of iodine becomes less than unity and the thermodynamic voltage falls correspondingly.

The observed load voltage from these batteries is quite different from the open-circuit voltage. In fact, the load voltage decreases monotonically throughout battery life. This is because the internal resistance of the battery continually increases with discharge as the LiI discharge product is precipitated between the anode and the cathode, forming an ever-thickening layer. These batteries may start out having a resistance on the order of 100 Ω, but may have a resistance of over 10,000 Ω by the end of their service. Thus, the principal limitation on lithium/iodine batteries is the range of currents they are capable of substaining. For most implantable cell configurations the absolute current is limited to about 200 μA.

Although all the lithium/iodine batteries commercially available are based on the same anode and cathode chemistry, there have been three distinct variations produced. The first is called the *uncoated* battery and consists of a lithium anode placed in direct contact with the iodine-based cathode material. Melted cathode depolarizer is poured into contact with the lithium surface and forms an insulating layer of lithium iodide by direct chemical reaction. This layer is electronically insulating but lithium ion conducting and serves as both the electrolyte and the separator. Batteries produced in this manner exhibit a linear decrease in load voltage at a given current density because of the planar lithium iodide discharge product grown between the anode and cathode.

The second cell variation is similar to that described above, but before the depolarizer is poured onto the lithium anode, the anode surface is covered with a coating of pure poly-2-vinylpyridine. When the cathode is poured onto the poly-2-vinylpyridine coating, a chemical reaction occurs that alters the performance of the battery dramatically. Discharge curves from batteries produced in this manner do not exhibit a linear decrease in slope with discharge. Rather, an exponential decay is observed. The internal resistance of these *coated* batteries is usually a factor of 10 or more less than for the corresponding uncoated battery. This decreased internal resistance helps to maintain a greater rate capability throughout discharge. This cell design has been used in the majority of the lithium/iodine batteries implanted.

A third variation in cell design is also commercially available. In the first two types of batteries discussed earlier, the iodine and poly 2 vinylpyridine are mixed and preheated for up to several days. However, in the third variety the iodine and poly-2-vinylpyridine are intimately mixed, pressed into a pellet and placed in the cell at ambient temperature. The batteries are then subjected to a short period of heating. These batteries are termed *pressed cathode* batteries and exhibit performance very similar to the poured cathode batteries utilizing the poly-2-vinylpyridine anode film.

The difference in behavior between the second two varieties of iodine cells and the first is very interesting. No one has put forth an explanation of why the pressed cathode batteries behave as they do. However, recent information has shed light on the effect of the poly-2-vinylpyridine anode coating on cell performance. It has recently been found that a low-temperature reaction takes placed between iodine, poly-2-vinylpyridine, and lithium. The product of this reaction is a very runny colored liquid.[14] This liquid wets lithium iodide and also dissolves lithium ions. This makes it an excellent liquid electrolyte and lowers the resistance of the lithium/iodine battery by providing an alternative path to solid-state conduction through the lithium iodide. This liquid electrolyte material has been experimentally observed using a microscope to view very thin Li/I_2 battery sections.[15] It is suspected that a similar mechanism is operating in the press cathode cells, although this has not been verified.

Most lithium/iodine batteries have proved to be extremely reliable. Projecting their operating characteristics was difficult during the first years of their use because the discharge curve is quite dependent on the load. Therefore, classical accelerated tests did not give accurate projected results. However, computer simulations and curve-fitting techniques have been applied to this cell system very successfully and now tests results from nearly 10 years of real-time testing have verified projected results to within 5%.[16,17] In addition, real-time experience in humans has proved this battery reliable for more than eight years of operation.[16,17] There are more than 800,000 lithium/iodine batteries currently implanted.

An offshoot of the lithium/iodine chemistry was an analogous system with bromine as the chemical cathode material. This system has a discharge reaction analogous to lithium/iodine, with a thermodynamic cell voltage of 3.50 V.

$$Li + \tfrac{1}{2}Br_2 \rightarrow LiBr \tag{4}$$

Some of these batteries were produced as an advanced prototype design around 1978, and one was clinically evaluated. However, bromine is much more reactive than iodine and this system proved to be erratic in performance. While some batteries at low rate showed good performance for several years, others showed variable performance with internal degradation of the battery occurring in relatively short periods of time. For this reason the Li/Br_2 battery never became commercially successful. There have been other attempts to develop a lithium/bromine system using various schemes to make the cathode material conductive and precipitate the very resistive lithium bromide discharge product in a location where it would not cause undue internal resistance. However, none of these attempts have been successful and there has always been the problem of the extreme corrosiveness of elemental bromine. For this reason attempts to produce higher-energy-density batteries have been channeled to other systems.

Both the lithium/iodine and lithium/bromine systems are often referred to as solid-state batteries; however, neither are truly solid state, since both cathode materials are really mixtures of solid plus liquid. There has been only one truly solid-state battery developed for implantable use. This battery system was based on lithium as an anode and a mixture of lead iodide and lead sulfide as the cathode. These materials discharge to form lithium iodide and lithium sulfide plus free lead. The reactions in Eqs. (5a) and (5b) occur simulataneously during discharge.

$$2Li + PbI_2 \rightarrow 2LiI + Pb \quad (5a)$$

$$2Li + PbS \rightarrow Li_2 + Pb \quad (5b)$$

The observed open-circuit voltage for this cell is around 1.9 V. The electrolyte/separator material used was a dispersion of lithium iodide and high surface area alumina. This mixture has an ionic conductivity several orders of magnitude larger than pure lithium iodide. Even so, this battery was useful only for very-low-rate applications. Energy densities approaching 0.6 Wh/cm^3 were achieved in pacemaker configurations. These cells were implanted prior to 1980, but they did not achieve wide acceptance. Self-discharge was exceedingly small, but early versions of this battery were subject to cracking and separating of the solid layers because of volume changes occurring on discharge.

Batteries based on the lithium/thionyl chloride chemistry system were also used to power pacemakers between 1975 and 1980. This battery system is capable of very high rates and inherently has a high energy density since the electrolyte solution is also the chemical cathode. The discharge reaction for this battery system is complicated but involves the formation of LiCl, SO_2, and various other sulfur products.

$$4Li + 2SOCl_2 \rightarrow 4LiCl + SO_2 + S \quad (6)$$

The thermodynamic voltage for this reaction is about 3.65 V. Observed values range from 3.63 to over 3.74 V. It is thought that the high open-circuit voltages are caused by low levels of very active impurities such as chlorine. However, it is also possible that solution equilibria and adducts between thionyl chloride and the lithium tetrachloroaluminate salt are responsible. The energy density of this system, in small cells, can be over 1 Wh/cm^3. The discharge curve is quitc flat, but careful engineering design is necessary in order to avoid an abrupt decline in voltage on cell depletion. For this reason considerable work has recently been done toward adding end-of-life indicators to this battery system.[18–24] Such indicators are designed to produce a voltage step[18–24] when approximately 10% of the battery capacity remains. This is a general problem for battery systems that have discharge voltages that are relatively independent of the state of discharge.

Early studies on this system indicated self-discharge losses on storage were reasonably small. Therefore, it was generally assumed that self-discharge would not be a problem even though thionyl chloride in direct contact with lithium is not thermodynamically stable. The observed stability was attributed to the formation of a passive lithium chloride film that served as electrolyte and separator. However, the estimates of self-discharge based on static experiments proved to be far too low for the dynamic operating state. The reason is that the passivating LiCl layer on the anode is disrupted by the passage of current. Thus, a lithium/thionyl chloride battery operating at a nonzero current has more self-discharge than one standing at open-circuit.

Microcalorimetry has been very useful in estimating the extent of this self-discharge. However, the results have been complicated by the fact that the heat flux varies in a complicated manner with the state of discharge[11] and leaves the question of whether all the observed heat is due to self-discharge or whether there are secondary reactions of the discharge product occurring. In any case, self-discharge, while not nearly as small as in the other lithium systems discussed earlier, can still be controlled well enough to allow the battery to maintain high energy densities at low rates. Work is continuing on this battery system to reduce the amount of self-discharge through purification of components and application of anode coatings that control the diffusion of thionyl chloride to the lithium surface.

Lithium/thionyl chloride batteries are not currently being used in cardiac pacemakers. There are two major reasons for this. The first is the poor experience with early lithium/thionyl chloride batteries. While these batteries functioned very well, their performance was extremely optimistically projected, with relatively little known about the details of discharge and self-discharge. In addition to this, the end-of-life was observed to be extremely abrupt. This abrupt end-of-life when combined with a very much higher than projected amount of self-discharge led to a great uncertainty as to when the batteries would cease to function. This became a large problem in their continued usage. The technical problems could have been overcome, but the overwhelming success of the lithium/iodine battery made it difficult to justify their usage for pacemaker applications.

However, there has recently been a resurgence of interest in this system. It is now being redesigned for the new, higher-current applications that are beginning to appear in new implantable medical products, including multichamber pacemakers and drug delivery pumps. The reason for this renewed interest lies in the fact that the lithium/iodine system probably cannot support the high currents needed by these next-generation devices.

One of the early cells to be qualified for implantable use was the lithium/silver chromate battery. Full discharge of this battery shows a two-step plateau. The open-circuit voltage for the first step is about 3.3 V and is due to the reduction of silver in the silver chromate according to the reaction.

$$2Li + Ag_2CrO_4 \rightarrow Li_2CrO_4 + 2Ag \quad (7)$$

This cell has low internal impedance and the discharge occurs at about 3.2 V at moderate rates. A second plateau with an open-circuit voltage of about 2.5 V is due to the further reduction of chromate ion. This second plateau appears about two thirds of the way through the useful discharge of the battery and is useful as an end-of-life indicator for the battery. The battery is made using propylene carbonate and lithium perchlorate as the solvent and electrolyte salt. Silver chromate is nonconductive and is therefore mixed with graphite to permit electron distribution to reduce the silver ion. Lithium perchlorate in propylene carbonate is a lithium ion conductor; therefore, the discharge product is formed in the region of the chemical cathode. Implantable cells were originally made with a crimp seal, but the seal was later changed to a welded hermetic design. Experience with both types of cells has been good.

The lithium/copper sulfide battery is another important implantable chemical system. This battery is a solid cathode system utilizing lithium perchlorate in a mixture of nonaqueous solvents, principally dimethoxyethane and dioxolane. Discharge involves transferring lithium ions to the cathode side of the cell and a two-step reduction of copper sulfide to copper plus lithium sulfide. At low rates the discharge curve shows two steps corresponding to reactions (8a) and (8b).

$$2\,Li + 2CuS \rightarrow Li_2S + Cu_2S \quad (8a)$$

$$2Li + Cu_2S \rightarrow Li_2S + 2Cu \quad (8b)$$

The first, with an open-circuit voltage of 2.15 V, reduces Cu(II) to Cu(I). The second discharge step has an open-circuit voltage of about 1.8 V and reduces Cu (I) to Cu (O). At fast discharge rates, no distinct plateau occurs, as the voltage is low enough to cause immediate reduction to copper metal. Energy densities in this system initially were of the order of 0.5 Wh/cm^3, but have been improved by advanced designs. The system is capable of fairly high current loads and has excellent reliability experience for implantable use. Self-discharge is very small.

A relatively new implantable battery system uses the lithium/manganese dioxide chemistry. This battery system is actually one of the three most widely commercialized lithium batteries for nonimplantable applications. The cell discharge reaction incorporates lithium ions into the MnO_2 lattice.

$$Li + MnO_2 \rightarrow Li{\cdot}MnO_2 \quad (9)$$

However, there is some controversy about the details of this reaction. The open-circuit voltage of the cell may be as high as 3.5 V, but the discharge voltage is usually between 3.0 and 3.1 V. The discharge curve is relatively flat and falls off sharply toward the end-of-life. For this reason, end-of-life indicators are advisable when using this system as an implantable power source. The energy density of the battery system may be as high as 0.8 Wh/cm^3, and the rate capabilities are certainly adequate for most implantable products. Self-discharge is low, but pretreatment of the cathode material is generally necessary in order to avoid initial gassing when the cell is constructed. This initial gassing appears due to the reaction of active cathode surface sites with the solvent. Implantable cells are hermetically sealed and usually in a can negative design.

The last chemistry to be discussed is lithium/vanadium oxide. This cell chemistry has been used to power prototype implantable heart defibrillators, since it can be designed for very-high-rate applications. The discharge curve has a lot of fine structure showing several discharge products are formed throughout its life. The discharge voltage lies between 3.4 and 3.2 V. The energy density of high-rate cells is about 0.3 Wh/cm^3.

A major concern with all lithium cells is their safety. Because of the very nature of the high-energy components in these batteries, a lot of energy is available in a relatively small volume. One of the major concerns of the battery design engineer is to package this energy in a way that precludes too quick a release. This involves design considerations such as controlling the anode and cathode areas, plus limiting the conductivity of the electrolyte. However, a lot of effort has also gone into determining the fundamental reactions associated with the chemistry of these cells. This is particularly true of the lithium/thionyl chloride battery, for which a number of mishaps with larger cells have been reported. While it is not possible to determine if a cell is completely safe under all circumstances, experience with the small lithium cells used for implantable applications has been extremely good and no mishaps have been reported. However, only continuing work and investigation will provide the information necessary to maintain this excellent record of reliability. It can certainly be said the advantages offered by the lithium systems for implantable uses have been one of the major reasons implantable medical devices have progressed so far in the last 10 years.

REFERENCES

1. K. R. Brennen, K. E. Fester, B. B. Owens, and D. F. Untereker, *J. Power Sources 5,* 25–34 (1980).
2. W. D. Helgeson, Medtronic, Inc. Minneapolis, Minnesota, private communication, 1984.
3. A DeHaan and H. Tataria, U.S. Patent 4388,380 (1983).
4. W. R. Brown, C. F. Holmes, and R. D. Stinebring, Corrosion Resistance of Lithium/Iodine Batteries Fabricated in an Extremely Dry Environment, *J. Electrochem. Soc. Extended Abstracts 81-2,* 228 (1981).
5. E. J. Prosen and J. C. Colbert, NBSIR 77-1310, National Bureau of Standards, Washington, D.C.
6. L. D. Hansen and R. M. Hart, Proceedings of the Reliability Technology for Cardiac Pacemakers Workshop III, NBSP 400–500 (H. A. Schafft, ed.), pp. 10, June, 1979, National Bureau of Standards, Washington, D.C.
7. L. D. Hansen and R. M. Hart, *J. Electrochem. Soc. 125,* 842 (1978).
8. W. Greatbatch, R. McLean, W. Holmes, and C. Holmes, *IEEE Trans. Biomed. Eng. BME-26*(5) (1979).
9. D. F. Untereker, *J. Electrochem. Soc. 125,* 1907 (1978).
10. K. R. Brennen and D. F. Untereker, Iodine utilization in Li/I_2 (poly-2-vinylpyridine) batteries, in: *Proceedings of the Symposia on Biomedical Implantable Applications and Ambient Temperature Lithium Batteries* (B. B. Owens and N. Margalit, eds.), Vols. 80–84. The Electrochemical Society, Princeton, New Jersey (1980).
11. R. Buchman, K. Fester, B. Patel, P. Skarstad, and D. F. Untereker, 164th National Electrochemical Society Meeting, Washington, D.C., October 1983, Abs. 42.
12. S. Ruben, U.S. Patent 2,422,045 (1947).
13. A. Senning, Cardiac pacing in retrospect, *Am. J. Surg. 145*(6), 733–739 (1983).
14. F. Gutmann, A. M. Herman, and A. Rembaum, *J. Electrochem. Soc. 114,* 323 (1967).
15. J. B. Phipps, T. G. Hayes, P. M. Skarstad, and D. F. Untereker, Lithium/Iodine Batteries with Poly-2-Vinylpyridine Coated Anodes: A Microstructural Investigation, 166th National ECS Meeting, New Orleans, October 1984, Abs. 175.
16. R. C. Stinebring, C. F. Holmes, and M. M. Safford, Determining the Reliability of Lithium/Iodine Pacemaker Cells Using a Life Test Sample, 163rd National Electrochemical Society Meeting, San Francisco, California, May 1983, Abs. 27.
17. W. D. Helgeson, Design Testing and Reliability of Lithium/Iodine Pacemaker Batteries, 163rd National Electrochemical Society Meeting, San Francisco, California, May 1983, Abs. 28.
18. K. Takeda, I. Kishi, K. K. D. Seikosha, U.S. Patent 4,144,382 (1979).
19. J. P. Gabano, U.S. Patent 4,371,592 (1983).
20. A. DeHaan and H. Tataria, U.S. Patent 4,388,380 (1983).
21. N. Marincic and J. Epstein, U.S. Patent 4,293,622 (1981).
22. F. Goebel and R. McDonald, U.S. Patent 4,416,957 (1983).
23. F. Geobel, U.S. Patent 4,418,129 (1983).
24. P. Keister, J. M. Greenwood, C. F. Holmes, and R. T. Mead, Performance of Li Alloy/Li and Ca/Li Anodes in $SOCl_2$ Cells, 2nd International Meeting on Lithium Batteries, Paris, France, April, 1984, paper 27.

4

Evaluation Methods

KEITH FESTER and SAMUEL C. LEVY

1. EVALUATION OBJECTIVES

Battery evaluation as applied to power sources for biomedical applications is the appraisal of both battery designs and individual units. The appraisals are necessary to determine a battery's worth for satisfying the needs of the biomedical device and, ultimately, the patient. The primary needs are (1) an adequate supply of energy to operate the device for the intended service life within the specified power requirements and (2) a highly reliable energy source. The battery evaluation activities have the objective of verifying how well the battery design satisfies these performance and reliability goals. Furthermore, when a particular design is qualified for a device and begins to be manufactured on a production line, it is necessary to evaluate individual units for quality assurance requirements.

This chapter briefly describes the aforementioned evaluation objectives and covers, in detail, some test methods that have been or could be applied. Special attention is given to the theory and methods of accelerated testing. Examples illustrate how the test data are acquired and processed for pacemaker batteries. Long-term battery reliability data are presented for some widely distributed pacemaker models. Current methods of statistical evaluation of battery data are described in detail.

1.1. Performance Data

The performance goals for a biomedical device battery are commonly established in a specification document prepared by the design engineers. The

KEITH FESTER • Medtronic, Inc., Minneapolis, Minnesota 55440. SAMUEL C. LEVY • Sandia National Laboratories, Albuquerque, New Mexico 87185.

parameters usually specified are battery capacity, open-circuit voltage, a specific load voltage, and the initial ac impedance at 1 kHz. All measurements are usually carried out at 37°C to simulate the body environment.

The capacity specification is usually set for operation between the initial load voltage and some lower voltage, at which point the battery can trigger an end-of-life indicator of some type. For most pacemakers, a change of the patient's pulse rate occurs when the battery reaches the triggering voltage. The load-voltage specification is usually a constant-load measurement that provides a current close to the typical discharge rate in the medical device. The ac impedance parameter is often used as an upper limit to control the uniformity of production batteries. In the case of Li/I_2 cells, the measurement is a fair indication of the amount of lithium iodide salt formed between the electrodes due to discharge during cell manufacturing and conditioning.

In addition to performance in the device, the battery design must withstand typical environmental exposures to high and low temperature extremes, and shock and vibration. These conditions occur prior to implant during the assembly and shipping of the batteries or medical device. Also, capacity losses of the batteries during the shelf time between the cell assembly and device implant must be low. Finally, implantable-grade lithium batteries must maintain a hermetic seal to prevent degradation of the lithium and leakage of the cell materials. All of these performance requirements have to be evaluated before a design goes into regular production to ensure that adequate device service life will be achieved.

1.2. Reliability Data

The other main objective of battery evaluation is the determination of the design's reliability. Here reliability is defined as the probability that the power source will perform as required during the normal service life of the medical device. In reporting reliability characteristics, it is common to present the random failure rate F which is the complement of the survival rate S ($F + S = 1$). This value is meaningful until the batteries begin to lose function because of chemical exhaustion of the cell electrodes.

A precedent was set for a pacemaker power source reliability standard when the nuclear pacemaker was introduced and came under evaluation by the U.S. Atomic Energy Commission in 1973.[1] The performance standard recommended to the AEC was based on two facts, the longevity claims for the existing chemical-battery-powered pacemakers (six years) and the claimed electronic components' failure rate of 0.15% per device month for the pacemaker models. In order to demonstrate a benefit from nuclear pacemakers, the condition was set that 90% of the devices avoid random failure in six years. This corresponded to the 0.15% per device month failure rate claimed for the electronics. Cumulative survival for such a device population would be 81% at 12 years and 50% at 38 years.

Based on this reliability goal the battery and other components must exceed a 99.85% per device-month survival rate during the specified battery life.

For conventional chemical batteries, no reliability standard has ever been formally established. Ideally, the goal is zero defects for medical device batteries, especially when the device provides a life-sustaining function. In practice, the battery development team works to produce and verify a design that is at least as reliable as the models currently in use. The reliability figures from some widely used batteries are far superior to the 1973 random failure rate value of 0.15%/month, and will be presented later on in this chapter.

1.3. Quality Assurance

In addition to the performance and reliability information needs, evaluation of batteries is conducted to satisfy quality assurance needs. A common practice of manufacturers is the partial discharge, or burn-in, of batteries at 37°C just after cell assembly. This type of burn-in is not like the MIL-STD (Military Standard) procedures commonly applied to electronic systems whereby they are subjected to failure-provoking temperature cycles prior to shipment.[2] Instead, the partial discharge is done to stabilize the electrical properties of the batteries and to identify units that perform outside the specified control limits. The inspection parameters for Li/I_2 cells are usually open-circuit and load-circuit voltage, and a 1-kHz impedance measurement after the burn-in.

Evaluation of pilot production samples is necessary to establish the quality control limits, using histograms and statistical analysis methods. Furthermore, follow-up tests can be run to determine the performance characteristics of batteries outside the control limits. Comparisons can be made with the acceptable units to determine the effectiveness of the quality criteria.

Before the batteries are ready for burn-in, many evaluation steps are applied to the battery materials and components both prior to and during the assembly process. Quality inspections run the gamut from chemical composition of the electrode materials to destructive analysis of completed cells. Intervening checks for such features as glass seal hermeticity, stainless steel microstructure, and weld closures have been described by Liang and Holmes.[3]

2. ACCELERATED TESTING

The purpose of accelerated testing of batteries is to assess their probable performance and reliability, particularly for long-life applications, in a much shorter time period than can be obtained by real-time testing. The usual approach to accelerated testing is to impose a stress (e.g., elevated temperature) on the battery at higher than normal levels to shorten the time to failure. Similar results can sometimes be obtained by the use of sensitive measurement techniques that are capable of detecting very small changes that occur within the battery at normal stress levels and that may lead to unsatisfactory performance. A good example of such a technique is microcalorimetry, used by the manufacturers of batteries for cardiac pacemakers to determine rates of self-discharge. Internal battery characteristics can also be measured with complex impedance analysis. While a complete understanding of the resultant data is not yet available, valuable information concerning the formation of resistive layers at various interfaces within the battery can be obtained. Combining this technique with microcalorimetry promises to yield even more detailed information about the self-discharge processes occurring in batteries under "normal" conditions. Use of these sensitive techniques for battery characterization has been reported in the literature.[4–16]

In an accelerated test, the expected battery lifetime is shorter than that under normal use conditions, because of the higher than normal stress levels. Most batteries can fail in a number of ways that are usually identified from real-time lab tests and field tests. In accelerated testing it is essential to know what these normal failure modes are. Only batteries failing by these modes should be included in the data analysis, as the acceleration process itself may induce additional, unrealistic failure modes.

Each failure mode is the result of a unique physical or chemical degradation process. Once a failure mechanism has been identified as being responsible for a certain mode of failure, it is usually possible to identify stresses that will increase the time rate of the degradation process. Accelerated tests are run at selected values of these "accelerating stresses". The acceleration factor f for the experiment is defined as the number of hours of normal operation required to produce the same amount of degradation as occurs in one hour of testing under the accelerated stress conditions. In order to minimize the impact of the test upon degradation mechanisms, valid acceleration factors are usually less than 10.[17]

Prior to starting an accelerated test, some of the samples must be analyzed to ensure that they are identical with respect to construction, quality, and initial performance. At least five samples of a reasonable size population should be analyzed to ensure homogeneity and establish a baseline for subsequent analyses. In addition, enough samples should be placed on test at each accelerating stress level to allow for periodic removal and destructive analysis. This is necessary, as nondestructive measurements alone frequently are not able to provide enough information on the degradation process as the sample ages. It is recommended that specimens be removed at five different times during the test period.[17]

There are three approaches to accelerated testing that may be employed: empirical, statistical, and physicochemical. In the absence of a detailed knowledge of the relationships between stress levels and aging, all three approaches should be applied. The following sections summarize the different approaches. A detailed summary of these approaches can be found in the literature.[18,19]

2.1. Empirical Approach

This approach regards the data as containing unknown relationships that may be imbedded in noise. Analysis of the data consists of searching for structure. A number of exploratory techniques may be applied in analyzing the data to reveal any significant patterns. To prevent suppression of the significant detail, the data are smoothed as little as possible. Upon identifying structure in the data, confirmatory statistical methods may be applied to assess the persistence of this structure in the background noise. A large number of exploratory techniques such as those described in Velleman and Hoaglin[20] are suitable for this testing approach. An empirical analysis is data dependent with no specified sequence to be followed. The search for structure in the data is by purely exploratory methods and the results need not explain, but just show these interactions. By analyzing the data in raw form, errors, inconsistencies, and suspicious outliers can be detected. After removing identified discrepancies, a search for more subtle features of the data may be initiated.

It is important to monitor the data by methods suitable for detecting errors and outliers. In this way suspicious behavior can be detected quickly and corrective measures taken if appropriate, in order to maintain a high level of quality of the data. The objective of an on-line exploratory approach is to identify typical behavior and "tail". The tails may be indicative of either poor or excellent performance. In either case, more measurements should be made on the tails in order to establish trends early in the test.

The advantages of the empirical approach to accelerated testing are several:

- Minimal prior assumptions need to be made.
- There is minimal smoothing of the raw data.
- A minimum amount of information is lost.
- There is an increased probability of discovering important, unsuspected relationships.

The main disadvantage is the large amount of data to be analyzed. Use of computers is a necessity and only partially alleviates the problem.

2.2. Statistical Approach

In a statistical analysis of accelerated test data, a model is generated. After assessing the noise component of the data, a smooth surface is fitted. Modeling the response surface coefficients as a function of time allows the future positions of the surface to be predicted. This is called a "Dynamic Response Surface" approach.[18]

Statistical methods are based on averages, since the standard deviation of an average divided by the standard deviation of a single observation is $\sqrt{1/n}$. For statistical stability, at least five matched cells are required at each stress level. According to the central limit theorem, the averages are approximately normally distributed, even if the individual readings are not. To determine whether or not there is a change in degradation mechanism with increased stress level, a minimum of five stress levels must be used. The lowest should be the intended use condition, while the highest should be as large as can be accommodated without changing the degradation mechanism. Usually, an educated guess is required for the high stress level, followed by experimentation to evaluate the validity of the value selected. For thermal stress testing, this can be accomplished by plotting the measured log of the degradation rate versus stress level ($1/T$). A linear relationship indicates a consistent degradation mechanism over all stress levels; a nonlinear relationship implies a change in mechanism.

Any number of statistical methods may be used to analyze the data. It is not the intent of this chapter to discuss statistical analysis, since there are many books on the subject. A good example of the use of statistics to analyze accelerated test data is found in reference 18.

2.3. Physicochemical Approach

To use the physicochemical approach to accelerated life testing of cells, the processes occurring in the cells must be known. The following assumptions are made concerning the degradation of cell performance:

- Every cell has a quality that can be derived from physical measurements.
- This quality varies with time and degrades after sufficiently long time.
- Higher degradation rates occur at higher levels of stress.
- The degradation will eventually lead to failure of the cell.
- The physicochemical mechanisms leading to degradation do not change with increasing level of stress, except in rate.

Measurement of the initial quality and its time rate of degradation are both necessary to an accelerated life test.

A number of physicochemical approaches to accelerated testing have been used.[19] The most widely used technique in testing all types of batteries involves the Arrhenius model. The Eyring model can be used similarly.

2.3.1. *Arrhenius Model*

If one assumes a normalized dimensionless measure of the quality $V(t)$, such that $V(0) = 1$ and a rate of degradation $r(T)$, that is dependent on the absolute temperature T and constant with time, then at any time t,

$$V(t) = V(0) + r(T)\,t \tag{1}$$

The Arrhenius equation may be written as

$$r(T) = e^{a-(b/T)} \tag{2}$$

or

$$\ln r(T) = a - (b/T) \tag{2a}$$

Using least squares estimates of the intercept, a, and slope, b, Eq. (1) may be rewritten as:

$$V(t) = 1 + [e^{a-(b/T)}]\, t \tag{3}$$

If t and T represent the time and absolute temperature at normal stress levels, and t' and T' the same parameters at a high stress level, and if one assumes the same amount of degradation has occurred at both stress levels, that is, $V(t) = V'(t')$, then:

$$1 + [e^{a-(b/T)}]t = 1 + [e^{a-(b/T')}]t' \tag{4}$$

$$t = [e^{-b(1/T'-1/T)}]\, t' \tag{5}$$

Since the acceleration factor f has already been defined as the number of hours of operations at normal stress required to equal one hour of operation at high stress, that is,

$$t = ft' \tag{6}$$

it follows that:

$$f = e^{-b(1/T'-1/T)} \tag{7}$$

Equation (7) states that the acceleration factor involves only the slope of the Arrhenius plot, that is, the acceleration factor is dependent on the activation energy of the degradation reaction.

If a degradation rate $r(T)$ is found to be valid over several levels of stress, the decay curves at the various stress levels are superimposable by a horizontal shift in the log time axis.[21] When superposition is not found, the accelerating condition did not result in a constant acceleration of the degradation process, making any extrapolated predictions highly speculative. This points out the advantage of following the degradation as a function of time, rather than depending upon one data point such as the time to failure. A shift factor for thermal stress s (the most commonly used stress in accelerated aging studies of batteries) may be defined by the following proportionality:

$$s \propto e^{-(E_a/RT)} \tag{8}$$

where E_a is the Arrhenius activation energy. A plot of ln s versus reciprocal temperature has a slope of $-E_a/R$ for degradations due to chemical reactions, since most chemical reactions exhibit Arrhenius behavior.

The preceding discussion assumes a single degradation process, that is, a single activation energy for all stress levels, resulting in a linear Arrhenius plot. If a change in the dominant degradation process occurs as a function of T, a nonlinear Arrhenius plot will result.

The degradation of a substance may be represented by the expression:

$$\frac{dC_A}{dT} = k_1(T)C_A + k_2(T)C_A \tag{9}$$

where C_A is the concentration of substance A, and there are two pathways for the degradation of A, both first order with respect to A, resulting in an overall degradation that is first order with respect to A. If we assume:

$$k_1(T) \propto e^{-(E_{a1}/RT)} \tag{10}$$

$$k_2(T) \propto e^{-(E_{a2}/RT)} \tag{11}$$

and

$$E_{a1} > E_{a2}$$

then the nonlinear plot of ln (dC_A/dt) versus $1/T$ has a segment in the higher-temperature region where the reaction with the steeper slope, E_{a1}/R, dominates the curve. At the lower temperatures where the process of activation energy E_{a2} is dominant, the curve resembles a straight line of slope E_{a2}/R.

2.3.2. *Eyring Model*

In the Eyring model,[19] the time rate r for the transition from an acceptable to a failed condition due to an increase in temperature stress is given by:

$$r = T\left(\frac{Ak}{h}\right)e^{-b/T} \tag{12}$$

where h = Planck's constant, A depends on the partition function, and b is the ratio of the activation energy to Boltzman's constant ($b = \Delta E/k$). The theoretical calculation of A and b is very difficult, but they can be estimated by taking logarithms of Eq. (12).

$$\ln (hr/kT) = \ln A - b/T \tag{13}$$

A plot of ln (hr/kT) versus $(1/T)$ yields a straight line with intercept ln A and slope $-b$, if degradation is controlled by a single mechanism.

The Eyring model differs from the Arrhenius model only in the fact that the coefficient of the exponential term is multiplied by a temperature-dependent factor kT/h. Usually this term is relatively small and can be neglected, which then reduces the Eyring model to the Arrhenius model.

An advantage of the Eyring model is that it can be used to consider stresses other than temperature (e.g., rate, pressure, mechanical stress, etc.).[19]

2.4. Accelerated Testing without Failure

A new approach to accelerated testing has been proposed by workers at Battelle.[19] In conventional accelerated tests, cells must fail at high stress levels. If, for some reason, no failures occur at the high stress levels, then nothing can be learned about failure at lower stress levels or at normal use levels. Suppose N samples are tested for a period of time D with an acceleration factor $f > 1$ and no failures occur. What can then be said about the life of the cells when $f = 1$? If one uses a Weibull distribution:

$$P_r\,[D>t] = e^{-(t/\tau)^{\beta}} \tag{14}$$

in which $P_r\,[D > t]$ is the probability a cell fails after time t, τ is the characteristic failure time, and β is a shape parameter, an exponential distribution will result when $\beta = 1$. Normally, however, β will vary between 1 and 4. If we assume β is independent of the accelerated stress level and τ is a function of the stress level, then the accelerated test can be expressed in a format analogous to the statistical sampling plans for lot acceptance (i.e., samples are drawn at random and subjected to specific tests). In a lot acceptance test, if fewer than k cells fail, the entire lot is accepted. The size of the sample to be tested n and the acceptance number k depend on γ, the manufacturer's risk of good lots being rejected, and σ, the consumer's risk of bad lots being accepted.

The fraction defective p, is the fraction of cells that would fail before a specified time at normal stress levels. The acceptance number, k, then represents the number of cells allowed to fail at the high stress conditions such that the fraction that will fail before some specified time under normal stress is $< p$. Depending on the acceptable risk level for γ and β, k can be zero. In this case, if no failures occur at the high stress levels, then the life will be acceptable at normal stress conditions. If one or more cells fail at high stress conditions, this implies unacceptably short lives at the normal stress or use condition.

The number of cells required for testing at high stress levels is:

$$n = N/f^{\beta} \tag{15}$$

where N is the number of cells required at normal stress levels (this is usually very large, 10^3–10^4). Using an acceleration factor $f = 10$ and a shape parameter $\beta = 2.5$, the number of cells required for testing is reduced by a factor of 1/316.

Equation (15) can be used to determine the feasibility of conducting an accelerated test, taking into account various limitations. Using the logarithmic form of (15):

$$\log n = \log N - \beta \log f \tag{16}$$

and plotting log n versus log f yields a straight line (Fig. 1). If the maximum number of cells that can be tested is $n_{\max}$ and the limiting acceleration factor for the particular cell to be tested is $f_{\max}$, the point ($\log f_{\max}$, $\log n_{\max}$) is plotted on the same graph. If this point lies on or above the line generated by Eq. (16), it is feasible to construct an accelerated test. If the point lies below the line, construction of an accelerated test is not feasible. This is illustrated in Figure 1.

If an accelerated test turns out to be feasible, segment AB of the line indicates the range of sample sizes and acceleration factors that can be used to determine whether performance will be acceptable at use conditions at the required risk levels. Therefore, an accelerated test can be designed using a variety of conditions described by line segment AB. The data obtained can then be reduced, using whichever model one chooses, to describe the anticipated dependence between the acceleration factor and the various stress levels chosen.

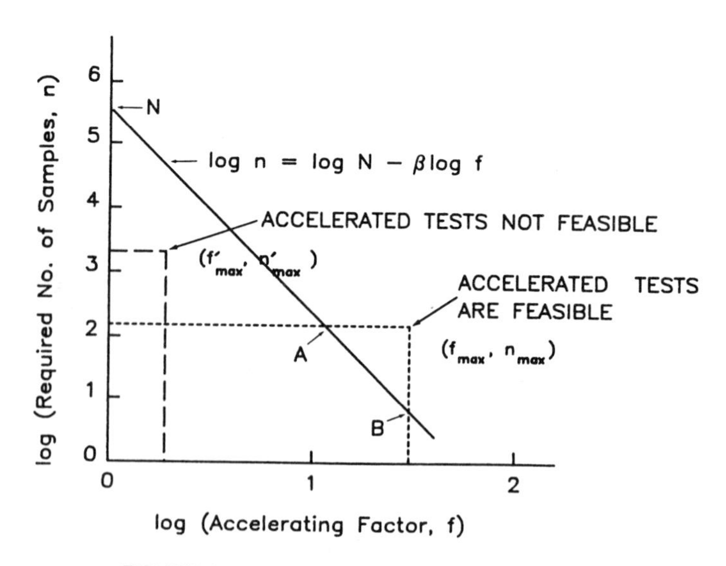

FIGURE 1. Feasibility Diagram for Accelerated Testing.

2.5. Designing an Accelerated Life Test

When designing an accelerated life test, a certain minimum number of experiments must be run to obtain meaningful information. A look at the three approaches to accelerated testing can give some insight as to what a minimal test matrix should consist of. In the empirical approach, sufficient data are required to identify typical behavior. Nonparametric statistical analysis includes a number of tests for significance of data, and these tests are described in the literature,[22–24] along with the minimum sample size for each. The majority of these tests require a minimum sample size of five. Therefore, from an empirical point of view, five cells should be tested at each stress level. Using similar logic, it can be concluded that a minimum of five measurements should be made on each cell during the time period of the test.

The statistical approach involves the use of parametric linear regression models that presuppose normally distributed errors. In order to estimate the parameters of the regression line (slope, intercept) with sufficient precision, we need to measure at least five cells seven times each.[25]

Using the physicochemical approach to accelerated testing, one can determine the minimum number of stress levels to be included from the Arrhenius and Eyring models. The data are plotted as the logarithm of the degradation rate versus reciprocal stress level (i.e., temperature), and a straight line is expected if only one degradation mechanism is occurring. Therefore, three stress levels should be adequate to validate the straight line. However, to extrapolate to normal use conditions, one must be certain that a change in mechanism has not occurred. If two degradation mechanisms operate over the stress levels of the accelerated test, a minimum of five stress levels is required to define both straight lines.

The three approaches to accelerated testing indicate that a minimal test matrix should consist of:

1. Five cells at each stress level
2. Seven measurements over the period of the test
3. Five levels of stress

Thus, 25 cells are required for a minimal accelerated life test, each cell having seven measurements made upon it. The 25 cells should be as nearly identical as possible. If different types of cells are to be tested (e.g., different manufacturer, various sizes) each cell type would require 25 samples. Thus, if one type of cell from two different manufacturers in four sizes were to be evaluated, a minimum of $25 \times 2 \times 4 = 200$ cells would be required to obtain data on every combination.

2.6. Other Acceleration Methods

In addition to the application of high temperatures to advance the aging of battery samples, other forms of accelerated stress testing can be applied to assess reliability. Two that have been used on pacemaker battery evaluation are high-rate electrical discharge and mechanical stress.

Pacemaker batteries typically operate at a 20–60-μA discharge rate, depending on the model, patient, and the output settings of the device. Real-time discharge of new battery designs would require 5–10 years of testing to determine design performance reliability. The accelerated discharge of batteries at rates 5–10 times faster has been used successfully to predict the design behavior in much less time. Several approaches have been developed along this line of evaluation.

2.6.1. *Fast Discharge*

One practice consists of initial cell discharge at a high rate and then stepping down the current when output voltage drops off due to the internal resistance. An example of a Li/I_2 cell test is shown in Figure 2. The last current is in the pacemaker application range. The capacity delivered at this rate to the end-of-life voltage level can be described as the accelerated application capacity.[26] Even with current acceleration, it can require over one year to reach the EOL of a Li/I_2 cell at application drain. Other cell chemistries that use more conductive cathodes and liquid electrolytes (e.g., MnO_2, $SOCl_2$) can be discharged at much higher rates (in the milliampere range).

An important performance feature determined by this step-down method is the rate of voltage decay due to normal cell exhaustion. This information is

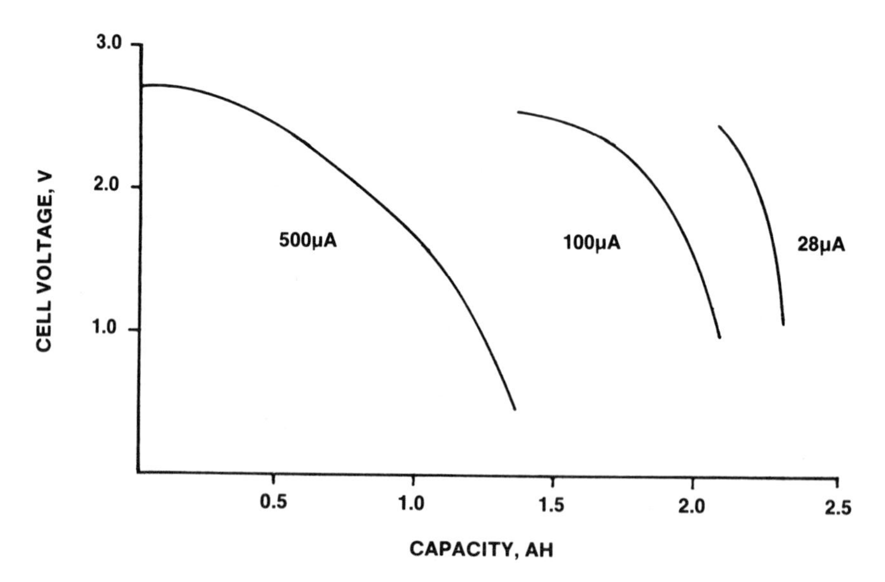

FIGURE 2. Accelerated Discharge of a Li/I_2 Cell, 2.2 AH Delivered to 2.0 V End-of-Life.

valuable for selecting and verifying the EOL circuit signals that indicate when a battery needs replacement. Also, the battery data can be used to estimate the remaining useful service life once the signal voltage has been reached.

Another variation of accelerated battery discharge has been described by Liang and Holmes.[3] Li/I_2 cells are discharged at high rates of 5- or 10-kΩ loads to predetermined levels and then switched periodically onto typical application loads, [e.g., 140 kΩ (or 20 μA) for a voltage measurement]. The low rate readings provide an estimated discharge curve for a normal device load that can subsequently be verified by continuous low rate discharge.

A third form of accelerated discharge testing consists of constant current tests that provide a family of curves as shown in Figure 3. These voltage–capacity curves are used to construct mathematical models for the purpose of making projections of performance at application rates. The solid lines show the extent of completed measurements. While this example shows two years of performance, high confidence projections can be prepared by this method on new Li/I_2 designs within six months of test experience. A more detailed discussion of this method can be found in Chapter 5 of this book.

2.6.2. *Mechanical Stress*

The second acceleration method in this section is mechanical stress. This technique has its roots in the MIL STD testing of electronic systems. Design samples are exposed to a mechanical shock simulation that represents dropping

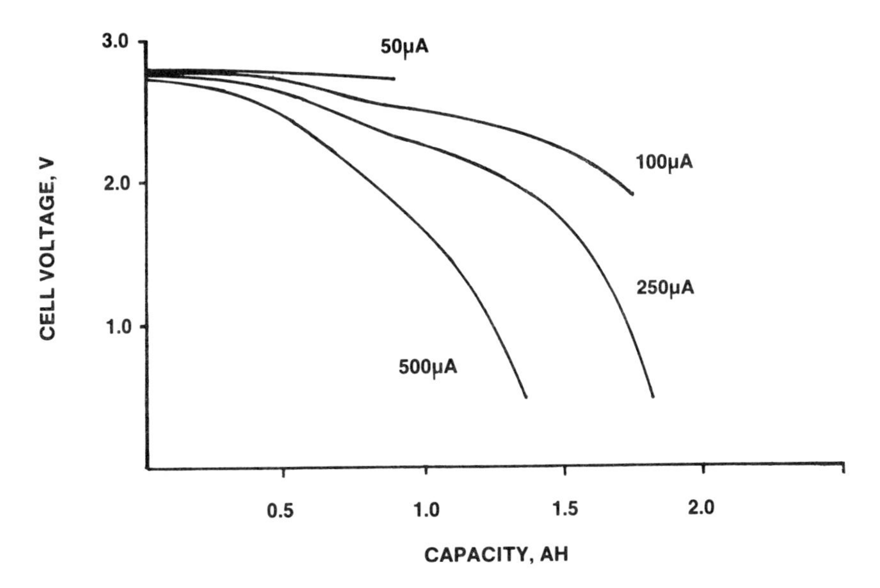

FIGURE 3. Family of Voltage-Capacity Curves, Li/I_2 Cells on Constant-Current Discharge.

the cell to a hard surface from a specified height.[27] Mechanical vibration is simulated by exposing samples to sinusoidal vibration sweeps along each of the cell's three axes.[28]

These test conditions do not accelerate the aging of the device. The goal here is to apply stresses much greater than expected in handling and transporting the batteries and to observe the results. Endurance testing (i.e., applying the stresses repeatedly to look for a wearout behavior) is not necessary for implanted batteries, since the body cushions the device. A mechanical shock strong enough to damage an implanted cell is also likely to significantly affect the nearby bones and tissue, rendering the battery condition a secondary concern.

3. NONACCELERATED TESTING

3.1. Real-Time Tests

In addition to accelerated performance testing, it is useful and often necessary to evaluate biomedical power sources by several other methods. Probably the most credible battery test is the real-time electrical discharge at application rate and temperature. While this test takes several years to complete for a pacemaker battery, there are no questions as to whether accelerated current or temperature is somehow distorting accurate performance measurement. Self-discharge losses, which accumulate with time, are fully reflected in the performance of the low-rate test samples. Reaction product conductance (e.g., LiI salt in Li/I_2 cells) sometimes varies directly with discharge current. This effect leads to an optimistic estimate of a battery's internal resistance from accelerated discharge data.

The low-rate, real-time battery discharge tests are often constant load or constant current. However, when high current pulses are required for the application (e.g., a drug pump, an internal defibrillator, etc), it is necessary to test the battery under actual pulsing conditions. The shifts from low standby currents at the microampere level to milliampere or higher rates can cause initial voltage delay in $Li/SOCl_2$ cells. The reasons for these varied behaviors (which cannot easily be estimated by equivalent circuit models) lie in the chemistry of the systems. The battery cannot be simulated accurately with an electronic power supply and decade box when the battery voltage exhibits time-dependent behavior during pulsing and load changes. It is best to set aside several battery samples connected to device circuits in order to obtain simulated operating data.

An early estimate of how well the battery will handle the circuit requirements can be obtained by testing samples that have reached various levels of advanced depletion on accelerated discharge. This approach must be used with the realization that the batteries will perform somewhat better in the circuit tests, since self-discharge and resistivity tend to be minimized by the shorter discharge times.

3.2. Materials Testing

Another area of battery evaluation is the verification of the reliability of the components and materials used in the battery. Most lithium batteries used in biomedical devices have an anode-centered design with the cathode contacting the walls of the container. The negative terminal from the lithium is usually a wire that exits the battery case through a glass hermetic feedthrough.

In Li/I_2 cells, reliability is ensured by an insulated feedthrough assembly (Fig. 4) that prevents the electronically conductive cathode material from reaching the anode lead and shorting the cell. The insulating material is often a polymer material, selected for its mechanical and chemical properties. The material must be resistant to degradation by the chemicals in the cell. Material stability is best evaluated by building actual cells and performing destructive analysis at selected exposure times.

Implantable battery samples are stored at 37°C. Upon disassembly the polymer component can be checked for electrical resistance, permeation by the cathode, erosion, fractures, weight change, hardness, tightness of the seal, or

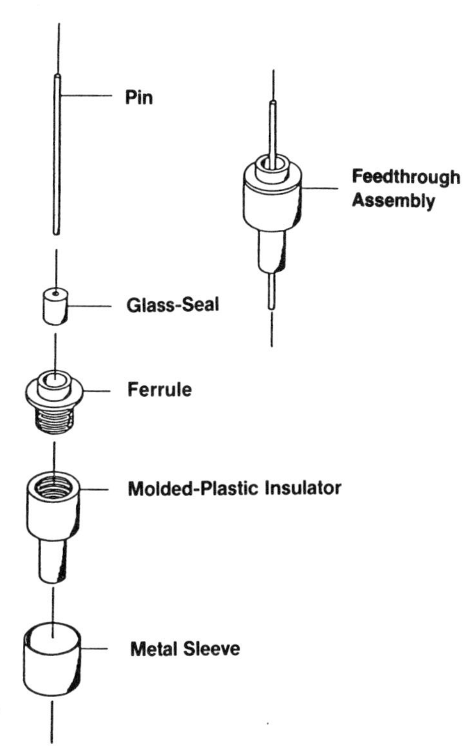

FIGURE 4. Cross Section of Li/I_2 Battery Terminal Feedthrough.

some other property. By sampling regularly over several months and years, the battery component changes can be used to predict real-time performance.

Using a similar methodology other components of the battery can be checked for performance and reliability. An early concern in Li/I_2 cells that received considerable attention was corrosion of the stainless steel battery case by the I_2-P2VP cathode.[29,30] Extensive materials testing showed that while pitting corrosion occurred upon cell assembly, the depth was less than 0.001 in. and did not progress further.

In recent years, attention has been given to glass seal degradation by the cathode/electrolyte in Li/SO_2 and $Li/SOCl_2$ batteries. Glass seal degradation occurs when lithium ions from the electrolyte either adsorb onto or ion exchange into the interior surface of the glass. These ions are reduced to lithium metal by the process of underpotential deposition. Since the glass is a poor electronic conductor, this process occurs at the interface between the glass and the negative terminal (source of electrons). Once formed, the lithium metal will react with the oxides of the glass (particularly SiO_2) to form Li_2O plus reaction products of the glass (e.g., Si). The Li_2O can react further with the matrix, forming oxides of the glass and compounds such as the lithium silicates. The resulting products have sufficient conductivity to allow electrons to reach the next layer of lithium ions and repeat the process. In this manner corrosion occurs across and through the glass from the negative to the positive terminal. Additional underpotential deposition of lithium ions in the corrosion product layer can result in a dispersion of lithium metal throughout the product, giving it a high electronic conductivity. This leads to self-discharge of the cell across the seal.[31] More recently, an alternative mechanism has been reported, involving lithium alloy formation rather than underpotential deposition.[32]

Research at Sandia National Laboratories has led to the formulation of a glass, TA-23, which is extremely stable to this type of degradation. Tests have shown TA-23 glass to be effective in preventing battery failure by glass seal degradation in a number of lithium systems, including Li/SO_2, $Li/SOCl_2$, Li/BCX, and Li/(CFx)n. Use of this glass in implantable batteries should lead to a more reliable product.

3.3. Microcalorimetry

Microcalorimetry is an evaluation technique covered thoroughly in Chapters 3 and 8. Self-discharge reactions in batteries, such as lithium corrosion by iodine vapor, are exothermic. The heat released by a cell on discharge can be measured by a highly sensitive microcalorimeter. The heat measurement, after subtraction of the entropic heat from the normal cell reaction and any heating from internal cell resistance, is indicative of the rate of self-discharge occurring within the battery.[33] This technique has become a very useful battery evaluation tool in recent years. Mathematical models of battery performance have been built from microcalorimetry data and accelerated discharge results and used to make accurate projections of medical device longevity.

4. QUALIFICATION PROTOCOL

The various evaluation methods applied to a biomedical battery design can be collected into a qualification plan. The following is an example of such a plan that could be used to verify the performance and reliability of small Li/I_2 cells capable of operating cardiac pacemaker circuits.

4.1. Sample Qualification Plan

4.1.1. Burn-in and Receiving Inspection

All the units (170) will be discharged at 2 kΩ for 16 h and then monitored on 100-kΩ loads for voltage and impedance per standard 2-kΩ burn-in procedure. OCV will be monitored on the units prior to burn-in and again 24 h after burn-in. After burn-in, cell weight and thickness will be measured on each cell and each unit will be X-rayed. For cells from the production line, a random selection method will be used to assign samples to the following tests.

4.1.2. Accelerated Discharge—64 Units

Sixteen units each will be placed on 400-, 200-, 100-, and 50-μA discharge test at 37°C. Units for the tests described in this test plan will be selected to give a select distribution of various physical and electrical characteristics. The selection procedures will be discussed in detail in the burn-in and receiving inspection report to be issued shortly after burn-in is complete.

Three units selected randomly from the 100-μA test group will be monitored for heat output on the microcalorimeter. The three units will be monitored on open circuit prior to being placed on 100-μA discharge and after three, six, and twelve months on discharge.

4.1.3. Application Discharge—15 Units

The cells will be placed on 100-KΩ load at 37°C, approximately 28 μA, to simulate application drain. If sample circuits become available, some of the cells will be switched to actual circuit drain. Six units from the application drain test will be monitored for heat output after burn-in, after three and six months on discharge, and every six months thereafter.

4.1.4. Destructive Analysis—10 Units

Initially, five units will be analyzed destructively for construction features, signs of corrosion, and any mechanical defects. If any problems are found, additional units will be destructively analyzed from the set of 10.

4.1.5. Environmental Test—16 Units

Environmental testing is done as a 2^3 factorial experiment, with the three independent variables being shock and vibration, exposure to high temperature, and exposure to low temperature.

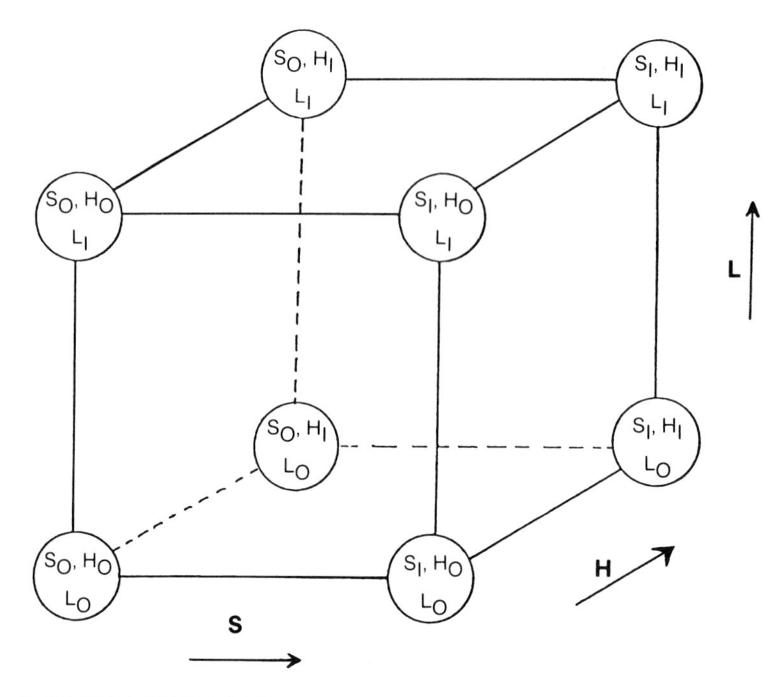

S: SHOCK AND VIBRATION
S_O — No Exposure
S_I — Shock and Vibration Treatments

H: HIGH TEMPERATURE EXPOSURE
H_O — No Exposure, Ambient
H_I — 60°C For 7 Days

L: LOW TEMPERATURE EXPOSURE
L_O — No Exposure, Ambient
L_I — 40°C for 7 Days

FIGURE 5. Environmental Test, 2^3 Factorial Experiment.

Details of the factorial experiment are presented in Figure 5. Two units will be tested at each of the eight conditions for a total of sixteen units. Visual inspection and hermeticity tests will be performed on the environmental test units prior to and after environmental testing. The order of exposure to the three environmental conditions will be as follows:

1. Shock and vibration
2. High temperature
3. Low temperature

Shock and vibration tests are to be run according to in-house procedures derived from MIL-STD-202E.

After the exposure to the environmental conditions, all the units will be placed on 400-μA, 37°C accelerated discharge test.

4.1.6. *Feedthrough Tests—65 Units*

Sixty-five cells will undergo routine receiving inspection and will be burned-in on a 2-kΩ load for 16 h. The electrical parameters will then be recorded before the cells are placed on the following test schedule.

No. of Cells	Constant Load, KΩ	Temp. °C	Number of Units Sampled, After Indicated Storage Time (months)					
			3	4.5	6	12	18	24
30	100	37	5	5	5	5	5	5
30	100	60	5	5	5	5	5	5
5			0-Time (Controls)					

Evaluation of the 65 units will consist of:

Electrical Data. Electrical parameters (voltage, impedance, etc.) will be monitored periodically for early indications of any feedthrough failures.

Visual Observations. Each feedthrough will be disassembled as per the test schedule and the individual components examined. One aspect of importance that will be closely examined is the extent of any cathode complex progression along the ferrule–Halar insulator interface as a function of time and temperature.

Metallographic Examination. Following visual observation, each feedthrough will be cross-sectioned parallel to the pin, polished, and examined using a high-powered microscope. Areas of interest will be:

a. Quality of cover to ferrule weld (depth of penetration and fusion, workmanship, etc.)
b. Glass to metal seal (ferrule to glass and glass to pin)
c. Cracks in the glass seal that could result from the extensive heat input during cover to ferrule welding operation
d. Compatibility of electrolyte with feedthrough components
e. Extent of any corrosion attack

Data analysis will commence with the zero-time controls. Periodic updates will be in the form of status reports after the indicated storage intervals, as per the referenced test plan. A final report will be issued after 24 months.

As the test results accumulate, they are compared with the design performance goals. The battery design may meet the performance and reliability criteria or it may need modification in order to be acceptable. The decision to manufacture the design for biomedical use can occur as early as six months after the initiation of qualification testing if it is a familiar chemistry with minor component changes. Changes involving new chemistries and major packaging design modification changes require longer evaluation times.

5. DATA ANALYSIS

Two important uses of battery evaluation results are (1) the projection of device service life and (2) the estimation of battery reliability statistics. Both of these needs require careful planning and operation of the evaluation tests and the acquisition and analysis of numerous measurements. A computer system and custom software is needed to acquire, store, and process the test data when sample sizes and reading frequencies overwhelm simple resources.

5.1. Longevity Projections

The development of a longevity model for a Li/I_2 pacemaker battery has been achieved by the analysis of accelerated discharge and microcalorimetry data.[34] Capacity projections were obtained by fitting an equation to curves of voltage versus delivered capacity from sets of batteries drained at several rates and then by generating projections of capacity to a depletion level of 2.0 V at 37°C. Figure 6 shows the family of curves. The solid sections are observed performance; the dashed lines are extrapolations by the model. The 2.0 V level is where device replacement indicators are activated, such as a pacemaker rate decline of several beats per minute.

Curve fitting and model projections can be completed with confidence for Li/I_2 batteries with only 6–12 months of discharge testing and calorimetric approximation of self-discharge. However, it takes several years to verify the results by real-time testing. Such verification has been observed for a two-cell, 5.6 V Li/I_2 battery that has been in laboratory tests at the pacemaker rate for eight

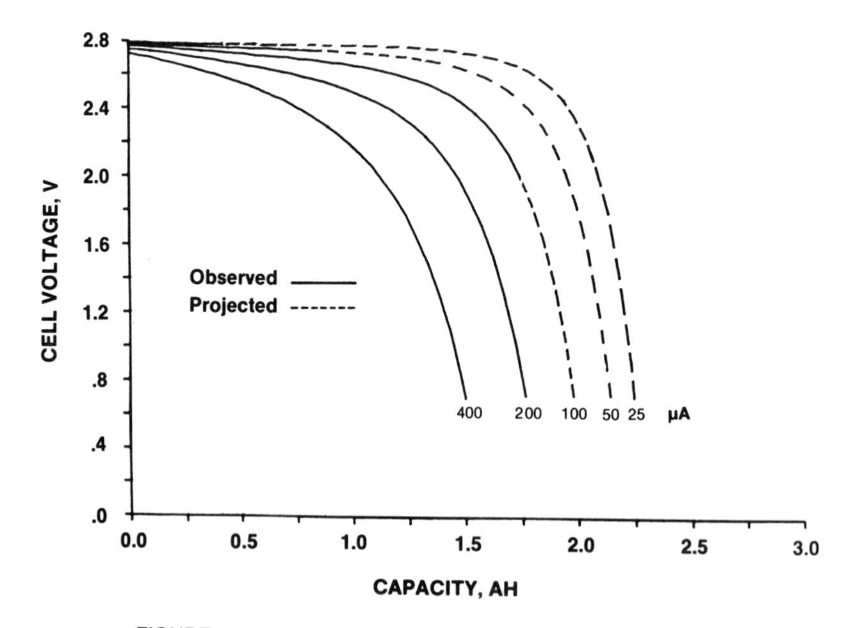

FIGURE 6. Family of Voltage–Capacity Curves—Real and Projected.

years.[35] The model projection of 8.1 years for average longevity compares favorably with the 50% survival of leading-edge test samples at 7.6 years. An even closer agreement of 8.0 years was observed as the rest of the test population (~2800) reached full depletion. More complete information on battery longevity projections is available in Chapter 5.

5.2. Statistical Evaluation of Battery Longevity

The reliability of most devices is divided into three time periods. Initially, some failures occur during burn-in (infant mortality). A long service period then follows during which the failure rate of the remaining cells is constant. Finally, toward end of life (EOL), a wearout period follows, where failure rate increases greatly because of exhaustion or wearout. A graph of such performance resembles a bathtub and is so labeled by reliability engineers. In pacemaker pulse generators and in implantable batteries, the burn-in period will be a few weeks or a day, while the service life may be four to eight years. The wearout period will vary greatly from battery to battery, depending on size, on the power requirements of ancillary functions such as programming or telemetry interrogation, and on the number of times the pacemaker was inhibited by the "demand" function and thus not required to deliver its usual energy pulse to the heart. However, a typical EOL wearout period might occur over a one- or two-year interval.

The improving reliability and longevity of pacemakers have emphasized the growing need for a standardized method of evaluation of pacemaker pulse generators and their power supplies. Two standard statistical techniques can be used to attain this objective: (1) Cumulative Survival Analysis, which gives an unconditional standardized measure of past performance, and (2) Random Linear Failure Analysis, which predicts future performance from past exposure, even if no failures have occurred, but requires an assumption that the failure rate is constant. This method is applicable only during the service life and not during burn-in or wearout.

5.2.1. *Cumulative Survival Analysis*

Berksow *et al.*[36] reported in 1950 the first medical use of Cumulative Survival Analysis at the Mayo Clinic to evaluate the survival of cancer patients. This technique has been long used in the insurance field and is known as the "Actuarial Method." Schaudig *et al.*[37] (1971) used it in Munich for pacemaker evaluation, and Green[38] (1974) independently intuitively developed some elements of this system. In this discussion, we use much of Schaudig's methodology, but with Green's terminology.

The usefulness of Cumulative Survival Analysis lies in the fact that it evaluates only the performance of exposed units. It does this in such a way that the results are relatively independent of any bias on the part of the data taker.

It is important, however, to define precisely each status class. Failure status in particular must be carefully defined. A practical definition by Greatbatch[39] interprets it from the patient's point of view as a malfunction, after wound closure, that requires surgical intervention. Others may well disagree with that definition, and perhaps rightly so. However, a pulse generator whose output has dropped 50% but still paces is still functional from the patient's point of view. A malfunction that requires surgical intervention, however, is quite another matter. In earlier work Greatbatch showed that three pacemaker series, from three different countries and three different doctors, using two different pacemakers, all showed the same Cumulative Survival: 12 to 15 months at 50% Cumulative Survival. The reason was that they all used the same battery, which was the principal cause of failure.

In Cumulative Survival Analysis individual quarterly failure rates and quarterly survival rates are calculated for exposed units only. The quarterly survival rates are then successively multiplied by each other, forming a cumulative survival. The output of the analysis is the number of months at which a specified percentage of the exposed units survive. Experience has shown that the quarterly failure rate breaks sharply upward somewhere between the two levels of 90 and 50% cumulative survival, and the point of the break is perhaps a good elective replacement time.

Using Green's terminology for identifying status:

1. Incomplete Lifetime: Units that have attained the particular age indicated and are still implanted and operating satisfactorily. The total value of the number of units, A, indicates the maturity of the series. When this equals zero, the series is completely matured (i.e., out of service).
2. Curtailed Lifetime: Units removed at the indicated age for reasons unrelated to unit malfunction. Such reasons might be unrelated death, electrode problem, pulse generator erosion, and so on.
3. Elective Lifetime: No system fault. (These are usually grouped with item 2, curtailed lifetime.)
4. Failed Lifetime: A unit removed sometime after wound closure because of unit malfunction. Note that the word *failure* is reserved for a clinical failure. The word *malfunction* is used to describe a unit that has gone outside of specification. The decision as to whether a malfunction is sufficiently severe to justify removal as a "failure" is a medical one that the physician should remain free to make.

As an example, the method is applied to Green's series of older, mercury-powered, pacemaker pulse generators from 1968 to 1970 (Table 1). Data are divided horizontally into quarterly periods and vertically into 10 columns.

M is the number of months.

N is the total number of units that entered the quarter (i.e., became that

TABLE 1 Cumulative Survival of 162 Medtronic 5870/5841 Mercury Pacemakers, 1968–1970

(*M*) Age (Months)	(*N*) Number of Units	(*A*) Incomplete Units	(*B*) Curtailed Lifetime	(*D*) Failed Units	(*T*) Total $A + B + D$	(*X*) Exposed $X = N - T/2$	(*F*) Fraction Failed $F = D/X$	(*S*) Fraction Survived $S = 1 - F$	(*S1*) Cumulative Survival $S1 = S1 \times S$
0–3	162	8	14	0	22	151	0	1.000	1.000
3–6	140	13	9	0	22	129	0	1.000	1.000
6–9	118	9	4	1	14	111.5	0.009	0.992	0.992
9–12	104	5	2	2	9	101.5	0.020	0.980	0.971
12–15	95	9	3	3	15	89	0.034	0.966	0.938
15–18	80	7	2	5	14	75.5	0.067	0.934	0.874
18–21	66	5	2	4	11	58.5	0.068	0.931	0.815
21–24	55	11	0	4	15	49.5	0.081	0.919	0.749
24–27	40	14	1	6	21	32.5	0.184	0.815	0.610
27–30	19	5	1	7	13	16	0.438	0.563	0.343
30–33	6	1	0	2	3	4	0.500	0.500	0.172
33–36	3	1	0	1	2	3	0.333	0.667	0.114
36–39	1	0	0	1	1	0.5			
39–42	0								
42–45									
45–48									
48–51									
51–54									
54–57									
57–60									
Total		88	38	36	162	(807.5)	× 3 = 2422.5 pacemaker months		

old). Obviously, any unit used entered the first quarter, so that N_1 is the total number of units in the series (162 in this case).

A is the number of units that have attained the indicated age and that are still working satisfactorily. Thus, $A_1 = 8$, indicating that eight units are between zero and three months old and are still working.

B is the number of units removed for reasons unrelated to pulse generator malfunction. $B_1 = 14$, so 14 unfailed units were removed in their first three months of life. Thus, a total of 22 units did not see a full three months of service. Therefore, only 140 units (N_2) entered the second quarter. Thus, $N_2 = N_1 - T_1$. Because the 22 units were not exposed a full three months; we average them and say that the exposed units (X) were $162 - 22/2 = 151$ units. Thus, $X_1 = N_1 - (A_1 + B_1)/2$, and when a failure occurs, $X_n = N_n - (T_n/2)$.

The first failure occurred in the third quarter. In this quarter, 111 units were exposed, so that the quarterly failure rate was $F_3 = 0.009$, or 0.9%/quarter. Quarterly survival rate is $S_3 = (1 - F)$, or 0.992. The next quarter two more units failed, giving a quarterly survival rate of 0.980. This, multiplied by the previous cumulative survival rate of 0.992, gives a cumulative survival of 0.971. Thus, the table grows until *N* goes to zero. In practice, the cumulative survival becomes of questionable significance when *X* drops under 30 units. At this point, an error of one unit will produce a 3% drop in cumulative survival for a single additional failure. Column A will move downward with age and will dissipate to the right into B or D. Existing entries in B and D, however, will stay there, increasing if A is not zero, but not decreasing.

Green's series is only half mature, with 88 A-type units still at risk out of a total of 162. Nevertheless, examination of column F is interesting. Up until 24 months, the quarterly failure rate ranged from zero to 8% per quarter, but in the ninth quarter it jumped to 18% and continued to increase. This sharp break upward could be regarded as a signal of the desirability of imminent elective replacement. Such a signal is evident in most series, and it always falls somewhere between 90% and 50% cumulative survival.

Cumulative survival analysis successfully gives an unconditional standardized measure of past device reliability, based on unit exposure. The measure is the number of months before a target survival rate is reached. This method is currently in use both in the Cardiac Data Corporation reports and in the "Performance of Cardiac Pacemaker Pulse Generators" reports by Bilitch et al., which have appeared in the PACE journal.[40]

5.2.2. *Random Linear Failure Analysis*

Table 2 represents a more modern series of 795 lithium-powered CPI pacemakers.[41] Cumulative Survival Analysis for Table 2 is unrewarding because of the immaturity of the series, both because of the short time the units have been implanted and because of the obvious superiority of the lithium batteries in this

TABLE 2 Cumulative Survival of 795 Lithium–Iodine Pacemakers, 1972–1974

(*M*) Age (Months)	(*N*) Number of Units	(*A*) Incomplete Units	(*B*) Curtailed Lifetime	(*D*) Failed Units	(*T*) Total $A + B + D$	(*X*) Exposed $X = N - T/2$	(*F*) Fraction Failed $F = D/X$	(*S*) Fraction Survived $S = 1 - F$	(*S1*) Cumulative Survival $S1 = S1 \times S$
0–3	795	3	17	5	25	786.0	0.006	0.994	0.994
3–6	770	22	10	1	33	754.0	0.001	0.999	0.992
6–9	737	165	6	0	171	651.5	0	1.000	0.992
9–12	566	240	0	1	241	446.0	0.002	0.998	0.990
12–15	325	198	2	0	200	225.0	0	1.000	0.990
15–18	125	74	0	0	74	88.0	0	1.000	0.990
18–21	51	32	0	1	33	35.0	0.02	0.971	0.961
21–24	18	16	0	0	16	10.0	0	1.000	0.961
24–27	2	2	0	0	2	1.0	0	1.000	0.961
27–30									
Totals		752	35	8	795				

series over the zinc-mercury batteries in Green's series. To analyze the lithium data, Random Linear Failure Analysis is a more relevant method.

Table 2 shows a pattern typical of newer pacemaker series. If early mortality (failures at one week or less) are ignored, the failure pattern looks relatively random. If such an assumption is valid, a new branch of mathematical analysis is open to us. Rosenbaum et al.[1] used the technique of Random Linear Failure Analysis to arrive at a longevity design target for nuclear-powered pacemakers. This method is particularly useful in situations where few or no failures exist and no failure pattern is discernible.

Furthermore, the confidence level (C) in a given maximum failure rate (P) for (R) failures in (X) pacemaker-months of exposure can be derived as:

$$C = 1 - \sum \frac{(PX)^r}{r!} (1-P)^{X-r} \tag{17}$$

summed iteratively over all the values of r from 9 to R. Rosenbaum suggested a P value of 0.15%/month. This would be equivalent to six failures out of 100 pacemakers over 40 months, or 94% survival.

Since that time, both lithium and nuclear pacemaker systems have demonstrated a far better reliability than this, but let us assume a design target of 0.1%/month failure rate and examine the 795 lithium pacemakers in Table 2. Of the first five units that failed in the first quarter, three failed in the first week and are eliminated from the calculation. Thus, the total number of failures to be counted are five rather than eight shown in the table. Thus, $R = 5, X = 8990$, and $P = 0.001$. Then:

$$C = 1 - \sum_{0}^{5} \frac{0.001\ (8990)^r}{r!} (.999)^{8990-r} \tag{18}$$

giving a confidence level of 88% for the assumed failure rate of 0.1%/month.

Random failure rate information has been reported for other implantable lithium batteries. Pacemakers using a lithium-silver chromate battery had a very low calculated failure rate, 1.4×10^{-4}% per month, over a five-year period going into 1979.[42] This data base, as with some other studies, relied on the customers to report battery failures to the battery manufacturer, which cannot always be guaranteed.

A more certain source of reliability statistics is controlled laboratory testing of production cells. These can be batteries set aside for testing at the beginning of a new model run or samples drawn regularly from the production line to provide an ongoing performance history. When operated at the low pacemaker currents, these laboratory test batteries accumulate very accurate reliability data.

One of the more recent reports on lithium-iodine battery reliability is sum-

TABLE 3 Reliability of Lithium–Iodine Pacemaker Batteries

Device	Initial Sample	Maximum Years	Device-hours	Random Failure Rate (%/month)
Battery A	4412	8.0	216 million	0.0023*
Battery B	1917	5.0	67	0.0025
Battery C	1332	4.4	27	0.0063
Silicon transistors				0.0028
Silicon diodes				0.0011

[a] Lowest RFR prior to natural end-of-life 90% confidence level.

marized in Table 3.[43] All of the battery device-hours in the listing are from laboratory test samples. Two of the models, A and B, have already reached failure rates on a par with high-reliability silicon transistors.[44]

ACKNOWLEDGMENT. Most of the preceding material regarding survival analysis methods appears in a previous publication by W. Greatbatch.[39]

REFERENCES

1. D. Rosenbaum, D. Kleitman, B. Singer, and A. Barnett, *USAEC Statistical Guidelines for Device Testing, Draft* 19, October 1973.
2. U.S. MIL-STD-781; *Reliability Qualification and Production Acceptance Tests—Exponential Distribution,* Issue C, available from National Technical Information Service, Springfield, Virginia.
3. C. C. Liang and C. F. Holmes, Performance and reliability of the lithium/iodine battery, *J. Power Sources, 5,* 3 (1980).
4. B. B. Owens and D. F. Untereker, Accelerated testing of long life primary cells, in: *Power Sources* Vol 7 (J. Thompson, ed.), Academic Press, New York (1979).
5. H. F. Gibbard, Heat generation in lithium thionyl chloride batteries, *Proc. Electrochem. Soc. 80-4* (1980).
6. W. S. Holmes, Pacemaker battery microcalorimetry, in: *Reliability Technology for Cardiac Pacemakers III,* NBS Special Publication 400-50, 1979.
7. L. D. Hansen and R. M. Hart, The characterization of internal power losses in pacemaker batteries by calorimetry, in: *Reliability Technology for Cardiac Pacemakers III,* NBS Special Publication 400-50, (1979).
8. D. F. Untereker and B. B. Owens, Microcalorimetry: A tool for the assessment of self-discharge processes in batteries, in: *Reliability Technology for Cardiac Pacemakers III,* NBS Special Publication 400-50, (1979).
9. E. J. Prosen and J. C. Colbert, A microcalorimeter for measuring characteristics of pacemaker batteries, in: *Reliability Technology for Cardiac Pacemakers III,* NBS Special Publication 400-50, (1979).
10. S. A. G. R. Karunathilaka, N. A. Hampson, R. Leek, and T. J. Sinclair, The impedance of the Leclanché cell. I. The treatment of experimental data and the behavior of a typical undischarged cell, *J. Appl. Electrochem. 10,* 357 (1980).

11. S. A. G. R. Karunathilaka, N. A. Hampson, R. Leek, and T. J. Sinclair, The impedance of the Leclanché cell. II. The impedence of individual cell components, *J. Appl. Electrochem. 10,* 603 (1980).
12. S. A. G. R. Karunathilaka, N. A. Hampson, R. Leek, and T. J. Sinclair, The impedance of the Leclanché cell. III. The impedance of the cell at different stages of discharge and state-of-charge indication by the impedance method, *J. Appl. Electrochem. 10,* 799 (1980).
13. S. A. G. R. Karunathilaka, N. A. Hampson, R. Leek, and T. J. Sinclair, The impedance of the alkaline zinc–manganese dioxide cell. II. An interpretation of the data, *J. Appl. Electrochem. 11,* 715 (1981).
14. M. Keddam, Z. Stoynov, and H. Takenouti, Impedance measurement on Pb/H_2SO_4 batteries, *J. Appl. Electrochem. 7,* 539 (1977).
15. R. J. Brodd and H. J. DeWane, Impedance of Lechanché cells and batteries, *J. Electrochem. Soc. 110,* 1091 (1963).
16. R. J. Brodd and H. J. DeWane, Impedance of sealed nickel–cadmium dry cells, *Electrochem. Tech. 3,* 12 (1965).
17. J. E. Clifford and R. E. Thomas, Study of Battery Accelerated Testing Techniques, Final Report, SAND82-7049, Sandia National Laboratories, Albuquerque, New Mexico (1982).
18. R. E. Thomas, E. W. Brooman, J. H. Waite, O. L. Linebrink, and J. McCallum, Study of Space Battery Accelerated Testing Techniques, Phase II Report, Ideal Approaches Towards Accelerated Tests and Analysis of Data, Contract No. NAS 5-11594, Battelle Memorial Institute, Columbus Laboratories, 1969.
19. J. McCallum, R. E. Thomas, and E. A. Roeger, Jr., Failure Mechanisms and Accelerated Life Tests for Batteries, Technical Report AFAPL-TR-68-83, Battelle Memorial Institute, Columbus Laboratories, 1968.
20. D. F. Velleman and D. C. Hoaglin, *Application, Basics and Computing of Expoloratory Data Analysis,* Oxbury Press, 1981.
21. K. T. Gillen and K. E. Mead, Predicting Life Expectancy and Simulating Age of Complex Equipment Using Accelerated Aging Techniques, SAND79-1561, Sandia National Laboratories, Albuquerque, New Mexico (1979).
22. S. Siegel, *Nonparametric Statistics for the Behavioral Sciences,* McGraw-Hill, New York (1956).
23. M. Hollander and D. A. Wolfe, *Nonparametric Statistical Methods,* Wiley, New York (1973).
24. E. I. Lehmann and H. J. M. D'Abrera, *Nonparametrics,* Holden-Day, Inc., San Francisco, California (1975).
25. H. Cramer, *Mathematical Methods of Statistics,* Princeton University Press, Princeton, New Jersey (1946).
26. K. R. Brennan, K. E. Fester, B. B. Owens, and D. F. Untereker, A capacity rating system for cardiac pacemaker batteries, *J. Power Sources 5,* 25–34 (1980).
27. U.S. MIL-STD-202; *Test Methods for Electronic and Electrical Component Parts, Issue E, Method 213B Shock (Specified Pulse)* available from National Technical Information Service, Springfield, Virginia.
28. U.S. MIL-STD-202; *Test Methods for Electronic and Electrical Component Parts, Issue E, Method 204C, Vibration, High Frequency,* available from National Technical Information Service, Springfield, Virginia.
29. F. E. Kraus, M. V. Tyler, and A. A. Schneider, in: *Reliability Technology for Cardiac Pacemakers III* (H. A. Schafft, ed.), NBS Special Publication 400-50, pp. 35–48, U.S. Dept. of Commerce, June, 1979.
30. W. R. Brown, An Examination of Wilson Greatbatch Limited Model 755 Cells After Load Testing, Calspan Corporation Report, April, 1978.
31. B. C. Bunker, C. J. Leedeke, S. C. Levy, and C. C. Crafts, in: *Power Sources 8* (J. Thompson, ed.), pp. 53–62, Academic Press, New York (1981).

32. N. S. Istephanous, K. Fester, D. R. Merritt, P. M. Skarstad, and D. F. Untereker, Glass Seal Corrosion in Liquid Lithium Electrolyte Batteries, *The Electrochemical Society, Extended Abstracts* No. 146, Fall (1984).
33. D. F. Untereker, The use of a microcalorimeter for analysis of load-dependent processes occurring in a primary battery, *J. Electrochem. Soc. 125,* 1907–1912 (1978).
34. J. S. Kim and K. R. Brennen, Mathematical modeling of lithium iodine discharge data, *Proc. Electrochem. Soc. 80-4,* 174–186, 1980.
35. W. D. Helgeson, Design Testing and Reliability of Lithium-Iodine Pacemaker Batteries, *The Electrochemical Soc., Extended Abstracts* No. 28, Spring (1983).
36. J. Berksow and A. Gage, Calculation of survival rates of cancer, *Proc. Staff Meeting, Mayo Clinic 25,* 270 (1950).
37. A Schaudig and M. Zimmerman, Comparison of function time of different pacemaker systems, *Ann. Cardiol. Angiol. 20*(4), 357 (1971).
38. G. Green, Progress in pacemaker technology, *J. Electrocardiol. 7*(4), 375 (1974).
39. W. Greatbatch, in: *Advances in Pacemaker Technology* (M. Schaldach and S. Furman, eds.), pp. 345–355, Springer-Verlag, New York (1975).
40. M. Bilitch, Performance of cardiac pacemaker pulse generators, a regular feature section of the *PACE* (Pacing and Clinical Electrophysiology), Futura, Mt. Kisco, New York.
41. W. Greatbatch, in: *Quantitative Cardiovascular Studies: Clinical and Research Applications of Engineering Principles* (N. Hwang, D. R. Gross, and D. J. Patel, eds.), pp. 581, University Park Press, Baltimore, Maryland (1979).
42. P. Lenfant, M. Broussely, J. P. Rivault, and M. Grimm, The lithium silver chromate cell: A five-year study, *Proc. Electrochem. Soc. 80-4,* 81–94 (1980).
43. K. Fester, W. D. Helgeson, B. B. Owens, and P. M. Skarstad, Long term performance of Li/I_2 batteries, *Solid State Ionics 9&10,* 107–110 (1983).
44. Reliability Prediction of Electronic Equipment, *Military Standardization Handbook, MIL-HDBK-217B,* September 1974.

5

Battery Performance Modeling

KENNETH R. BRENNEN and JOHN S. KIM

1. DESCRIPTION OF THE PROBLEM

The function of a battery is to supply electrical energy at a rate sufficient to power the device it drives. Since power is the product of current and terminal voltage, the ideal battery performance model will project these quantities mathematically over the entire range of conditions likely to be encountered in the lifetime of the battery. The first use of a model then is to project how long a new battery will last under a known set of application conditions. If the model is very complete, it serves a second purpose: to project the amount of useful life left in a battery after it has been partially discharged. For an implanted medical device, this information could be a key element in planning for surgical replacement procedures.

The difficulty of battery modeling resides in the fact that lifetime depends on an array of interrelated physical factors: temperature, discharge rate, form of the load current (i.e., constant, stepped, increasing, decreasing), length of storage or standby intervals, and location of storage intervals in the history of the battery. The way in which each of these affects performance depends, in turn, upon the design of the particular battery (i.e., the chemical system and the mechanical design factors).

Statistics also play a role in battery modeling, since even batteries free of gross defects often show statistical variations in performance. These are a reflection of variations associated with cell construction: anode weight, cathode weight and composition, electrolyte composition, separator homogeneity, and reproducibility of assembly operations. These statistical variations should be recognized so that the modeling is not done from a data base that is too small,

KENNETH R. BRENNEN and JOHN S. KIM • Medtronic, Inc., Minneapolis, Minnesota 55440.

yielding highly skewed data and resulting in an inaccurate model. Statistical variations inherent in the battery, combined with statistical variations inherent in the devices themselves and in the use of the devices in medical practice, yield a statistical longevity distribution for any population of real devices. The contribution of the battery is required for the most meaningful calculation of this distribution.

2. IMPORTANCE OF THE SOLUTION

Emphasis on knowing the exact service life of a battery generally increases as the price of the battery rises. It also increases as the expense and complexity of the replacement process increases. Thus, very few consumers are vitally concerned with the real service life of a flashlight battery, but a single battery failure in a space vehicle or an implanted cardiac pacemaker will not only draw public attention, but may also involve large financial costs, ranging from expensive warranty settlements to possible lawsuits.

Aside from the business rationale, there is a great deal of professional satisfaction for the scientist or engineer working on a project where performance modeling is given a high level of importance. Battery modeling challenges mathematical creativity and tests one's knowledge of the detailed inner workings of very complex chemical systems. This, in turn, often leads to further physicochemical insights and enhances our knowledge of battery systems.

3. DESCRIPTION OF THE VARIABLES AND RELATIONSHIPS

The battery model is always expected to provide an answer to a question about time. How long will the device last under the specified conditions? How much longer will a device operate before the voltage falls to level V?

The classical variables for describing battery function are current, I, and voltage, V. The mathematical bridge required between these and the time variable, t, is simply:

$$dt = dQ/I(t) \quad (1)$$

If Eq. (1) is solved for dQ and integrated over the lifetime of the battery, the final value of charge is the capacity, Q, associated with the discharge conditions considered. A classical approach consists of obtaining the capacity for one set of discharge conditions and then assuming it will be a constant over a wide range of discharge conditions. The approximation is often a good one for consumer and industrial applications. It may not be accurate enough for medical applications with a device lifetime targeted to be five to 10 years. Consider the example of a battery drained at 463 μA that delivered 2.0 Ah of useful charge during a six-month discharge test. Suppose this battery has internal loss processes that consume the reactants at a constant rate equivalent to 10 μA of discharge current. The capacity delivered during the six-month test will be 98 percent of the capacity initially available (2.04 Ah). If the external drain current is halved, this battery will run not for a full year, but for about 350 days and will deliver about 1.95 Ah to the load. This is still 97% of the service life based on the six-month discharge test and is not a significant deviation from the full year that one might expect.

But now suppose the external load is set at 23 μA with the expectation that the device will have a 10-year service life, based on the six-month test. When the (previously insignificant) loss rate is accounted for, this device will actually last 6.9 years, during which time the battery will deliver 1.39 Ah, or about 70% of its presumed capacity, to the device. This example illustrates the problem that can arise if one assumes deliverable battery capacity is a constant when considering long-term applications.

The integration of Eq. (1), therefore, is not necessarily simple in the case of a long-lived battery operating at very low current drain. To make the complexity more explicit, one should write the differential of charge with subscript L to indicate the charge delivered to the load. The current may also be rewritten as the ratio of battery terminal voltage V_T, and the effective resistance of the load R_L. But the terminal voltage is a function of both the device impedance and the internal impedance of the battery, since these two factors determine the internal voltage loss or polarization. Internal impedance is, in turn, determined to a large extent by the depth of discharge, Q, the amount of charge already taken from the battery. Thus V_T should be written as $V_T(Q,R_L)$. Equation (1) then becomes:

$$dt = [R_L/V_T(Q,R_L)]\ dQ_L \tag{2}$$

Finally, to provide an upper limit to the integration and to emphasize the loss processes occurring within the battery, the charge delivered to the load, dQ_L, can be written as the difference between total charge lost, dQ_T, and charge lost to internal self-discharge processes, dQ_s. Equation (1) then becomes:

$$dt = [R_L/V_T(Q,R_L)](dQ_T - dQ_s). \tag{3}$$

Equation (3) is not a particularly useful form for integration, but it does illustrate the relationships between the physical variables that must be known before an integration can be achieved. In particular, the required voltage terms can be obtained from a plot of voltage versus total depth of discharge, Q_T, for a fixed load. An additional relationship is needed between Q_T and Q_s to specify completely the solution of Eq. (3).

Figure 1 is a simple equivalent circuit for the battery/load combination described in Eq. (3). Here V_O is the electrode voltage and V_T is the battery terminal voltage. R_i is the internal resistance of the electrodes, electrolyte, separators, and any slow chemical reactions directly involved in battery discharge. R_s is the equivalent resistance of self-discharge pathway.

The variables associated with the resistances indicate that both are functions of time or depth of discharge. To keep the discussion general and to include the most complex case, R_i and R_s are shown having a common component. However, the extent to which the common component contributes to R_i and R_s is also a variable and may be discharge rate dependent. The processes that contribute to the internal and self-discharge resistances may be classified as electrochemical reaction, bulk conductance, and surface conductance. The resistance arising from electrochemical reaction rates is generally known to be strongly rate dependent. But the value of the surface and bulk contributions to the internal resistance can also depend strongly on rate because the size and shape of crystallites formed by the discharge process will vary with the local current density.[1] This, in turn, affects the ratio of bulk to surface conductance within the battery. Thus, two initially identical batteries discharged to the same depth at different rates may exhibit quite different values of internal resistance, and hence polarization. The situation just described produces a problem only when a model is required to project performance for a battery subject to step-wise changes of load conditions. The battery data must then be integrated piece-wise over each successive set of discharge conditions if an accurate projection is needed.

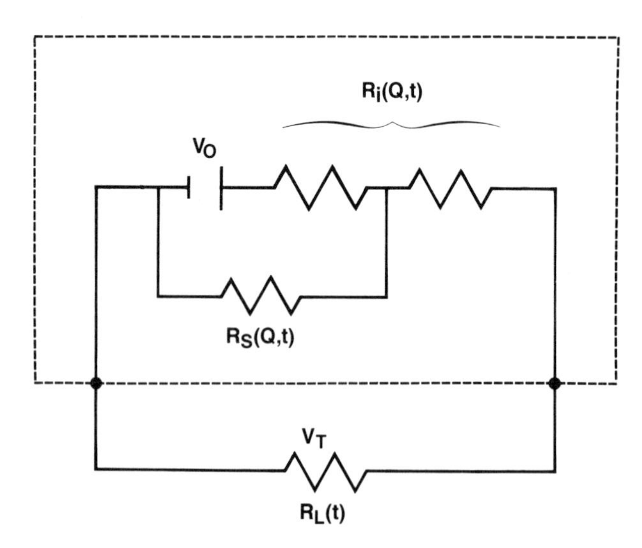

FIGURE 1. An Equivalent Circuit for a Battery Emphasizing Common Elements of the Discharge and Self-Discharge Pathways.

Verification of the validity of the battery performance model described previously requires a time span equal to the projected service life of the battery. Lithium/iodine batteries put on test in 1975 are showing depletion at application drain at the time of this waiting. Fester *et al.*[2] have shown that laboratory results conform to projections by the model described below. The measured average lifetime of the earliest lithium/iodine batteries tested with this model is within 7% of the 8.4-year projected service life.

4. CLASSIFICATION OF MODELS

Mathematical models derived strictly from physical laws are properly referred to as physical models. Such models are most easily constructed in the region of linear behavior. Physical models often provide a good first approximation to the behavior of real systems composed of nonlinear elements. Physical models represent one end of a continuum of descriptive techniques.

At the opposite end are purely empirical models constructed solely by the curve fitting of experimental data. Such models serve as convenient computers but implicitly contain no physical insights. The use of such a model to make projections outside the range of the original data is generally not reliable and frequently impossible.

Models may also be classified as static, steady state, or dynamic. Open-circuit storage might be thought of as a static condition, but as soon as self-discharge is considered, the model becomes dynamic (i.e., the model must account for changing conditions). Steady-state models in which processes are imagined to occur at constant rate throughout time are of limited use in battery modeling, since the internal state of a battery tends to change constantly during discharge.

Battery performance is usually best treated with a hybrid model because battery materials are usually condensed phases that exhibit nonlinear relationships between the relevant properties such as current and voltage drop. Physical insights into battery processes provide a rationale for combining nonlinear elements in certain ways and also provide limits that can be built into a model. The final result, however, generally incorporates a statistical approach superimposed on the framework dictated by the physical model.

5. STATISTICAL METHODS

Statistical methods are also useful for the extrapolation of battery performance when complete physical data are not available because of time and resource constraints. Statistical modeling consists of three general steps. The first step is to formulate the mathematical relationship between the variables of interest based on preliminary knowledge and data. The success of this step depends on how well one develops a model to explain the chemical reactions taking place during the discharge life cycle. For instance, the preliminary overview should be able generally to describe self-discharge, polarization, and voltage delay effects over the entire range of application rates.

The second step in the modeling process is to collect appropriate data. At this stage, experiments are designed such that measurements can be obtained for a wide range of discharge rates. Typically, the discharge rates chosen at the early development stages of a new battery design range from the application level discharge rate to significantly higher accelerated rates. Since the data collected from these tests will be used to fit the model and since time constraints will dictate that accelerated testing data will comprise the majority of the data points, one should examine whether it is appropriate to make statistical inferences from these accelerated test data.

The last step in this modeling process is fitting the data into the previously discussed mathematical model to estimate the parameters of the model and the error associated with these estimates. Various types of statistical methods are available for such purposes under the general heading of linear/nonlinear regression analysis. These methods not only estimate parameters of interest, but also examine the appropriateness of the proposed mathematical model and select the best model from the family of possible models.

5.1. Self-Discharge

The key to accurate modeling for long-lived battery systems is the accurate description of the self-discharge rate. The self-discharge rate may be time dependent, rate dependent, depth-of-discharge dependent, or a combination of these.

Earlier it was shown that an accurate model for a long-lived device requires the modeling of extremely small self-discharge losses. Before the advent of microcalorimetry to access self-discharge rates, there was no method to obtain data in a short time for low-level self-discharge. The lithium/iodine model discussed later is based on microcalorimetry data that indicate that the self-discharge rate decreases as the depth of discharge increases. Figure 2 shows the relationship obtained from some early data by Untereker.[3] These data fit an exponential decay equation where division of the preexponential factor by the heat of formation of LiI will yield the molar flux of iodine as a function of Q_L. Division of the preexponential factor by the battery voltage yields an equivalent self-discharge current.

This model is intuitively satisfactory because one can imagine that as discharge proceeds, the depletion of iodine near the anode and the thickening of the solid electrolyte layer both tend to reduce self-discharge. In addition, the

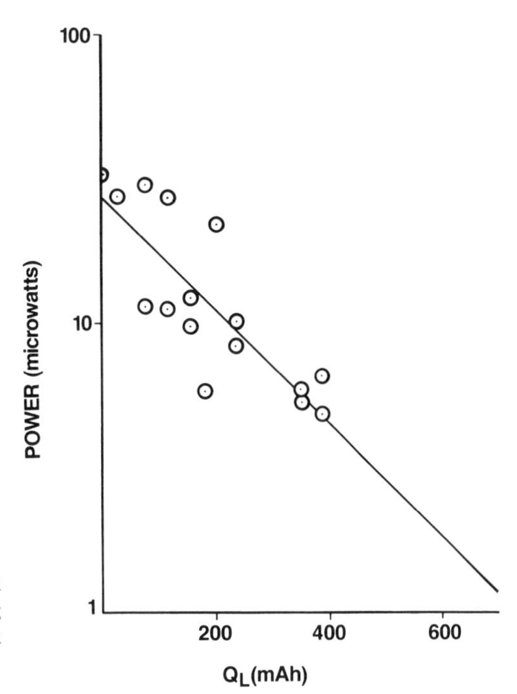

FIGURE 2. Microcalorimetric Heat Flux for a Lithium/Iodine Battery Illustrating the Reduction of Self-Discharge Rate with Increasing Depth of Discharge.

exponential form is mathematically convenient because it is a smooth, continuous function that can be readily integrated in many equations, including those used here. An interesting point to note is that under constant current discharge conditions, a time-dependent self-discharge rate is equivalent to a depth-of-discharge dependent rate. This fact can be useful in reformulating and simplifying solutions under constant current conditions.

A more general approach in those cases where a simple function does not fit the self-discharge rate would be the use of:

$$I_S = A(Q_T) + I_L B(Q_T) \tag{4}$$

where the terms A and B are functions only of depth of discharge and can be determined by microcalorimetric experiments. Early indications are that lithium/thionyl chloride systems fit this model.[4] However, the complete characterization of $A(Q_T)$ and $B(Q_T)$ has not been published to date. The experimental values of $A(Q_T)$ and $B(Q_T)$ can be determined by microcalorimetrically determining the rate dependence of self-discharge at an array of points spread over the depth of discharge axis from zero to battery depletion.

5.2. Polarization

Polarization, a rate-dependent phenomenon, tends to be nonlinear. The mathematical description of polarization will depend very much on cell chemistry and design. Cells with fast electrode kinetics and highly conducting electrolyte and cathode layers show very little polarization until depletion is imminent. The lithium/thionyl chloride cell is a prime example of such a system. Conversely, systems with slow electrode kinetics or sluggish transport processes will exhibit appreciable polarization even at low current drains. In the latter case, the polarization voltage can be determined experimentally as a function of I_L and Q_T and can then be fit to an empirical model, which, in turn, can be used to project when a system will reach the cut-off voltage for a given current.

In the following example for lithium/iodine, the charge loss to polarization was modeled directly from capacity versus current drain data for a region of high rate discharge. Separate microcalorimetric experiments had already shown self-discharge to be negligible in this region. Therefore, observed capacity losses could be regarded as a pure polarization effect. The data could be extrapolated back to the region where self-discharge was not negligible (i.e., at low current drains). In the region of intermediate current levels, where self-discharge and polarization cause about equal capacity losses, the following check is useful. From independent measurements of self-discharge one may calculate the self-discharge loss and then add this to the observed capacity. The result should equal the extrapolated capacity projected from high rate discharge using the polarization correction alone. This assumes a negligible interaction between self-discharge and polarization.

6. MODELING OF THE LITHIUM/IODINE PACEMAKER BATTERY

Unlike mercury/zinc power cells, which maintain a relative small and constant source impedance (R_{DC}) throughout service life, lithium/iodine batteries tend to exhibit a continuously rising impedance. The level depends not only on the state of battery discharge (Q), but also on the current drain (I) or sequence of drains employed to achieve the given discharge state. A longevity model for a lithium/iodine-powered pacemaker must therefore characterize the relationship between R_{DC}, I, and Q before it can properly incorporate the effect of pacemaker circuitry on current drain and, hence, on generator life.

Accelerated test data for one model of a lithium/iodine pacemaker battery, EnerTec® Alpha 332, are summarized in Table 1. The data, obtained at 37°C at a series of constant current loads, were generated by the Battery Evaluation Department of Medtronic, Inc. From these data a mathematical computer model was derived to estimate source impedance increments and hence loaded battery

TABLE 1 Alpha-332 Battery Test Data[a]

Test Lot	Constant Current (μA)	Number of Points	Number of Cells	Capacity Delivered to Date (A Hr)
DO47PQ.021	400	382	16	1.487
DO47PQ.031	200	421	16	1.764
DO47PQ.041	100	629	16	1.859
DO47PQ.051	50	492	16	1.015
DO47PQ.061	28	461	15	0.566
DO47PQ.022	100	244	16	2.002
DO47PQ.032	100	175	16	1.977
DO47PQ.023	28	239	16	2.253
DO47PQ.033	28	236	16	2.233

[a] Data for the curves for Figure 3. Test lot suffix sequences 021, 022, 023, and 031, 032, 033 represent continuation tests on two lots of cells with current drains stepped down for each test in the sequence. Two tests are incomplete but are continuing at the time of writing.

voltage under actual pulse generator operating conditions. Under these conditions neither resistive load nor current drain remain constant.

Figure 3 shows the plots of mean battery voltage (V_L) versus capacity delivered (Q_L) and nominal current drain (I_L) observed in the tests of Table 1. For the higher current drain tests, the curves suggest that three different states occur in the life of the cell. Beginning from a relatively linear decrease, a period of gently decreasing exponential form is encountered, terminating in a sharp downward swing. This pattern is amplified in Figure 4, where $\ln \Delta V$ is shown in relation to Q and I. Voltage drop (ΔV) is defined as open-circuit voltage (V_{oc}) minus the observed battery voltage (V_L).

The three stages suggested by Figures 3 and 4 are postulated to be:

1. A period of incubation or stabilization
2. A period of electrolyte columnar/planar growth
3. A period of severe iodine depletion within the cathode

These states, it should be emphasized, are a simplification of the actual complex changes occurring within the cell. However, they may be shown to have some basis in theory and offer something more than a purely empirical rationale for this mathematical model selection. With respect to the effect of current drain, the pattern of curves as shown suggests a log-linear relationship between ΔV and I. Independent current-voltage scans of other lithium/iodine cells substantiate this behavior within the range of accelerated tests shown.

From these considerations, the following mathematical form was selected:

$$\ln \Delta V = \beta_1 \ln I_L + (\beta_2 \ln I_L)Q + P_3(Q), \tag{5}$$

where

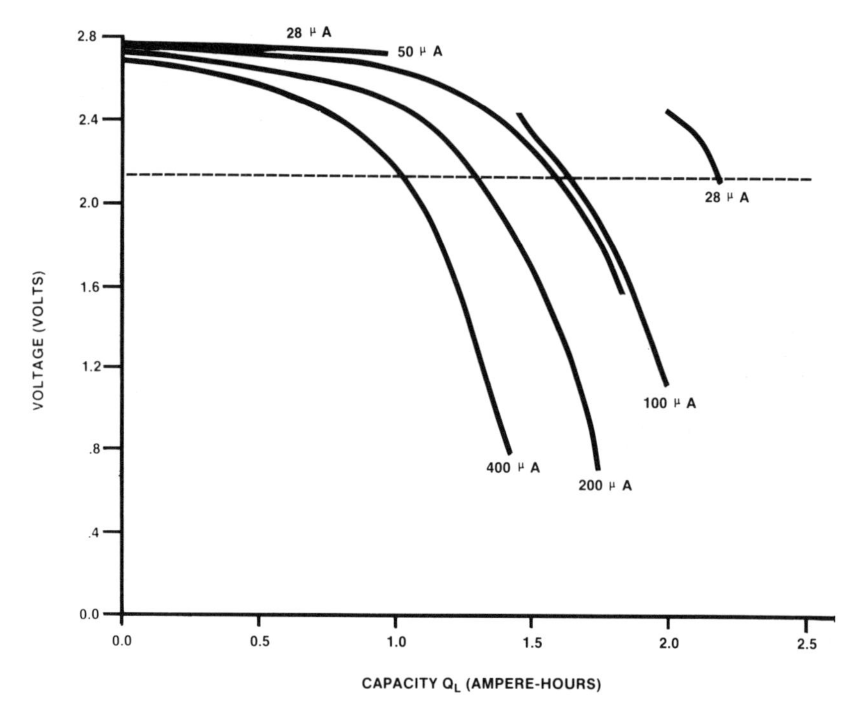

FIGURE 3. Battery Terminal Voltage as a Function of Charge Delivered to Load for the EnerTec™ Alpha 332 Lithium/Iodine Battery Discharged at Five Different Rates. The Five Curves on the Left are Constant-Current Discharges at the Indicated Current Drains; the 200- and 400-μA Test Cells Were Subsequently Discharged at 100 μA, and then 28 μA, Resulting in the Two Curves on the Right.

$$P_3(Q) = \alpha + \gamma_1 Q + \gamma_2 Q^2 + \gamma_3 Q^3 \tag{6}$$

This model was found adequate for modeling data taken at and above the application current drains of the batteries of interest. However, it was not able to account for self-discharge, since it only considered current and charge delivered to the load. As self-discharge data on lithium iodine batteries became available, a need for a more rigorous model became apparent. Such a model had to account for the self-discharge process so that the effects on the battery capacity of open-circuit storage and periods of low-drain standby operation could be calculated.

Since the polarization, or voltage loss under load, is a function of the internal state of the battery, and since the internal state is a reflection of the total depth of discharge, whether achieved by self-discharge or delivery of load current, it seems reasonable to select the total amount of discharge product formed as one independent variable. This quantity, designated Q_T and expressed in mAh, is

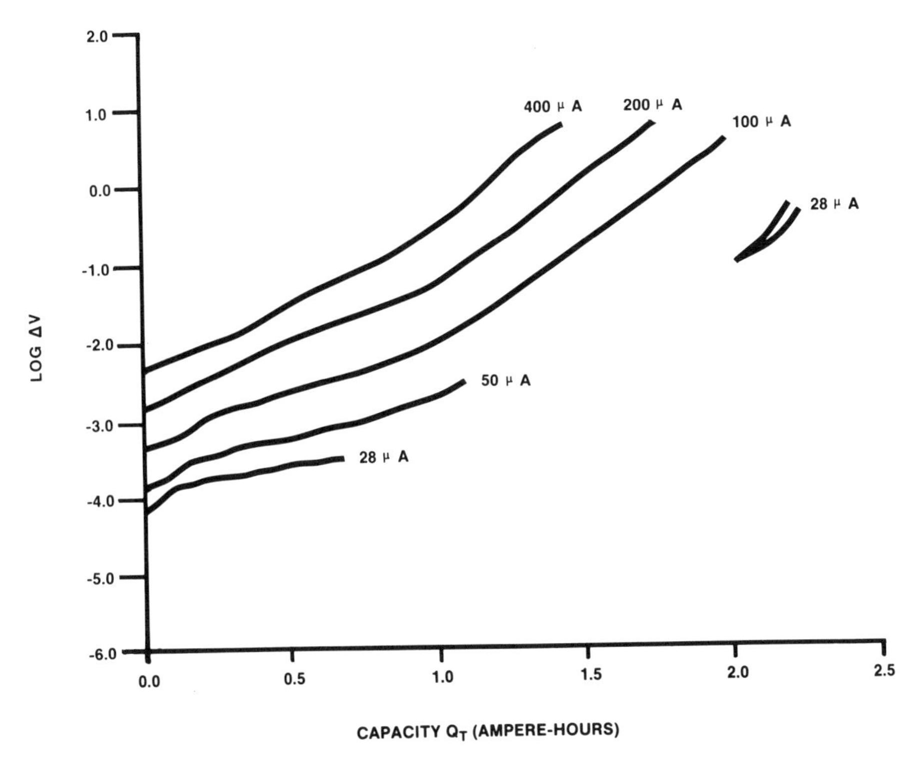

FIGURE 4. Logarithm of Polarization Voltage Versus Charge Delivered to Load for the EnerTec™ Alpha 332 Lithium/Iodine Battery Discharged at Five Different Rates. See Table 1 for Data.

written explicitly as the sum of the charge delivered to load (Q_L) and the charge lost to the nonelectrochemical combination of lithium and iodine (Q_S). Load current becomes the second independent variable, and time is treated as an implicit variable, relating current and charge. The equation relating the discharge quantities is then:

$$Q_T = Q_S + Q_L \tag{7}$$

Since Q_S and Q_T are not readily observable, it is necessary to relate them to the observable quantity Q_L. Studies of open-circuit heat fluxes from batteries discharged to various depths at several currents indicated that the self-discharge rate was related primarily to Q_L.

Since introspection shows self-discharge losses will be relatively low, it seems reasonable to replace Q_L with Q_T in the exponential self-discharge relationship. This was done with the assumption that the exponential form and the value of the exponent were invariant under the transformation. After two numerical iterations to adjust the Q axis for the addition of self-discharge, the relationship between the equivalent self-discharge current and the total discharge product for the model Alpha 332 was found to be:

$$I_S = I_0 \exp(-k^* Q_T) \mu\text{A} \tag{8}$$

where $I_0 = 10$ and $k = 0.0016$

The reciprocal of the exponential constant in Equation (8) is 625 mAh. This is the interval of Q_T for which I_S is decremented $1/k$. If the load current is high and Q_T is built up quickly, the self-discharge loss will be small. Small load currents increment Q_T more slowly and allow Q_S to reach higher values.

To quantitatively relate I_S and I_L, Eq. (7) was differentiated with respect to time and Eq. (8) was substituted for I_S while I_L was held constant. The result of integrating this new expression is:

$$Q_S = \frac{1}{k} \ln \frac{I_L + I_0 \exp - kQ_T^1}{I_L + I_0 \exp - kQ_T^2} \tag{9}$$

This equation combined with Eq. (7) yields values of Q_L and Q_S associated with ΔQ_T between the lower limit of Q_T^1 and the upper limit of Q_T^2. This expression may easily be applied to stepwise changes in battery drain rate by merely setting Q_T^1 equal to the previous value of Q_T^2 in any load sequence.

The remaining problem, to incorporate this self-discharge segment of the model with a segment that would yield a voltage versus Q_L curve, was solved by fitting Eq. (5) to the experimental data, using the least squares regression method.

The complete procedure for generating a voltage-Q_L curve now consists of placing a value for Q_T in Eq. (5) to obtain ΔV and in Eq. (9) to obtain Q_S. Eq. (7) then yields a value of Q_L appropriate to ΔV, while the terminal voltage can be obtained by subtracting ΔV from the open-circuit battery voltage V_{oc}.

Figure 5 shows the projected V_L versus Q_L curves generated by the fitted model. The coefficients for the model are $\alpha = -6.2976$; $\beta_1 = 0.6281$; $\beta_2 = 0.4397$; $\gamma_1 = 0.2388$; $\gamma_2 = -1.5151$; $\gamma_3 = 0.6893$. Excellent agreement of the fitted model with the data is indicated by the multiple correlation coefficient of 0.994 and by visual inspection of the fitted curves. Application of this same modeling procedure to other battery designs has shown that these coefficients are a function of battery geometry and cathode composition. Based on these coefficients, one can project Q_L, Q_S, and expected battery longevity for any given load condition. Table 2 summarizes the projected Q_L, Q_S, and longevity for the model Alpha 332 battery under constant current discharge with the assumption that the useful life of the battery ends when load voltage of the battery drops to 2.0 V.

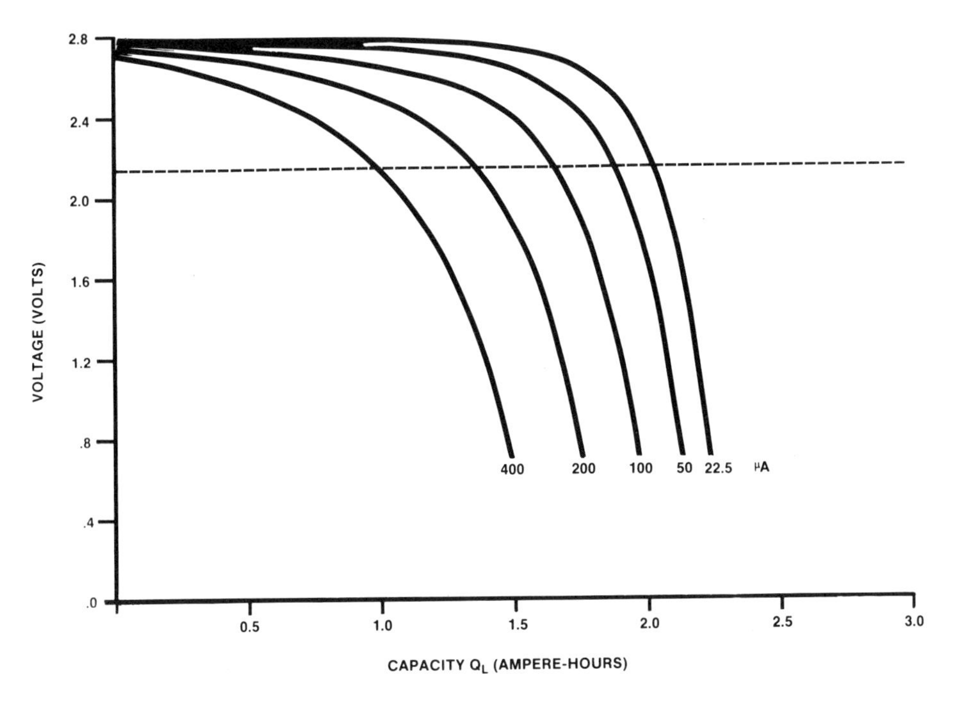

FIGURE 5. Battery Terminal Voltage Versus Charge Delivered to Load, for Five Current Drains, as Projected by the Model Described in the Text.

TABLE 2 Projected Deliverable Capacity, Self-discharge Losses, and Battery Longevity for the Enertec™ Alpha 332 Lithium/Iodine Battery at Five Constant-Current Drain Rates

Load Current Drain (μA)	Capacity Delivered (Q_L) (Ah)	Self-discharge Loss (Q_S) (Ah)	Projected Longevity (years)
10	1.12	0.41	24.1
25	2.08	0.20	9.5
50	1.95	0.11	4.5
100	1.74	0.05	2.0
200	1.46	0.03	0.8

7. DEVICE LONGEVITY

Device longevity should be simple to calculate if one knows the battery capacity Q (in ampere-hours) and the average current drain I (in amperes). The longevity of the device is simply equal to Q/I. This expression gives the time in hours and can be readily converted to months or years. In reality the problem is not so readily solved and the prediction as device longevity for implantable devices requires an extensive effort. Consider the two factors in the preceding mathematical equation. I is taken to be the nominal current drain that the device requires when it is operating at a certain factory preset rate in a typical patient. However, modern implanted devices are interactive with the patient and do not always operate at the preset rate. There will be a spread in the real longevities as a result of the variation in this nominal current.

In addition, the battery capacity (Q) is not a rigorous single value for a given battery. Generally, the device manufacturer receives a battery capacity value from the battery manufacturer. Since the battery manufacturer seldom knows the precise conditions under which the battery is going to be discharged (including storage or shelf life, actual current drain, end-of-life voltage, and other parameters), the battery manufacturer may be able to give a capacity term that is based only on the battery design rather than on performance testing simulating the actual application. This can lead to highly significant errors for some battery designs.

The factors affecting the uncertainty relating to the longevity of the device can be partitioned into three general categories: (1) the device characteristics, (2) physiological factors; and (3) the physician's medical management practices. Figure 6 illustrates the interrelationship among these three groups of variables. It is the inherent uncertainty that exists within each of these variables that gives rise to the uncertainty regarding device longevity. The discussion in this section will focus on pacemaker longevity to illustrate the impact of each of these variables on the total variation in device longevity. For the convenience of discussion, a specific multiprogrammable pacemaker, the SPECTRAX SXT manufactured by Medtronic, Inc. and powered by a lithium/iodine cell (model Alpha 332) will be used for illustration.

7.1. Pulse Generator Hardware

7.1.1. Battery

The battery is the most critical variable in the estimation of pacemaker longevity. The deliverable capacity[5] of the ENERTEC® model Alpha 332, was projected to be 2.0 Ah. At the nominal factory settings and 500-Ω load impedance, it was estimated to give 10.2 years of serviceable life.[6] This so-

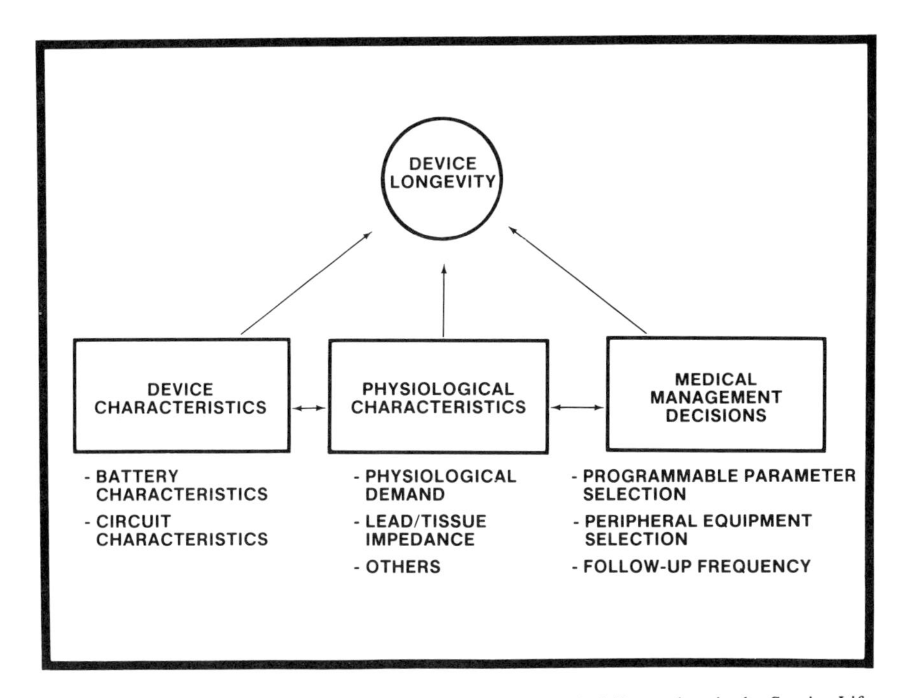

FIGURE 6. The Array of Factors that Contribute to the Statistical Uncertainty in the Service Life of a Cardiac Pacemaker and that Need to be Considered in Projecting Device Lifetime from Battery Performance Models.

called projected nominal longevity takes into account the self-discharge losses in the battery during both its service life and its shelf storage.

The performance of individual batteries, however, will differ because of (1) variations in the rate and efficiencies of chemical reactions within the battery and (2) variations arising from the manufacturing materials and processes such as quantitites of electrode chemicals and hardware dimensions. These variations were first determined from battery discharge data using the statistical model discussed earlier. The result indicated that the nominal variation in battery performance alone would cause an uncertainty of $\pm 10\%$ in device longevity, even if the load current was constant and identical for each individual pacemaker.

7.1.2. *Circuit*

The current consumption by the pacemaker circuit varies because of manufacturing tolerances. This variation is the product of variations in the performance of the electrical components. In the SPECTRAX SXT™, such a variation is relatively small in its impact on battery drain, relative to other variables to

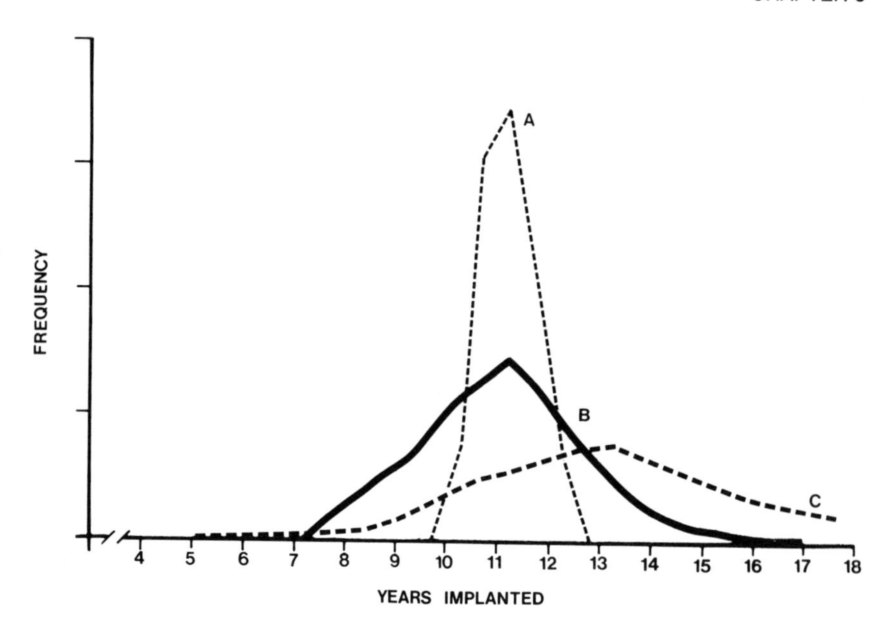

FIGURE 7. The Uncertainty in Pacemaker Longevity Results from Statistical Variations in A, Battery and Circuit Parameters; B, Lead–Tissue Impedance Values that Affect Current Drain; C, Therapeutic Requirements that Determine the Output Settings of a Programmable Pacemaker and Hence the Fraction of Time that a Patient is Paced.

be mentioned. In Figure 7 the cumulative effect of variations in the battery and circuit is shown by the dotted line (*A*). It indicates that the variation in the battery and circuit results in about ± 1 year of uncertainty in the projected nominal longevity of the operating pacemaker.

7.1.3. Patient/Lead Interface

The current delivered to the heart can vary from one patient to another because the net load impedance resulting from the electrode/tissue interface is variable. As shown in Figure 8, clinical studies indicate that the lead/tissue impedance of the Medtronic Model 6961 lead in humans follows a log-normal distribution with a mean of 600 Ω and a standard deviation of 170 Ω.[7] This means that the majority of the patients with this lead will have a load impedance in the range of 400–800 Ω, with some as high as 1000 Ω.

The output current delivered to the heart through a 300-Ω load is twice that delivered through a 600-Ω load. Conversely, with the 1000-Ω load this current is about one-half the nominal value. The impact of load impedance on longevity is less than one might at first anticipate because the stimulating output current only flows during the short output pulse time period—typically 0.6% of the time.

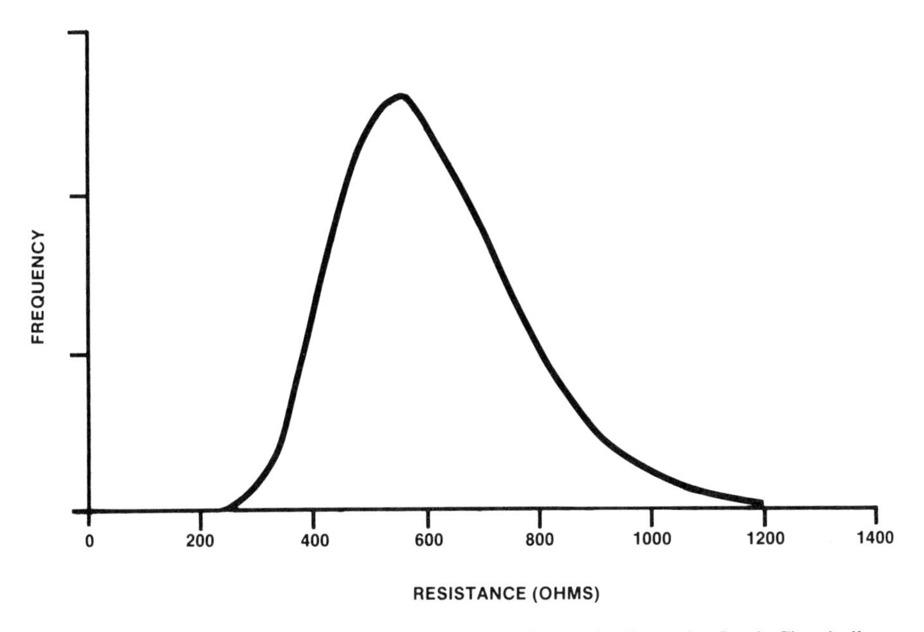

FIGURE 8. The Statistical Distribution of Electrical Resistance for Pacemaker Leads Chronically Implanted in a Typical Patient Population.

With the 300-Ω load, the projected longevity is reduced by 30% from the nominal value, assuming no compensating adjustments were made to the pulse width or amplitude. In Figure 7 the solid line (*B*) represents the cumulative effect of variations due to the load impedance, the battery, and the circuit. It indicates that this pulse generator, which has a nominal longevity of 10.2 years, may not exhibit end-of-life characteristics for up to 14 years in some patients with high load impedance.

7.1.4. Medical Management Practices

Variation in patient condition results in different medical management practices from patient to patient and further contributes to the variability of device lifetime. The choice of unipolar or bipolar pacing mode, and the type of lead, can be major factors here. In demand (VVI) pacemakers, the pacemaker draws far less current when not pacing. Thus, the pacemaker's service life is directly affected by the percent of time it is pacing. A survey by Goldman and Parsonnet[8] showed that only 28% of pacemaker implants in the United States were for complete heart block (CHB) patients. Typically, these are the only patients requiring pacing 100% of the time. A later study by Kim[9] further indicated that there are some patients who are rarely dependent on the pacemaker. The average percent pacing in the general patient population is unknown but was estimated to be near 60%. With nominal parameter settings, the longevity of the SPECTRAX SXT unit at 60% pacing would be three years longer than at 100% pacing. Therefore, this variability in percent pacing adds significant uncertainty to pacemaker longevity.

A more important factor of medical management, however, is the growing trend toward pacemaker programmability. Programmable parameters that affect longevity include rate, pulse width, pulse amplitude, hysteresis, and pacing mode. For example, longevity can be improved for patients who do not require high stimulation currents if the pacemaker's output is reduced by reprogramming. The effect of these variables on pacemaker longevity is quantified through the computer simulation method. The dashed line (*C*) in Figure 7 shows the net result of all variations, including percent pacing, pacemaker programming, and the previously mentioned variables of battery, circuit, and load impedance. It shows that the majority (83%) of pacemakers in this study will last longer than 10.2 years.

8. CONCLUSION

In the preceding section it was pointed out that the battery, circuit, and patient variables, and medical decisions all contribute to the uncertainty in device longevity. When all these variables are considered simultaneously their effects may range from halving to doubling the nominal longevity of a device performing normally in all regards. To reduce this variability as much as possible, the battery should be made as reproducible as possible and should be supported by an adequate mathematical model. However, even this would not provide an adequate basis for projecting device depletion and scheduling replacement surgery. This illustrates the importance of developing reliable indicators of impending battery depletion (end-of-life or depletion indicators) as part of the device. The usefulness of these indicators depends largely on the reproducibility of the battery performance characteristics as verified by mathematical modeling. The topic of depletion indicators is treated elsewhere in this book.

REFERENCES

1. J. B. Phipps, T. S. Hayes, P. M. Skarstad, and D. F. Untereker, *Extended Abstracts 84-2*, Abstract 175, pp. 258, The Electrochemical Society, Pennington, New Jersey (1984).
2. K. Fester, W. Helgeson, B. Owens, and P. Skarstad, *Solid State Ionics 9&10*, 107 (1983).
3. D. F. Untereker, *J. Electrochem. Soc. 125*(12), 1097 (1978).
4. H. F. Gibbard, in: *Power Sources for Biomedical Implantable Applications and Ambient Temperature Lithium Batteries* (B. B. Owens and N. Margalit, eds.), Vol. 80-4, pp. 510, The Electrochemical Society, Princeton, New Jersey (1980).

5. K. R. Brennen, K. E. Fester, B. B. Owens, D. F. Untereker, *J. Power Sources 5,* 25 (1980).
6. B. B. Owens, *Stimeucoeur Med. 9,* 161 (1981).
7. R. Dutcher, Medtronic, Inc., internal communication, unpublished.
8. B. D. Goldman and V. Parsonnet, *PACE 2*(5), (1979).
9. J. S. Kim, Medtronic, Inc., internal communication, unpublished.

6

Lithium/Halogen Batteries

CURTIS F. HOLMES

1. INTRODUCTION

The Gibbs free energies of formation for lithium halides vary from -64.5 kcal/mol for lithium iodide to -140.7 kcal/mol for lithium fluoride. Therefore, electrochemical power sources based on the reaction between lithium and halogens offer potentially very attractive energies. In fact, lithium/halogen systems, specifically the lithium/iodine couple, have seen extensive application as the most widely used power sources for cardiac pacemakers from 1976 through the present time (1984). The basic idea behind lithium/halogen systems is rather straightforward. The anode is elemental lithium. The cathode/depolarizer is a pure or mixed halogen whose electronic conductivity has been enhanced with inert additives or reactive species. The electrolyte/separator is the reaction product formed *in situ* between anode and cathode as the cell is discharged. During discharge this electrolyte layer grows as the reaction products are formed and the anode and cathode materials are depleted.

In 1967 Gutmann *et al.* reported cells using metal anodes and cathodes fabricated from "charge transfer complexes" between iodine and organic donors.[1] Metal anodes used included magnesium, indium, cadmium, barium, calcium, and a magnesium/lithium alloy. In 1972 Moser reported a lithium/iodine battery.[2] Further refinements were disclosed by Schneider in 1972.[3] In 1973 Mead reported a lithium/iodine cell whose cathode was prepared from a mixture of iodine and poly-2-vinylpyridine (PVP) that had been thermally reacted.[4]

As early as 1971 it was recognized that the lithium/iodine–PVP cell showed promise in fulfilling the low-current, long-life, high-reliability requirements demanded of a power source for implantable cardiac pacemakers. Greatbatch and

CURTIS F. HOLMES • Research and Development, Wilson Greatbatch Limited, Clarence, New York 14031.

co-workers reported a battery specifically designed for cardiac pacemakers that offered several advantages over the then currently used zinc/mercuric oxide cells.[5] In April 1972 the first human implant of a pacemaker with a lithium/iodine power source took place in Italy.[6] Since then the lithium/iodine–PVP battery has seen a considerable amount of development and design evolution. A threefold increase in energy density has been achieved since 1973, and improvements in cell design, current-delivery capability, and reliability have been made. Today it is estimated that well over 1 million lithium/iodine–PVP batteries have been implanted, and nearly 80% of all pacemakers manufactured in 1984 were powered by such batteries.

Attempts to develop batteries with bromine[7] and chlorine[8] cathodes have been made since 1975. Though none of these efforts have resulted in commercially successful implantable batteries, some interesting results have been obtained. Some work on mixed halogen and interhalogen cells has also occurred.[9]

The following sections will present general thermodynamic, kinetic, and structural considerations for lithium/halogen cells. Next, a brief survey of the lithium/bromine system will be presented. The lithium/iodine–PVP system will then be reviewed in considerable detail. The cell chemistry, cell construction, discharge characteristics, and cell performance in implantable applications will be described. Finally, a look at the current state of the art and some speculation about future trends in lithium/halogen battery systems will be presented.

2. GENERAL FEATURES OF LITHIUM/HALOGEN SOLID ELECTROLYTE BATTERIES

2.1. Thermodynamic Considerations

All of the halogens react with elemental lithium to form lithium halides according to the reaction

$$\mathrm{Li} + \tfrac{1}{2}\,\mathrm{X_2} \rightarrow \mathrm{LiX} \tag{1}$$

The enthalpies, entropies, and Gibbs free energies of these reactions are well characterized.[10] Table 1 summarizes these functions for LiI, LiBr, LiCl, and LiF, together with the theoretical open-circuit voltages, calculated from the Nernst equation, for cells resulting from these reactions (at 27°C).

The obvious conclusion that can be drawn from the data presented in Table 1 is that batteries whose cell reactions involve the direct reaction of elemental lithium with elemental halogens offer rather high voltages and energy densities, based on thermodynamic considerations alone.

TABLE 1 Thermodynamic Functions for Formation of Lithium Halides (300°K)

Halide	ΔH (kcal/mol)	$T\Delta S$ (kcal/mol)	ΔG (kcal/mol)	OCV (V)
LiF	−147.450	−6.754	−140.696	6.100
LiCl	− 97.578	−5.792	− 91.786	3.980
LiBr	− 83.870	−2.220	− 81.650	3.540
LiI	− 64.551	−0.101	− 64.450	2.795

2.2. Kinetic Considerations

Three fundamental kinetic limitations present potential barriers to the successful realization of lithium/halogen batteries. These are (1) self-diffusion of lithium atoms and its effect on the anode/electrolyte interface, (2) the conductivity of the lithium halide solid electrolyte formed *in situ,* and (3) the conductivity of the cathode material.

The first of these limitations can have a potentially serious effect on the anode/electrolyte interface. As the cell is discharged, lithium atoms at the anode/electrolyte interface are oxidized and the resultant ions migrate through the lithium halide to the electrolyte/cathode interface. The resultant vacant sites left at the anode surface by this migration must be filled by diffusion of lithium atoms to the interface. Thus, self-diffusion of lithium atoms presents a kinetic limitation to the discharge rate of the cell because if the cell is discharged at rates higher than this self-diffusion rate, vacancies can coalesce into voids, resulting in loss of interface contact.[11]

A second and more formidable kinetic limitation is present in the ionic conductivity of the solid electrolyte. Lithium ions must travel from the anode/electrolyte interface to the cathode/electrolyte interface during the discharge process. Moreover, the thickness of the electrolyte increases as cell discharge proceeds. Consequently, the internal resistance of the cell constantly increases as the cell is discharged.

Table 2 presents the ionic conductivities of the four lithium halides at 25°C.[12] Lithium iodide shows the highest conductivity ($10^{-7}\ \Omega^{-1}\ \mathrm{cm}^{-1}$) at 37°C, implying that even this couple will show significant current-delivery limitations and that such limitations will be more severe in batteries using the more reactive halogens. In practice, techniques have been introduced that have alleviated this problem for the lithium/iodine and lithium/bromine couple. These techniques, which will be discussed in detail in the foregoing sections, have resulted in the development of cells whose current-carrying capability, although limited in comparison with liquid-electrolyte systems, is considerably higher than would be predicted from the conductivity values shown in Table 2.

TABLE 2 Ionic Conductivities of Lithium Halides (25°C)

Salt	Conductivity (Ω^{-1} cm^{-1})
LiF	10^{-12}
LiCl	10^{-10}
LiBr	10^{-9}
LiI	10^{-7}

The third kinetic consideration is the conductiviy of the cathode material. In lithium/halogen systems, electrons must migrate from the cathode current collector to the cathode/electrolyte interface where the cell reaction takes place. The electronic conductivities of pure halogens are much too low to be used as such for cathode materials. Accordingly, the electronic conductivity must be enhanced considerably in order to produce useful batteries. This has been accomplished in the cases of iodine and bromine by the introduction of charge transfer complexing agents that render the material electronically conductive[(1–4)] or by mixing inert conductive material such as carbon with the halogen.[(13)] These techniques, which will be discussed in the following sections, have resulted in the formulation of cathode materials of sufficient conductivity to be useful in practical low-rate battery designs.

3. THE LITHIUM/BROMINE SYSTEM

3.1. General Considerations

The lithium/bromine couple offers several theoretical advantages as a long-life, low-power cell for implantable devices. Among these are high volumetric and gravimetric energy densities and a high open-circuit voltage (3.5 V). Compared with the lithium/iodine system, the lithium/bromine system offers a 94% increase in gravimetric energy density and a 25% increase in volumetric energy density. It offers no theoretical advantage in capacity per unit volume. Even so, the theoretical advantages of the Li/Br_2 couple have made it an attractive candidate system as a power source for implantable medical devices such as cardiac pacemakers, and the system has been the subject of research and development in several laboratories.[(7,14,15)] To date, none of these efforts have produced viable commercial cells for implantable devices. There are fundamental limitations that suggest that the theoretical advantages of the Li/Br_2 system are unlikely to be attained in practice. Moreover, practical problems such as materials compatibility and the production of cells meeting the high standards of reliability and repro ducibility demanded of implantable power sources have led to the conclusion that, given the satisfactory performance of the lithium/iodine system, the lithium/ bromine system is not competitive.[16] However, the efforts have met with some success, and the research has led to a greater understanding of lithium/ halogen batteries in general.

This discussion is restricted to those systems that have been reported as candidate implantable batteries. That is, systems in which the bromine-containing cathode material is placed in direct contact with the lithium anode and the discharge product, lithium bromide, is formed *in situ* as the reaction progresses. The fundamental kinetic problems discussed in Section 2 present themselves in considering this approach to battery design. The first is the low conductivity of pure bromine. The second is the current-carrying limitation imposed by the conductivity of LiBr.

The conductivity of pure bromine has been reported as $1.5 \times 10^{-11}\ \Omega^{-1}\ \text{cm}^{-1}$.[14] Therefore, no practical battery could be made using pure bromine as a cathode material. By adding salts,[14] charge-transfer complexing agents rendering the solution electronically conductive,[7] or inert conductive material such as carbon,[13] the conductivity of the cathode material can be raised to levels useful in practical battery design. The second of these approaches has been the subject of most of the effort in development of lithium/bromine cells. Howard *et al.* have reported nine candidate additives for bromine cathodes.[14] These compounds, six monomeric pyridine derivatives, two quaternary ammonium bromides, and PVP, result in cathode conductivities ranging from $1.3 \times 10^{-3}\ \Omega^{-1}\ \text{cm}^{-1}$ to $5.5 \times 10^{-2}\ \Omega^{-1}\ \text{cm}^{-1}$ at room temperature. Such cathode materials are, in principle, useful in the construction of low-rate lithium/bromine cells.

A more formidable limitation is that of the conductivity of the lithium bromide formed *in situ* as the cell is discharged. The conductivity of pure lithium bromide is around $10^{-9}\ \Omega^{-1}\ \text{cm}^{-1}$ at 37°C. Since the electrolyte thickness increases as the cell is discharged, a significant current-delivery limitation is imposed that worsens as cell discharge progresses. Measurements on actual cells, however, indicate that the internal resistance of Li/Br_2 batteries is two to three orders of magnitude smaller than that predicted by the theoretical value.[14] The rate of resistance buildup is affected by the cathode additive employed in the cell and by the rate of cell discharge. Additional steps can be taken, such as precoating the anode with PVP, to reduce cell resistance further.[17]

Cell design considerations for lithium/bromine batteries share many features with the corresponding iodine cells. A typical design approach utilizes a stainless steel (304L) prismatic case, a central lithium anode, a feed-through design that protects the anode current collector and lead from contact with the conductive cathode mixture, and a sealable fill port through which the liquid cathode material is poured into the cell. The case itself acts as the cathode current collector, and the cell is hermetically sealed.

3.2. The Li/Br_2–PVP Cell

The first approach in the development of an implantable lithium/bromine cell was the design of a cell that was essentially analogous to the lithium/iodine–PVP system. The cathode material was the product of the reaction between bromine and PVP. The anode was precoated with PVP. In constructing the cell, pure PVP, in either powder or pellet form, was introduced in equal amounts on either side of the central anode. The lid and anode assembly were then welded to the cell case and pure bromine was introduced through the fill port.[18]

Cells constructed in this manner have been subjected to a variety of discharge tests. One group of these cells has been on constant 100 kΩ load test at 37°C for six years. Of an original group of 16 batteries, with stoichiometric capacity of approximately 4.6 Ah, eight were still operating six years later. Their average voltage versus time is shown in Figure 1. These data indicate that any of these cells would have powered a typical cardiac pacemaker for six years. However, other batteries in the original group of 16 attained an arbitrary end-of-life voltage of 2 V in times varying from 15 months to 70 months, indicating a wide variability in real-time performance that is unacceptable for current standards of implantable cell performance.

3.3. Other Cathode Formulations

As stated earlier, Howard *et al.* have reported monomer additives that gave various results in cell performance testing.[14] They found that the rate of resistance increase as a function of cell discharge was always at least one order of magnitude greater than that in corresponding lithium/I_2–PVP cells. The best delivered capacity obtained in their discharge experiments was approximately 50% of the stoichiometric bromine capacity under 34 kΩ constant load.

Holmes and Mueller have reported bromine batteries using as a cathode additive the solid reagent pyridinium bromide perbromide.[19] This compound can be pelletized and incorporated into the battery before filling with pure bromine. The compound dissolves in bromine and renders the cathode material conductive. A group of cells made in this manner was tested for over 60 months under a cyclic discharge regime of two weeks at 30 kΩ, three days open circuit, and three days at 140 kΩ, with cycles repeated until cell depletion. The cells delivered an average capacity of around 1.7 Ah, which is 57% of the capacity delivered by a Li/I_2–PVP cell of identical shape.

3.4. Summary

Though much investigation has been conducted on the lithium/bromine system, no commercial utilization of such cells in implantable applications is occurring today. Given the excellent performance of the lithium/iodine–PVP

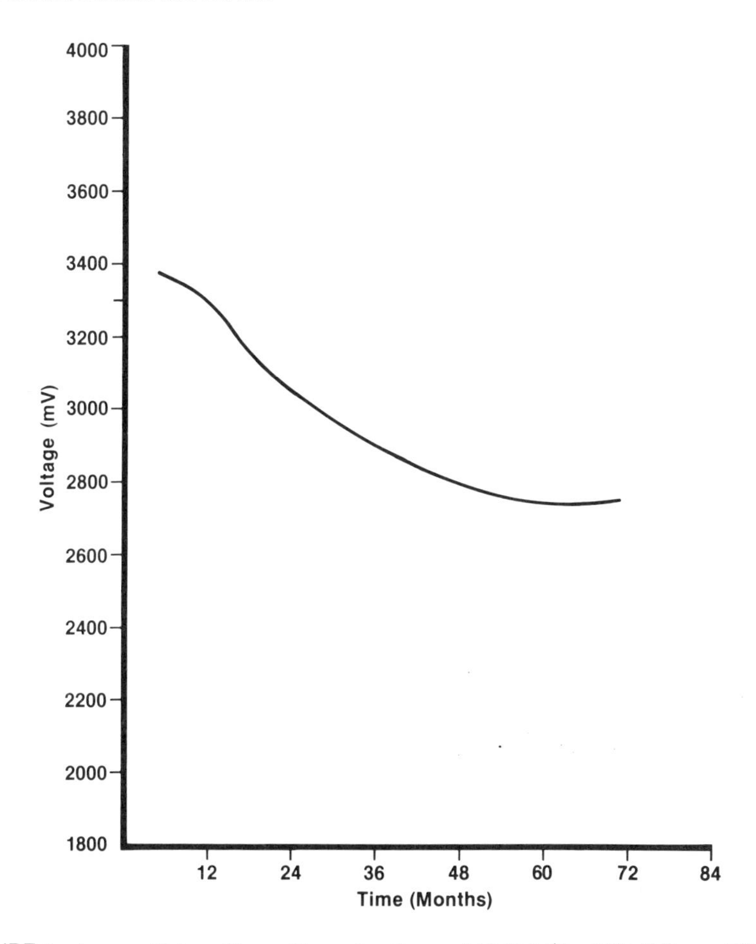

FIGURE 1. Average Voltage Versus Time of a Group of Eight Li/Br_2–PVP Cells on 100 kΩ Constant Load.

system as well as other systems such lithium/silver chromate and lithium/cupric sulfide, the lithium/bromine system does not appear to offer sufficient advantages to justify its commercial use in implantable applications.

4. CHEMISTRY OF THE LITHIUM/IODINE–POLYVINYLPYRIDINE SYSTEM

4.1. Cell Reaction

In the previous sections of this chapter some general features of lithium/halogen systems have been presented and the lithium/bromine system has been discussed. Since no commercially viable cells have been developed using halogens other than iodine, and since the only iodine system to achieve practical importance in an implantable battery design is the iodine–PVP system, the remainder of this chapter will deal exclusively with this particular battery chemistry.

The basic cell reaction in this battery is quite simple, namely,

$$Li + \tfrac{1}{2}I_2 \rightarrow LiI \qquad (2)$$

The Gibbs free energy of this reaction is −64.5 kcal/mol; thus, the open-circuit voltage is 2.8 V. The cathode material includes the compound poly-2-vinylpyridine, which reacts with a portion of the iodine to render the mixture conductive. It is generally accepted that the PVP plays no part in the actual cell reaction, at least at voltages useful for applications discussed here.

In the following paragraphs the properties of the anode material, the cathode material, and the reaction product will be discussed in detail.

4.2. The Lithium Anode

Lithium, the third element in the periodic chart, is an attractive anode material for high energy-density batteries. Its density is 0.534 g/cm^3 at room temperature. The first ionization potential of lithium is 5.39 eV, and the electrochemical equivalent is 3.86 Ah/g. It reacts readily with halogens at room temperature. Its physical properties are such that it is easily formed into sheets that can be cut to specific anode sizes, and it is easily pressed into specific anode shapes, such as the corrugated-anode design introduced into more recent versions of the Li/I_2–PVP cell.

The high reactivity of lithium presents some practical problems in battery design and construction. In the presence of water vapor lithium will react with oxygen to form the oxide, carbon dioxide to form the carbonate, nitrogen to form the nitride, and water itself to form the hydroxide. Since all these compounds will lead to deleterious effects on battery performance, care must be taken to avoid their formation. In the absence of moisture these reactions proceed very slowly or not at all. Thus, lithium batteries must be constructed in a dry or inert atmosphere. Implantable-grade cells are typically assembled in dry rooms where the dew point is kept below −40°C.

The melting point of lithium is 180.5°C. This fact is important because many lithium battery systems, including the Li/I_2–PVP system may present serious explosion hazards if heated to temperatures above the melting point of lithium.

In many versions of the Li/I_2–PVP battery the lithium anode is coated with a solution of PVP dissolved in a volatile solvent such as tetrahydrofuran. The solvent is evaporated, leaving a contiguous film of pure PVP on the entire anode surface. As will be discussed later, this precoating of the anode has a profound effect on the nature of the discharge product and leads to lower cell impedance and generally enhanced performance.

4.3. The Cathode Material

4.3.1. *General Remarks*

Iodine, element number 53 in the periodic chart, has an electron affinity of 3.24 eV and an electrochemical equivalent of 0.211 Ah/g. It is a black crystalline solid that sublimes at atmospheric pressure. The melting point of iodine is 113.7°C, and its density is 4.94 g/cm^3. The electronic conductivity of iodine is insignificant, rendering the pure material of no use in practical lithium/iodine battery systems.

In their initial report on batteries made from metal anodes and halogen-containing cathodes in 1967, Gutmann *et al.* described the cathode materials as "charge-transfer complexes".[(1)] This they defined as a "system involving the partial transfer of one or more electrons from a donor (typically an aromatic hydrocarbon) to an acceptor (typically a halogen or an organic compound of high electron affinity)". The authors pointed out that one significant property of such materials is a relatively high electronic conductivity. The cathodes in their study were made by mixing certain organic compounds with iodine to form these "complexes". One such organic compound was polyvinylpyridine mixed with carbon black. Moser [(2)] and Schneider[(3)] later reported lithium anode batteries wherein the cathode material was a mixture of iodine and a charge transfer complex of iodine and an organic donor component such as polyvinylquinoline or poly-2-vinylpyridine. It was recognized by Mead[(4)] in 1973 that heating the iodine–polyvinylpyridine mixture to temperatures in excess of 100°C produced a cathode material of relatively high electronic conductivity. In further development, batteries were designed wherein this material was poured into the battery in the molten state and allowed to freeze into the final shape. From 1973 to 1979 virtually all Li/I_2–PVP cells were made using this thermally reacted depolarizer material, and the majority of cells produced today use this same concept. In 1979 Schneider *et al.* reported lithium/iodine cells containing a pelletized particulate mixture of iodine, polyvinylpyridine (or polyvinylquinoline), and a charge transfer complex consisting of the polymer and iodine.[(20)] This material is not thermally treated above 100°C. Batteries using this cathode material are also produced today.

4.3.2. *Thermally Reacted Cathode Material*

The reaction between iodine and polyvinylpyridine has been the subject of much study over the past few years. Although the chemistry of this reaction is quite complicated, and the number of products obtained by the thermal reaction is relatively high, considerable progress has been made in understanding the thermally reacted cathode materials.

The first commercially available cells made with iodine and PVP contained 10 parts of iodine by weight to one part polyvinylpyridine. Since 1975, the weight ratio of iodine to PVP has increased from 10:1 to 15:1, 20:1, 30:1, and today some cells are being produced with a 50:1 weight ratio.

In 1977 Greatbatch *et al.* studied the reaction between iodine and polyvinylpyridine using differential thermal analysis.(21) They found that the reaction between I_2 and PVP is exothermic and begins as low as 25°C. At 105°C the reaction is intensely exothermic, though at higher ratios of iodine to PVP the exotherm is masked by concurrent melting of iodine at 113.7°C. Subsequent cooling and heating indicated the presence of at least two species at higher I_2/PVP ratios, one of which is elemental iodine.

Brennen and Untereker studied the iodine activity in the cathode material as a function of the iodine/PVP ratio (expressed as R, the ratio of moles of iodine to moles of monomer unit in the PVP).(22) They employed both an isopiestic method and EMF measurements to determine iodine activity. It was concluded that at values of R above 1.25 all iodine in the cathode mixture is at unit activity.

In later studies these authors studied the chemistry of the iodine–PVP complex further.(23) The replacement of hydrogen atoms in the polymer by iodine was studied using nuclear magnetic resonance spectroscopy. The authors found evidence that the alpha hydrogen on the polymer "backbone" was replaced by iodine but that neither the beta hydrogen nor the aromatic hydrogen atoms were replaced.

The reaction between halogens and PVP was also studied by McLean and Bleecher.(24) By using high-performance liquid chromatography and intrinsic viscosity measurements they demonstrated that the chain length of the original PVP was considerably reduced during the reaction with iodine. Electron spin resonance spectroscopy of the reaction product showed a single narrow signal with a g value of 2.002, indicating the presence of free electrons.

The chemical picture that emerges from these studies of the reaction between iodine and polyvinylpyridine is that one molecule of iodine is bound through a donor–acceptor-type bond to the nitrogen of the pyridine ring and another atom of iodine is bound to the alpha carbon of the much-shortened aliphatic chain. Figure 2 shows this structure.

The phase diagram of the iodine/PVP system has been studied by Phillips and Untereker.(25) The technique of differential scanning calorimetry was used to study the phase relationship of the iodine–PVP system in the temperature range 150°–400°K. Particular attention was paid to the mole fraction range $0.5 < X_{I_2} < 1.0$, where X_{I_2} is the mole fraction of iodine. (Here mole fraction refers to molecular iodine and the monomer unit of PVP.) This mole fraction range is of interest because it includes the concentration range of the cathode material during discharge. Figure 3 shows the phase diagram determined in this

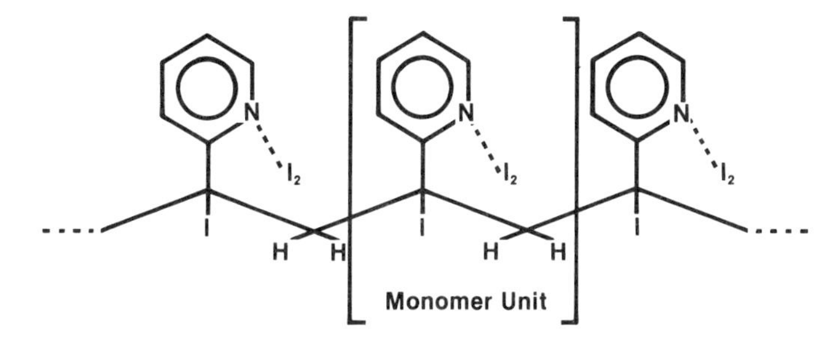

FIGURE 2. Proposed Structure of Iodine–PVP Reaction Product.

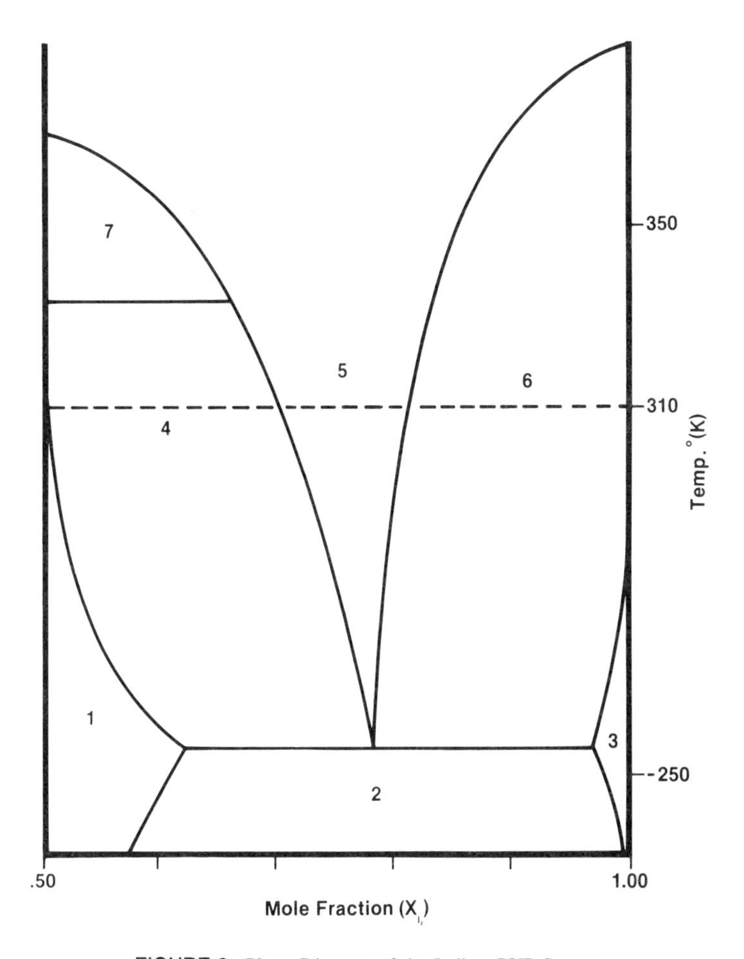

FIGURE 3. Phase Diagram of the Iodine–PVP System.

study. Regions 1, 3, and 7 were only partially characterized and are of little interest in discussing behavior of the cathode material during discharge. Region 2, a two-phase solid region, is also of little interest. Regions 4, 5, and 6 are, however, all encountered as the cell is discharged. The dotted line at 310°K (37°C) shows the temperature at which the cathode operates. As the cell is discharged, the cathode composition passes through region 6, region 5, and region 4. At $X_{I_2} = 0.5$ is the adduct phase wherein 1 mole of iodine has reacted with 1 mole of PVP monomer unit. At $X_{I_2} = 0.78$ there exists a eutectic between this adduct phase and pure iodine. Region 5, above the eutectic point, contains a single liquid phase. Region 4 is a two-phase system wherein the one-to-one iodine/monomer unit adduct phase coexists with the melt. Region 6 is also a two-phase system containing the melt and pure iodine. The discharge at 310°K thus begins in the two-phase iodine/melt region, passes through the single-phase liquid region, and ends in the two-phase adduct/melt region.

The viscosity of the system is a minimum at the eutectic composition. As discharge begins, the viscosity is fairly high, as the two-phase initial composition has a tarlike consistency at 310°K. As the eutectic composition is approached the cathode material becomes quite liquid. Near the end of discharge the viscosity again increases, and the cathode material in nearly depleted cells is quite hard and glasslike. These observations have been confirmed through destructive analysis of actual cells in various stages of discharge.

The conductivity of the cathode material has been the object of much study. Early workers asserted that the conductivity was electronic.[2] In 1976 Kraus and Schneider determined the resistivity of the cathode mixture as a function of iodine concentration.[26] The resistivity exhibits a minimum at an iodine–PVP mole ratio of about 3.4 (the eutectic point) and rises steeply at lower iodine concentration. Phillips and Untereker also measured the conductivity of the material as a function of the mole feactoin of iodine.[25] Figure 4 shows their results. The conductivity exhibited a maximum of about $10^{-3}\ \Omega^{-1}\ \text{cm}^{-1}$ at the eutectic point. It was suggested that the conductivity had a predominantly ionic component. However, in later work these authors studied anionic transport in the system.[27] By measuring the amount of iodine moved from one position to another in a test cell through which a known amount of charge was passed, these workers were able to determine the transference number of iodine. They found that ionic conductivity contributed no more than 10% of the total conductivity of the system and that the majority of the conductivity of the system was in fact electronic.

4.3.3. *Pelletized Cathode Material*

The pelletized particulate cathode material reported in 1979 exhibits characteristics considerably different from those of the thermally reacted cathode described previously.[20,28] This material is formed by mixing iodine, ground to

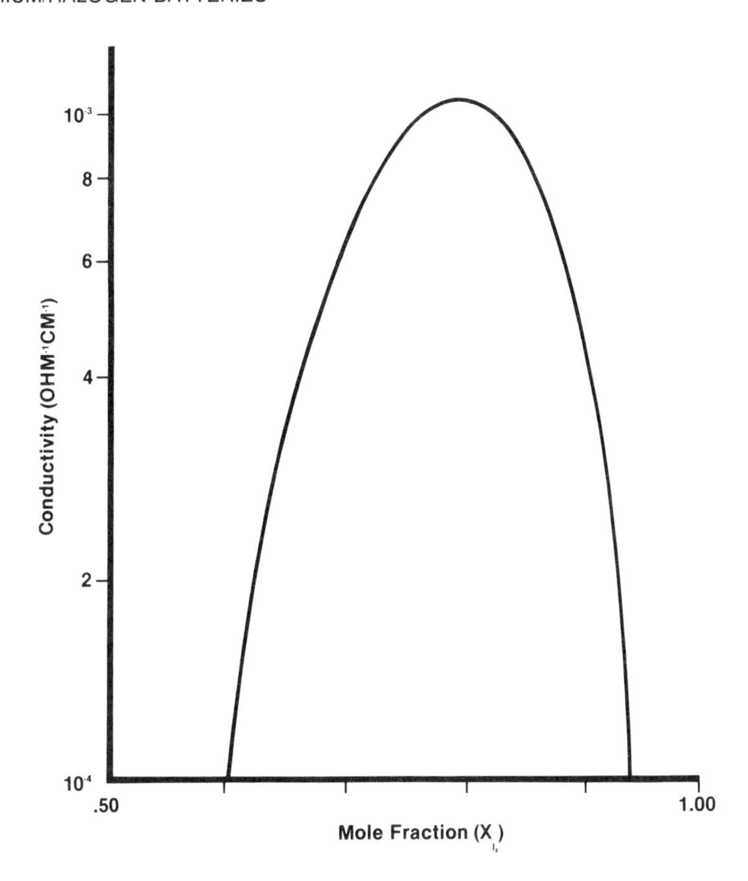

FIGURE 4. Conductivity of the Iodine–PVP Cathode Material as a Function of Mole Ratio of Iodine.

a particulate having a size less than 0.19 mm, with a polymer. The polymer, which may be either poly-2-vinylpyridine or poly-2-vinylquinoline having a weight average molecular weight greater than 1000, is ground to a particle size of less than 0.15 mm and preferably within the range 0.074–0.1 mm. From 18 to 26 parts of iodine by weight and one part polymer are mixed together. The mixture is placed in a die and pressed into a pellet under a pressure of 5000 to 15,000 lb/in.^{2} It is reported that the conductivity of the material is initially quite low but soon increases to an acceptable level. This increase in conductivity can be accelerated by mild heating of batteries made from this material to temperatures less than 100°C. The pelletized cathode retains its pressed shape but exhibits a slight tackiness that enhances electrical contact. Unlike the thermally reacted cathode material, which exhibits phase changes during discharge, it is reported that these cathode pellets remain solid throughout cell discharge.[29] Less iodine is apparently bound to the polymer chain in this cathode formulation than in the case of thermally reacted cathode mixtures. The solid pellet cathodes also affect the nature of the discharge product, as will be discussed.

4.4. The Electrolyte/Separator

4.4.1. General Properties

Lithium iodide is a crystalline solid with a rock salt type of crystalline structure. The lithium and iodide ions are arranged in a face-centered cubic structure. The lithium ion conductivity of lithium iodide is on the order of 10^{-7} Ω^{-1} cm^{-1}. The electronic conductivity is negligible compared to the current capability of the cell. The density of the material is 3.494 g/cm^3. The compound is relatively inert, so the electrolyte is stable in contact with both lithium and the iodine–PVP cathode material.

The unique feature of the eletrolyte/separator in this system is the fact that it is self-forming. When the cell is first formed the iodine–PVP material is either poured in the molten state around the anode or pressed against it. The initial separator thickness is zero, but it rapidly forms by direct reaction between lithium and iodine. As the cell is discharged the thickness of the lithium iodide increases, causing a corresponding rise in cell impedance. The self-forming nature of the electrolyte is a very appealing feature from the standpoint of reliability. The risks associated with separator failure, which is not an uncommon phenomenon in long-life batteries, are minimized in this system. If the separator were to crack, allowing cathode material to migrate to the bare anode, a new layer of lithium iodide will immediately form, healing the temporary short circuit caused by anode/cathode contact. This self-healing feature of the electrolyte enhances the inherent reliability of the system.

4.4.2. Enhancement of Electrolyte Performance

An ionic conductivity of 10^{-7} suggests that the cell is severely current-limited. Indeed, the first lithium/iodine cells were limited to low-current applications. As designs evolved, it was found that the manner in which the lithium iodide was formed could be modified by various techniques. These modifications led to a decrease in the resistance of the lithium iodide and a corresponding increase in current-carrying capability.[30]

4.4.2a. Modifications Induced by Anode Precoating. In 1976 Mead and co-workers reported a lithium/iodine–PVP cell in which the anode was coated with pure PVP dissolved in a volatile solvent.[31] Batteries made with such precoated anodes show lower internal resistance and higher voltages during discharge. The apparent ionic conductivity of lithium iodide formed in coated-anode cells is between one and two orders of magnitude higher than that of nominally pure polycrystalline lithium iodide. This phenomenon is illustrated in Figure 5, which shows the resistance rise of two otherwise identical cells with coated and uncoated anodes. The resistance versus capacity is plotted for the two cells, which were discharged under a 9.5-kΩ constant load at 37°C. It should be noted that the cell resistance as plotted also contains a component due to the resistance change of the cathode material, but this component should be equal for the two cells at any given level of discharge.

Holmes and Brown studied the macro- and microstructural differences in depleted cells made with precoated and uncoated anodes in an attempt to relate

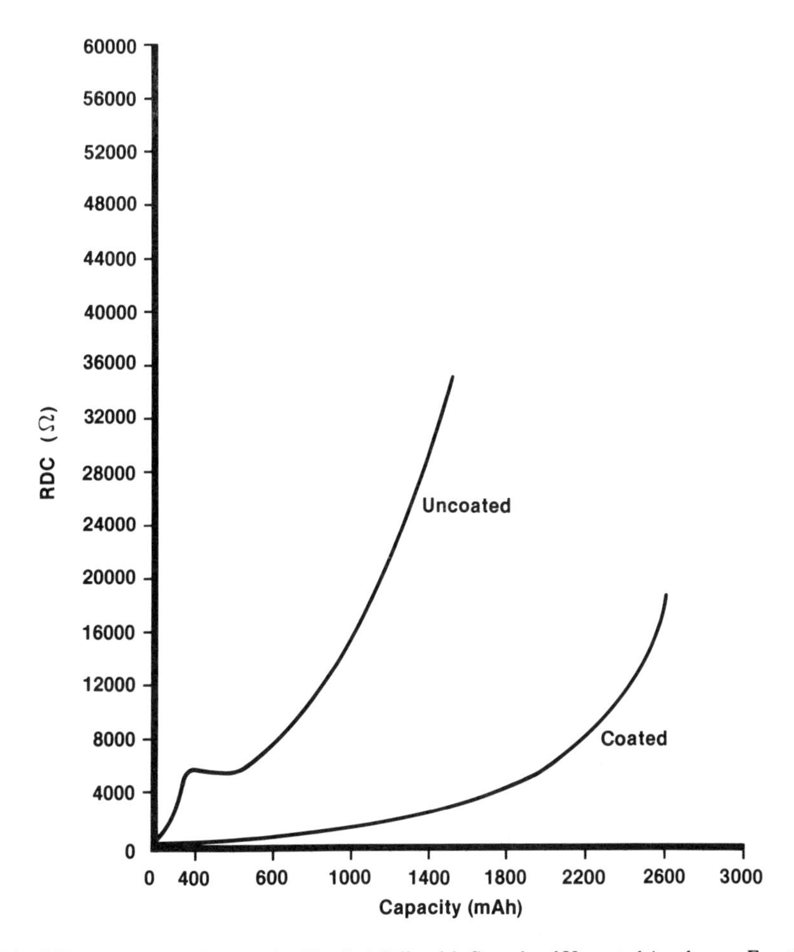

FIGURE 5. Resistance of Otherwise Identical Cells with Coated and Uncoated Anodes as a Function of Capacity.

this enhanced electrical performance to electrolyte structure.[32] They found striking differences in both the microstructure of the lithium iodide formed during discharge and the macrostructure of depleted cells. Figures 6 and 7 show scanning electron micrographs of lithium iodide formed during discharge of otherwise identical cells made with coated and uncoated anodes. Lithium iodide from the coated-anode cell is formed of apparently individual crystallites that are equiaxed and well formed. The general appearance is that of a granular mass. The electrolyte formed from the uncoated-anode cell is of the form of a bonded mass of much smaller granules in a matrix of amorphous or ultrafine lithium iodide. The material appears to be of higher density than that of the coated-anode electrolyte.

On a macroscopic scale (i.e., 40× light microscopy), dramatic differences between cells made with and without anode precoating are evident. In cells with flat anodes the coating feature leads to completely different discharge profiles. Uncoated-anode cells exhibit a flat interface between the lithium anode and the lithium iodide formed during discharge. The lithium iodide layer appears to be hard, regular, and completely flat. Well-defined and clearly delineated lithium, lithium iodide, and cathode layers exist. Figure 8 shows a cross-sectional view of such a cell.

The situation is much different in the case of the coated-anode cells. The originally flat lithium anode is macroscopically distorted during discharge. Col-

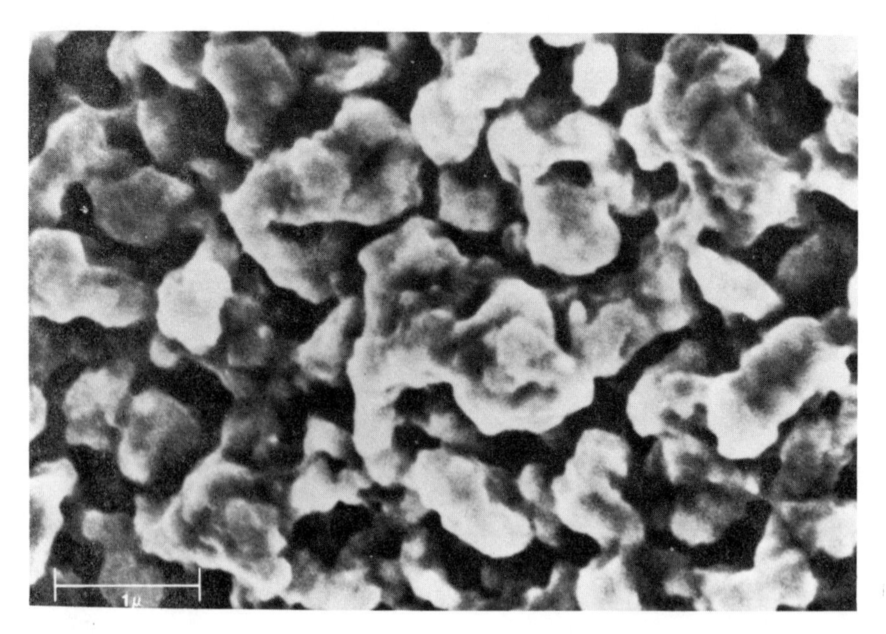

FIGURE 6. Scanning Electron Micrograph of Lithium Iodide Formed During Discharge of Coated-Aode Cell.

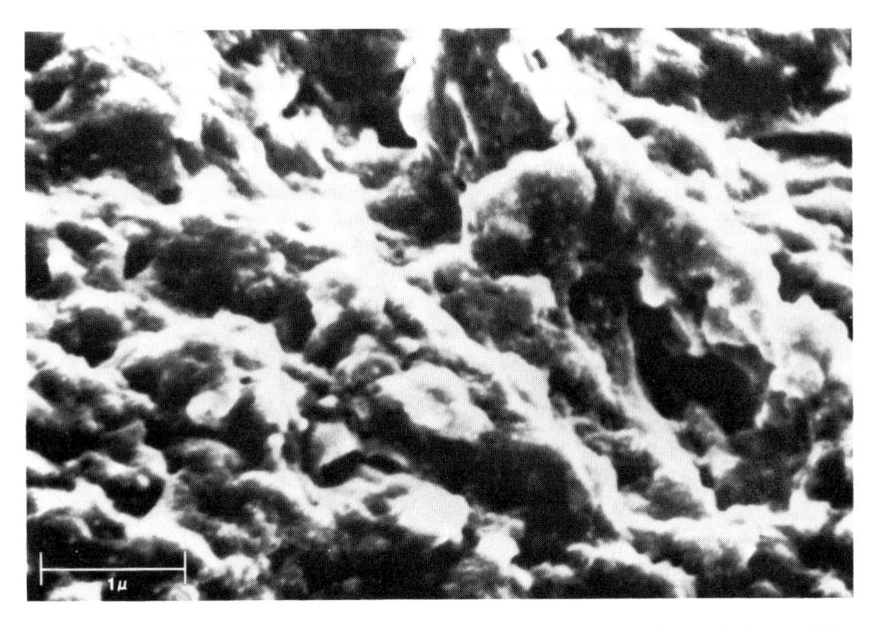

FIGURE 7. Scanning Electron Micrograph of Lithium Iodide Formed During Discharge of Uncoated-Anode Cell.

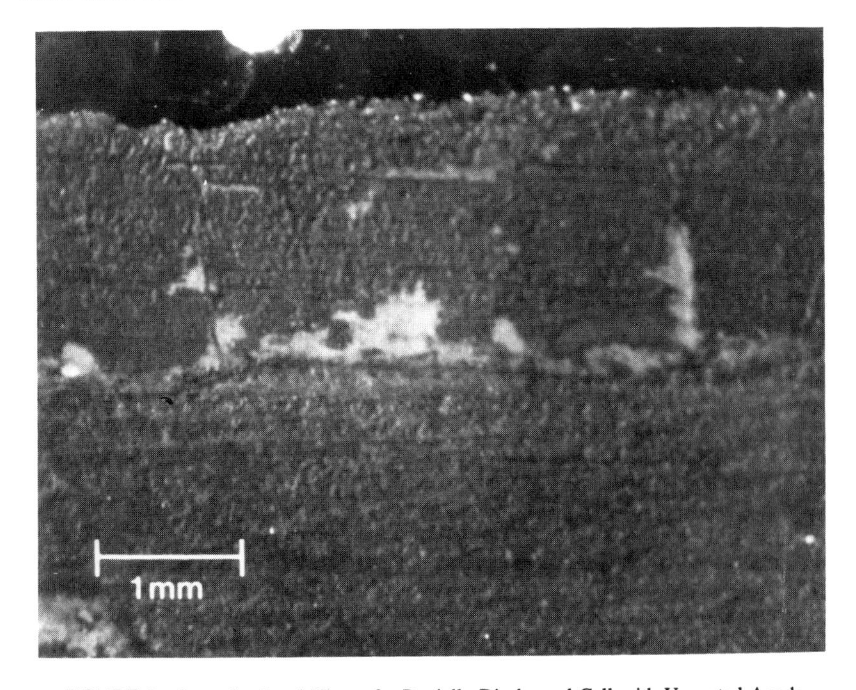

FIGURE 8. Cross-Sectional View of a Partially Discharged Cell with Uncoated Anode.

umns of iodine-rich zones protrude into an irregular lithium iodide area. Likewise, lithium-rich zones are present. Figure 9 shows a cross-sectional view of a discharged cell that exhibits these features. Two important resistance-reducing effects are evident from this structure. First, the effective surface area of the lithium/lithium iodide interface is substantially increased from its originally flat configuration; second, the distance that a lithium ion must travel through the lithium iodide is reduced.

These macroscopic differences in discharge profiles are even more evident in cells with precorrugated anodes. In such cells with coated anodes there exist columns of lithium-rich areas intermingled with iodine-rich columns, between which has formed a band of lithium iodide. Otherwise identical cells made with uncoated anodes show a very regular, hard lithium iodide layer that has retained the original corrugated shape. The interface between lithium and lithium iodide has been partially destroyed by the creation of voids in the "peaks" of the original anode corrugations.

The reasons for these rather striking differences in the formation of the lithium iodide and the integrity of the anode/electrolyte interface have been the subject of several investigations. Skarstad and Owens have suggested that the resistance-reducing effect is less related to the form of the lithium iodide itself than to an effective increase in the area of the anode/electrolyte interface.[33,34]

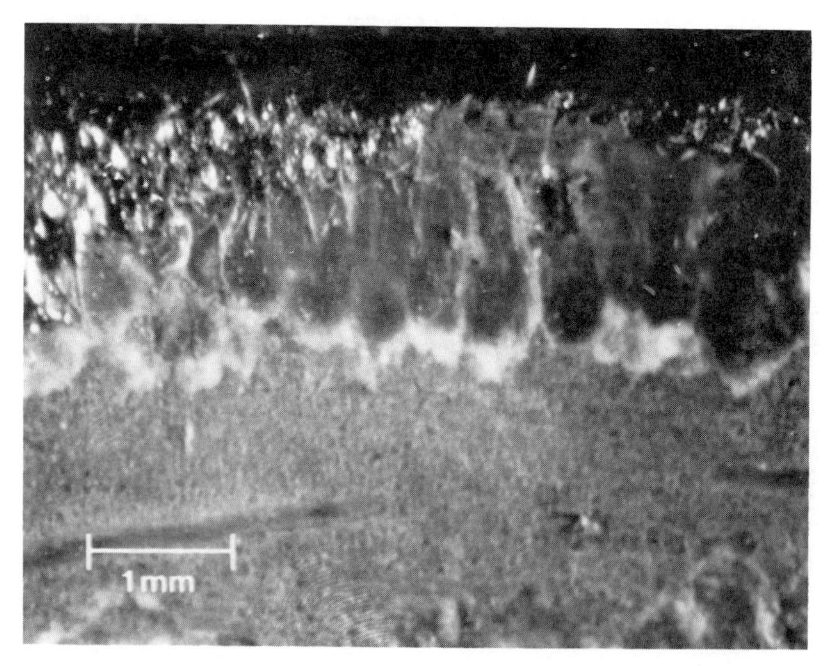

FIGURE 9. Cross-Sectional View of Partially Discharged Cell with Coated Anode.

Brown and coworkers have described the lithium iodide in the PVP-rich interfacial region as being of greater linear dimension than the lithium metal substrate on which it is formed.[35] This leads to a buckling of the reactive area into convolutions, increasing the contact of cathode material with the area nearest the anode. These workers suggest that the electrolyte adheres to the lithium so strongly that the anode is plastically deformed into ridges and peaks, causing the observed sinusoidal convolution of the anode/electrolyte/cathode area into alternate iodine-rich and electrolyte-rich columns perpendicular to the anode.

Phipps and co-workers have reported results of a light-microscopic study of the discharge of small glass cells using time-lapse photography.[36] They have discovered the formation of a highly conductive liquid electrolyte phase in the battery. They suggest that this phase is formed from an electrochemical reaction between the PVP coating, lithium, and iodine. A time-lapse motion picture film of the discharge of a small, coated-anode cell revealed this liquid phase penetrating the porous microstructure of the lithium iodide and participating in the formation of additional discharge product. Preliminary chemical analysis of this liquid material suggests that the material contains no aromatic character, which is surprising in view of the fact that the only organic component in the system is the aromatic PVP.

4.4.2b. Modifications Induced by Pelletized Cathodes. Cells made with pelletized, unheated cathode materials also show a lower-resistance rise than do uncoated-anode cells made with the thermally treated cathode material.[20] This decreased resistance is again attributed to the way in which the lithium iodide is formed during discharge.[28] Since the electrolyte is formed in contact with a solid pellet material, fracturing results and very small crystallites are created. This constant disruption of the lithium iodide layer produces a material of higher ionic conductivity.[37] The result is a system that exhibits a lower internal resistance during life than does an otherwise identical cell with an uncoated anode and the thermally treated, softer cathode material.

5. CONSTRUCTION OF LITHIUM/IODINE–PVP CELLS

5.1. Principles of Cell Design

Several options are available to the designer of lithium/iodine–PVP cells. These options show certain constraints imposed by the chemical and physical properties of the reactants and reaction products. Inherent limitations such as the conductivity of the electrolyte must be considered in the optimization of cells. Because of the critical nature of the application and long-life objective, materials compatibility is an important problem to be dealt with in cell design.

The physical properties of lithium metal permit it to be formed into a variety of shapes. This permits the design of cells with optimized surface area. The electrolyte/separator thickness is thus minimized and current-delivery capability is enhanced.

Cells made with preheated cathode material must be designed with the viscous properties of this material in mind. In particular, care must be taken to prevent this material from contacting the anode current collector or lead. If migrating cathode material should contact these components, a low-resistance short circuit will result because of the electronic conductivity of the material. The thickness of the lithium between the anode current collector and the anode/electrolyte interface must be sufficiently thick to insure that lithium depletion does not occur, since the bare current collector could then be contacted by cathode material migrating through microcracks in the electrolyte. Likewise, the anode lead must be insulated adequately with an iodine-resistant material to prevent contact with the cathode material. This is particularly critical in the area of the glass-to-metal seal. The problem is less troublesome in cells made with pelletized cathode material that retains its solid-state nature throughout discharge. However, since this material is not fluid, care must be taken to insure the integrity of the interface contact between cathode and cathode current collector and between cathode and electrolyte.

Lithium reacts with many materials often considered to be "inert" (e.g., Teflon(TM) and glass). Thus, insulating materials in contact with lithium must be chosen carefully to avoid long-term degradation. Materials such as Halar(TM) and high-purity alumina can form acceptable insulating components. Metals used as anode current collectors must also be carefully chosen, as some can form intermetallic compounds with lithium. For example, early experimental cells made with platinum anode screens showed significant degradation because of a reaction between platinum and lithium.[38] Nickel, however, makes an ideal anode collector material.

The potentially corrosive nature of the cathode material limits the choice of cathode current collectors and case materials. Titanium, for example, is very reactive with iodine,[39] but both 304L stainless steel[40] and nickel[28] show excellent corrosion resistance to iodine under water-free conditions.

Water, of course, is very deleterious to the performance of lithium/iodine–PVP cells, since both active components react with it. Care must be taken both during construction and throughout the life of the cell to prevent introduction of water into the system. This generally means true hermetic sealing, with electrode leads brought out of the cell with ceramic or glass-to-metal seals.

The most straightforward approach to the design of a prismatic cell is to place lithium on one side of an insulated case and to place the cathode mixture over the lithium. This design was, in fact, the first to be proposed and tried for a lithium/iodine–PVP battery.[41] This concept does not maximize lithium surface area and was abandoned in favor of a central lithium anode with cathode material on either side. This design, which reduces cell impedance by a factor of 4, is still in use today. A third alternative is to encase the cathode material with lithium so that the surface between cathode and anode is large and the iodine material is encapsulated by both the electrolyte formed *in situ* and the anode itself. This approach has also seen extensive use in commercial cells. The active components may be totally encapsulated by inert insulating material before being sealed into metal cases, producing a "case-neutral" battery with two hermetic feedthroughs, or the case itself can act as the sole encapsulant, producing a "case-grounded" battery with a single feedthrough. In the following paragraphs the various designs that have seen significant use in implantable applications will be reviewed.

5.2. The Central Anode/Case-Neutral Design

The first commercially successful lithium/iodine–PVP battery consisted of a single, flat, central lithium anode surrounded on either side by the iodine–PVP cathode material. Figure 10 shows a cutaway view of this cell. Because of initial uncertainty regarding the reactivity of iodine with the metal case material, the battery was enclosed in a fluoropolymer support and then potted in a polyester casing. Internal leads were brought out of this casing for attachment to glass hermetic feedthroughs in the stainless steel lid. The structure was then welded into a case made of 304 stainless steel. Because of the redundancy built into the

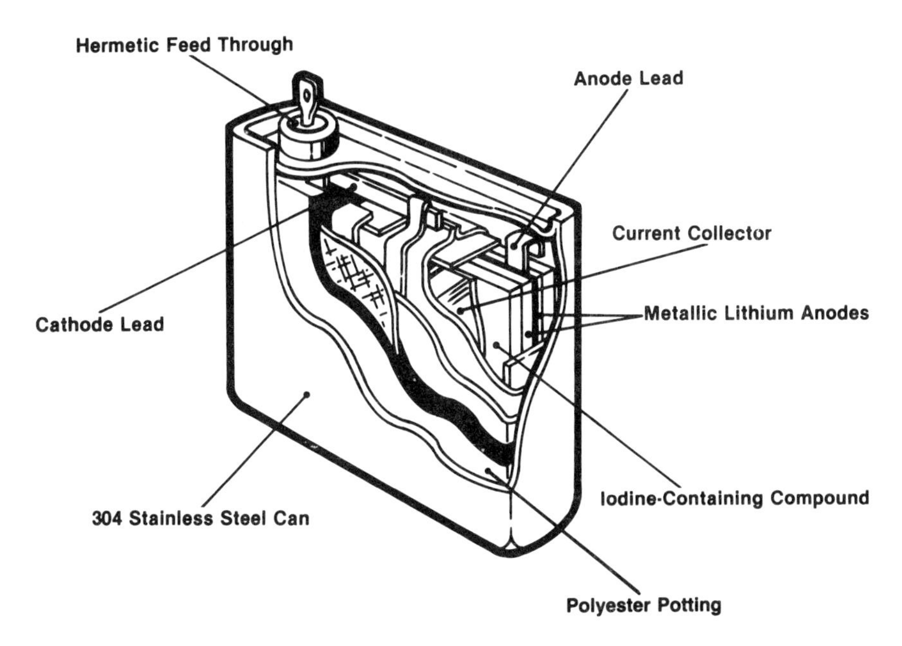

FIGURE 10. Cutaway View of Central Anode/Case-Neutral Design.

design and the desire for a long-life power source, this battery was rather large. Its nominal dimensions were 14 × 45 × 52 mm and its weight was 80 g.

5.3. The Central Cathode/Case-Neutral Design

The desire to provide more efficient encapsulation of cathode material led to a concept wherein this material was completely encapsulated by the lithium anode. The anode material was then insulated from the case by either a fluorocarbon insulator or a polyester potting material. Anode and cathode leads were brought out of the cell through glass-to-metal seals.

A cutaway view of one such design is shown in Figure 11. The design is a bipolar configuration in which two cup-shaped lithium anodes pressed into a fluorocarbon mold hold the cathode material. The two subassemblies were joined together by heat-sealing the fluorocarbon holders, forming a battery in which the cathode material was surrounded by lithium. The lithium anodes were precoated with polyvinylpyridine before addition of the cathode material. The battery was potted into a polyester material and then hermetically sealed into a stainless steel case.[42]

This concept was also used in a two-cell battery. Two such cells connected in series were potted in polyester and hermetically sealed into a stainless steel case.[43] This battery saw extensive use in a pacemaker design that did not need voltage-doubling circuitry.

A second central-cathode design, known as the "lithium envelope" design, eliminated the polyester potting material, thus improving energy density.[26,28] An exploded view of this design is shown in Figure 12. The lithium anode is formed into a Halar™ "cup", which fits snugly to the contours of the case. The

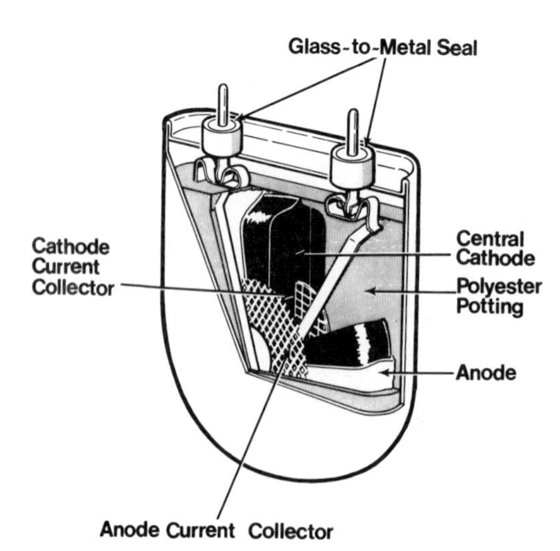

FIGURE 11. Cutaway View of Central Cathode Cell Encased in Polyester.

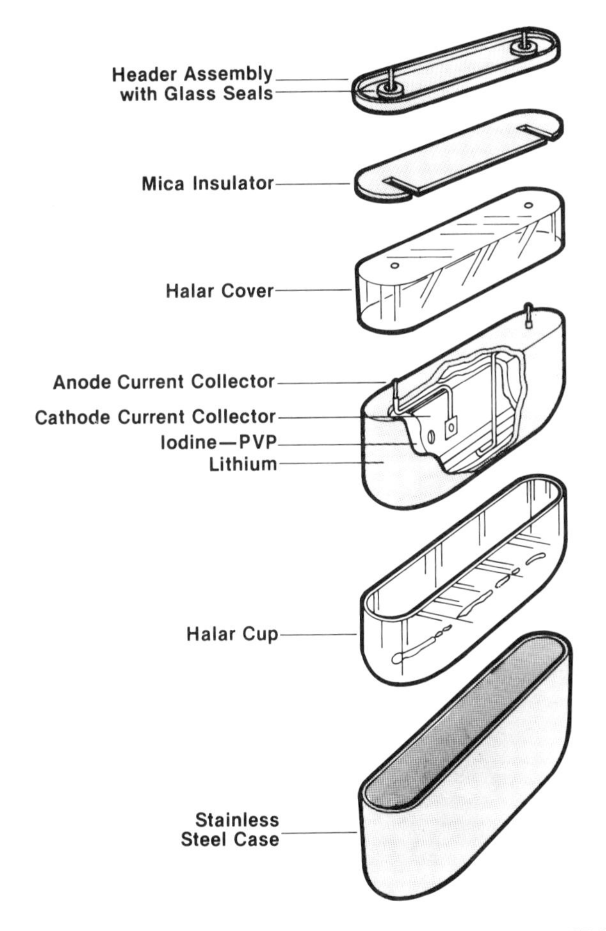

FIGURE 12. Exploded View of "Lithium Envelope" Cell (Courtesy CRC).

molten cathode material is poured into the lithium "container" and covered on top with additional lithium totally encapsulating the iodine material. A second Halar™ insulator covers the lithium, and the cell is hermetically sealed into a stainless steel case. In this design the lithium anode was not precoated with polyvinylpyridine.

5.4. The Central Anode/Case-Grounded Design

As more information was accumulated about the reactivity of cell components it became apparent that much of the inert material present in early versions of the lithium/iodine–PVP battery was unneeded and that the 304L stainless steel case itself was an adequate container of the cathode material. This conclusion led to the development of the so-called case-grounded battery. In this design a central lithium anode is surrounded by the cathode material, which is poured into the battery in the molten state. The case itself serves as the encapsulant of the cathode material and as the cathode current collector.[44] The anode is precoated with polyvinylpyridine. In early versions of this design concept, the lithium anode was flat and was surrounded by a Halar™ strap designed to insulate it from the case. The lid was welded onto the battery case after the molten iodine material was poured into the cell. Later versions of this concept employed an anode that was corrugated or "ribbed" during the pressing process to increase surface area. The lid was welded to the case before filling the cell with cathode material, and the cathode material was introduced into the cell through a fill-ferrule, which was then redundantly sealed by two separate sealing mechanisms.[45,46] Later versions of this design eliminated the Halar™ strap to allow utilization of iodine material, which may have been blocked by the strap. Figure 13 shows a cutaway view of a typical cell designed in this manner.

5.5. Central Anode/Case-Grounded Pelletized Cathode Cells

The development of cathode materials made by pelletizing particulate iodine and polyvinylpyridine led to cell designs in which a central, flat, uncoated lithium anode is surrounded by such cathode pellets. Figure 14 shows an exploded view of a typical design of this type. The central anode is in contact with the solid cathode material on the sides and is covered by an insulator on top. The anode current lead is brought out through a ceramic seal brazed into the cell. The case itself is the cathode current collector. Cells of this design have been produced using both nickel and 316L stainless steel as case material.

Jolson and Schneider have noted that the pelletized-cathode concept lends itself well to the design of a multicell battery in which alternate anodes and cathodes are stacked in a parallel arrangement.[47] The increased surface area of

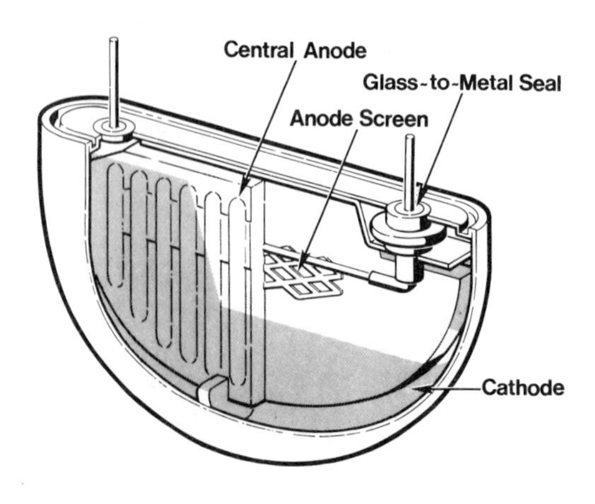

FIGURE 13. Cutaway View of Case-Grounded/Central Anode Cell with Corrugated Anode.

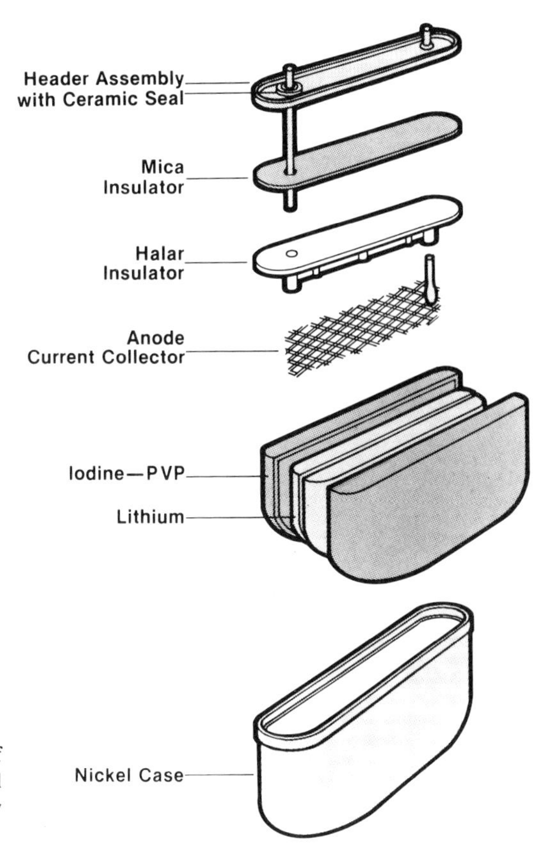

FIGURE 14. Exploded View of Central Anode/Case-Grounded Cell with Pelletized Cathode (Courtesy CRC).

such batteries enhances their current-delivery capabilities. Schneider and Goldman reported a pacemaker battery with two parallel anode plates surrounded by pelletized cathodes and electrically connected.[48] Such a configuration results in an anode surface area double that of an otherwise-identical single-anode cell. The theoretical impedance of such a cell is one fourth that of the corresponding single-anode cell.

6. DISCHARGE CHARACTERISTICS OF THE Li/I_2–PVP BATTERY

6.1. General Considerations

The application of the lithium/iodine–PVP battery system as a power source for cardiac pacemakers requires an average current drain of from 10 to perhaps 40 μA, although more recent dual chamber pacemakers may require currents approaching 100 μA. Accordingly, many years are required before meaningful performance data are available for these batteries. During the last 10 years a very large data base has been accumulated because of the large "life-testing" activities performed by manufacturers and users of these batteries. Literally thousands of batteries have been discharged under very controlled conditions, providing a very well-characterized set of discharge data. Tests are typically conducted at 37°C under constant load or constant current conditions. Batteries are also tested in pacemakers, providing valuable "real-time" data for application conditions. During the last two years some early, smaller versions of the battery have completed their discharge in real time, providing a very useful set of data that allow for the first time a comparison of real-time performance to projections based on accelerated testing and modeling.

The basic form of the discharge curve can be deduced from the thermodynamic, kinetic, and chemical considerations presented earlier. Under a current drain of 10–40 μA the discharge voltage begins near the open-circuit voltage of 2.8 V. As the layer of lithium iodide increases in thickness, the cell resistance increases and the voltage decreases. In the case of cells containing thermally reacted cathode material and uncoated anodes, this voltage decrease is linear and rather pronounced. In coated-anode cells and cells with pelletized cathode material, the voltage decrease is less dramatic as discharge proceeds. As iodine is depleted from the cathode material the cathode conductivity follows the curve shown in Figure 4. This causes the discharge curve to become nonlinear and exhibit a higher rate of voltage drop near the end of the discharge curve. The result is a discharge curve that exhibits a clinically desirable gradual approach to the "elective replacement voltage", a value selected by the manufacturers of pacemakers to allow elective replacement of the pacemaker before it ceases to function. Typical elective replacement voltages range from 2.3 to 1.9 V.

In the following paragraphs both projected and real-time data will be presented. That data shown in this section will be based on manufacturers' published data or the results of in-house testing of groups of cells. The statistical aspects of cell performance will be discussed in Section 7. Only nominal or average cell discharge characteristics are considered here, so that discharge curves representative of the four basic cell designs discussed in Section 5 can be seen. All discharge curves shown below are from experiments conducted at 37°C.

6.2. Discharge Characteristics at Application Current Drain

6.2.1. *The Central Anode/Case-Neutral Discharge Characteristics*

The first Li/I_2–PVP cell to see widespread commercial implantable use was known as the Model 702E. The cell was produced by both the Catalyst Research Corporation (CRC) and Wilson Greatbatch Limited (WGL). The cell had a stoichiometric capacity in excess of 6 Ah. Thus, cells on test for more than 8 years have not reached the elective replacement voltage. Figure 15 shows the average voltage of a group of 47 such cells on test for over 8 years. The curve shows the typical linear voltage decrease characteristic of the uncoated-anode technology. Figure 15 also shows the impedance rise during discharge.

6.2.2. *The Central Cathode/Case-Neutral Design*

6.2.2a. Coated-Anode Technology. Figure 16 shows 100 kΩ constant load discharge data for a group of 55 Model 752 batteries manufactured by WGL.

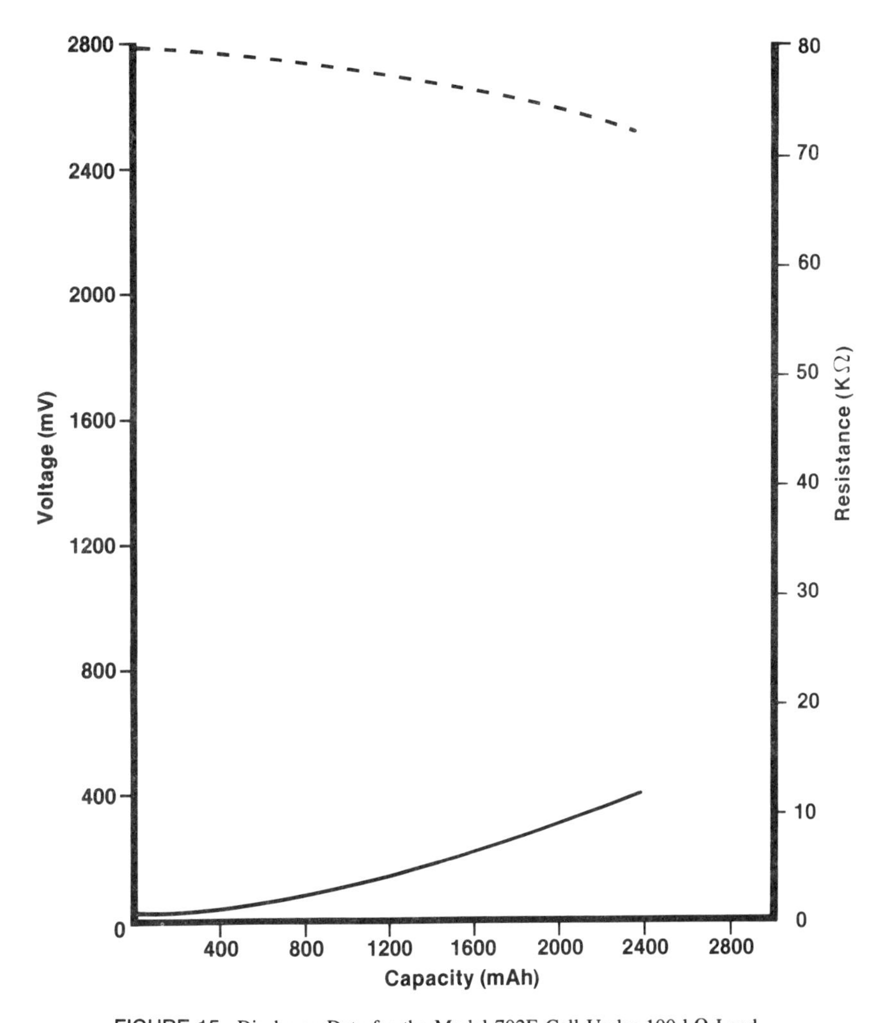

FIGURE 15. Discharge Data for the Model 702E Cell Under 100 kΩ Load.

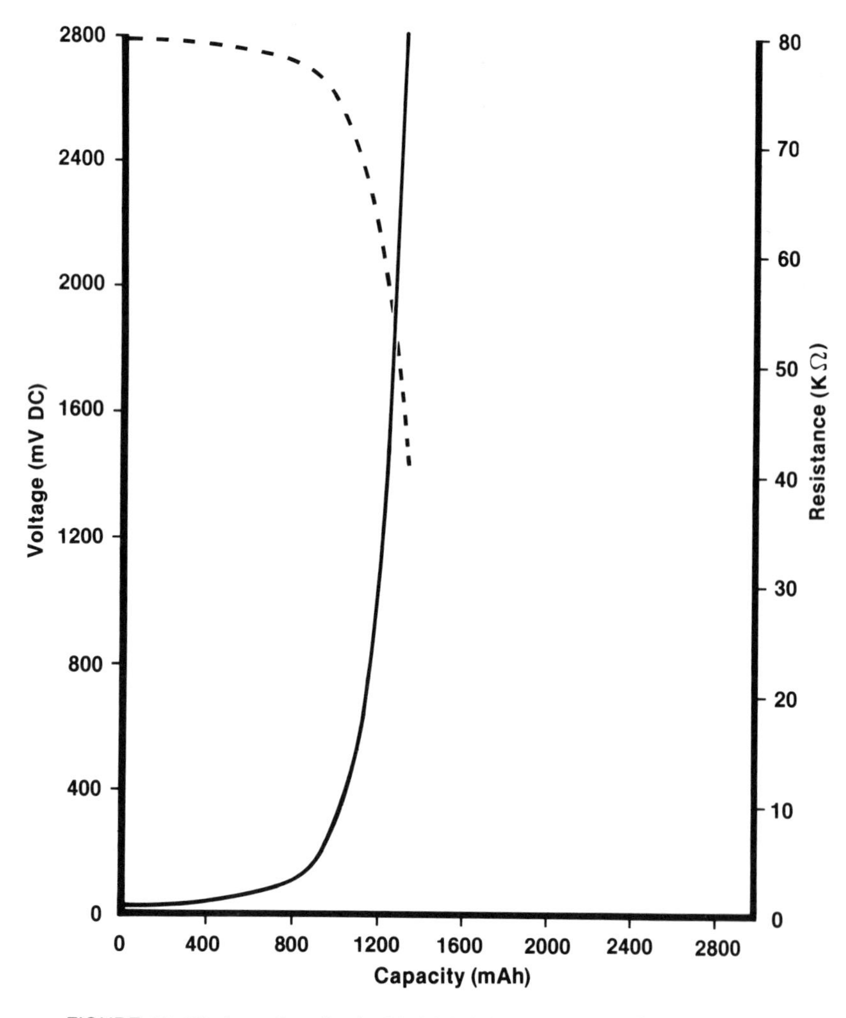

FIGURE 16. Discharge Data for the Model 752 Cell Under 100 kΩ Constant Load.

The cell was rated at 1.5 Ah less 10% for anticipated self-discharge losses. The average performance shown in Figure 16 is typical for coated-anode cells.

6.2.2b. The "Lithium Envelope" Uncoated-Anode Technology. A variety of cell sizes and shapes in the lithium envelope design were produced by CRC. Projected discharge curves for three such designs are shown in Figure 17. Voltage curves under 100 kΩ load are shown for the CRC Models 810B/16, 804B/23, and 810B/23.[28]

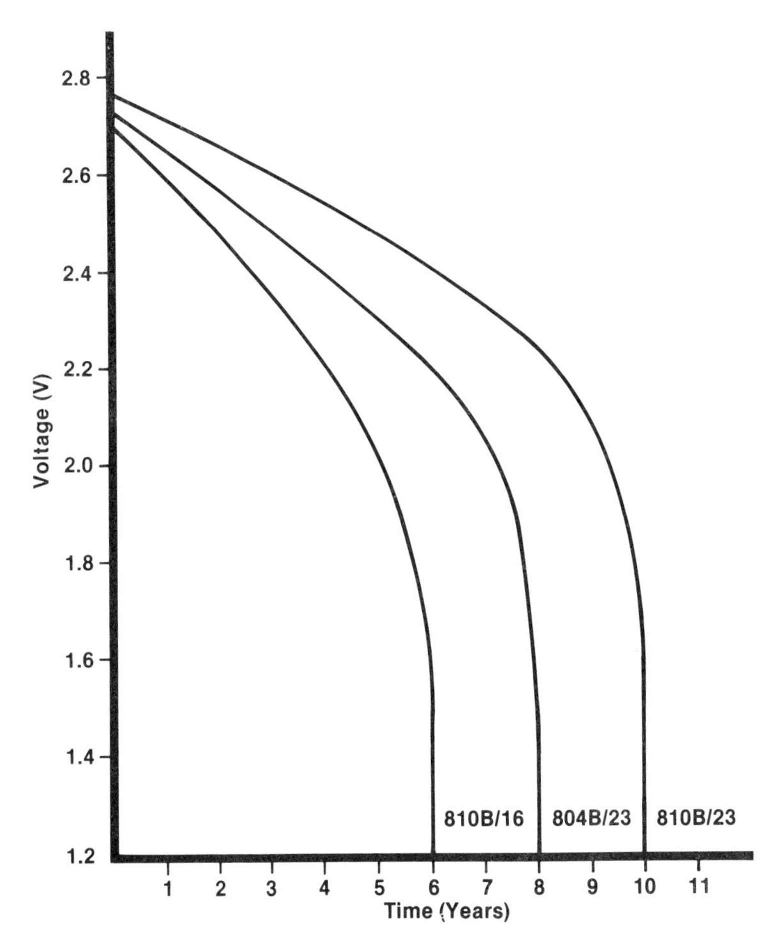

FIGURE 17. Projected Discharge Curves for the Models 810B/16, 804B/23, and 810B/23 Cells Under 100 kΩ Load.

6.2.3. *The Central Anode/Case-Grounded Design*

A variety of cells employing the central coated anode, case grounded concept are produced by WGL, Cardiac Pacemakers, Incorporated, and Medtronic Energy Technology. Later versions of this cell design included a corrugated anode to increase surface area. The various shapes and sizes of this design show generally similar discharge characteristics. An early group of one such model of the cell, the WGL Model 761/15 battery, has completed its discharge under 100 kΩ constant load. Figure 18 shows the average voltage and impedance of a group of 20 of these cells that has been on test under 100 kΩ constant load for over

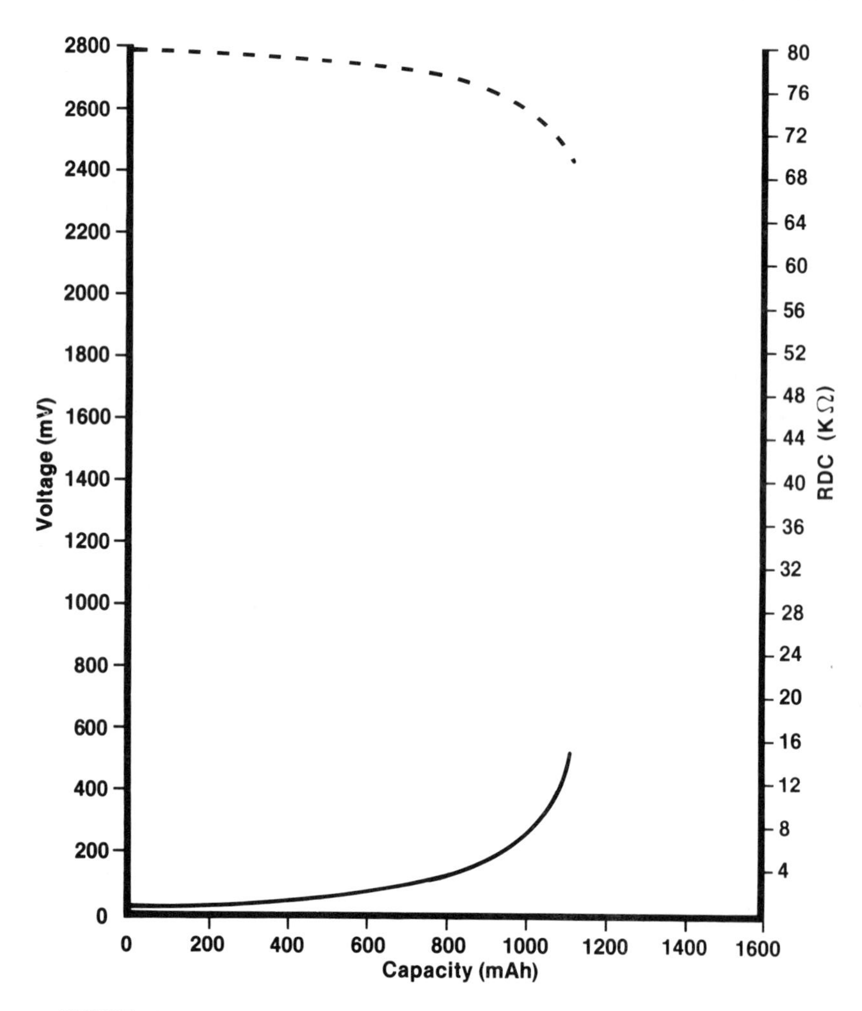

FIGURE 18. Discharge Data for the Model 761/15 Cell Under 100 kΩ Constant Load.

five years. The cell was originally rated at 1.3 Ah. These tests are not yet completed, but after 60 months on discharge these cells have delivered an average capacity of approximately 1.2 Ah to an average voltage of 2.4 V.

6.2.4. The Central Anode/Case-Grounded Pelletized Cathode Design

A variety of different cell shapes are produced by CRC using this design concept. Figure 19 shows both real-time data and projected discharge data for

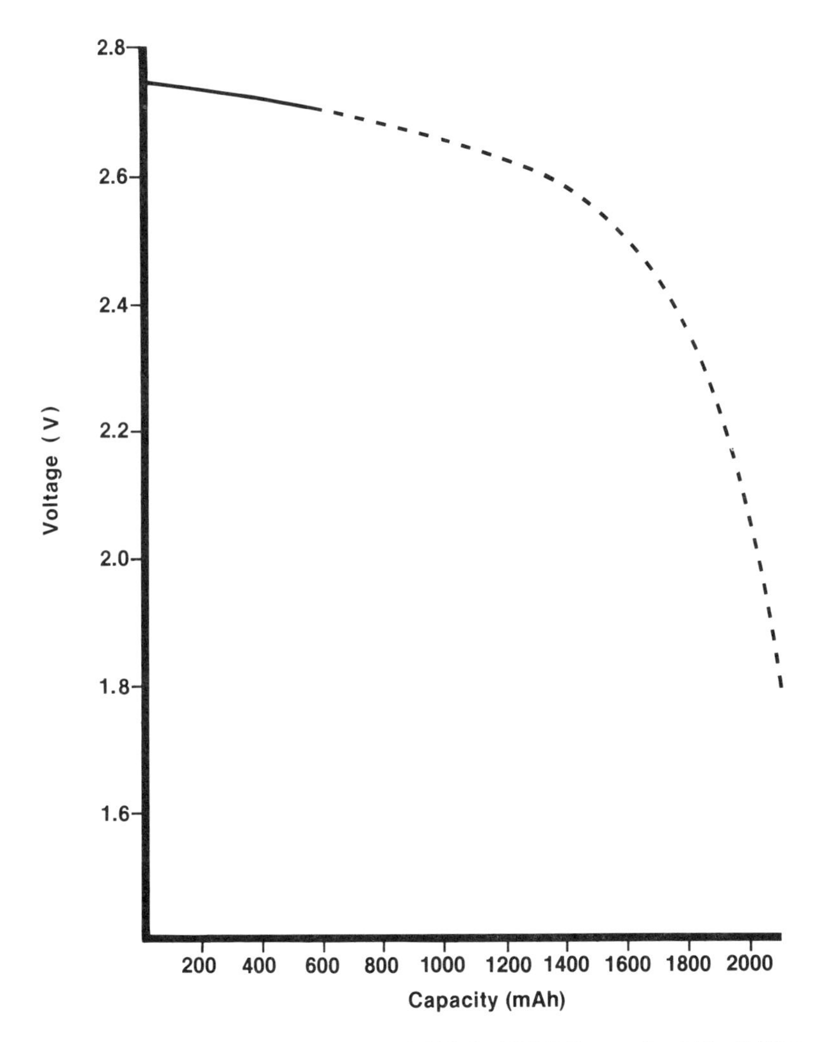

FIGURE 19. Discharge Data for the Model 902 Cell Under 100 kΩ Constant Load. The Solid Line is Real-Time Data, the Dotted Line Represents Projected Performance. (Data Supplied Courtesy CRC.)

one such version of this cell, the CRC Model 902. Shown on the graph are actual data amassed through 600 mAh and the projected curve based on accelerated test data.

Figure 20 shows the discharge characteristics of a double-anode cell of nominal dimensions 9 × 23 × 45 mm discharged under a constant load of 5 kΩ. Shown in the figure are the average voltage and average resistance of a group of 10 such cells.

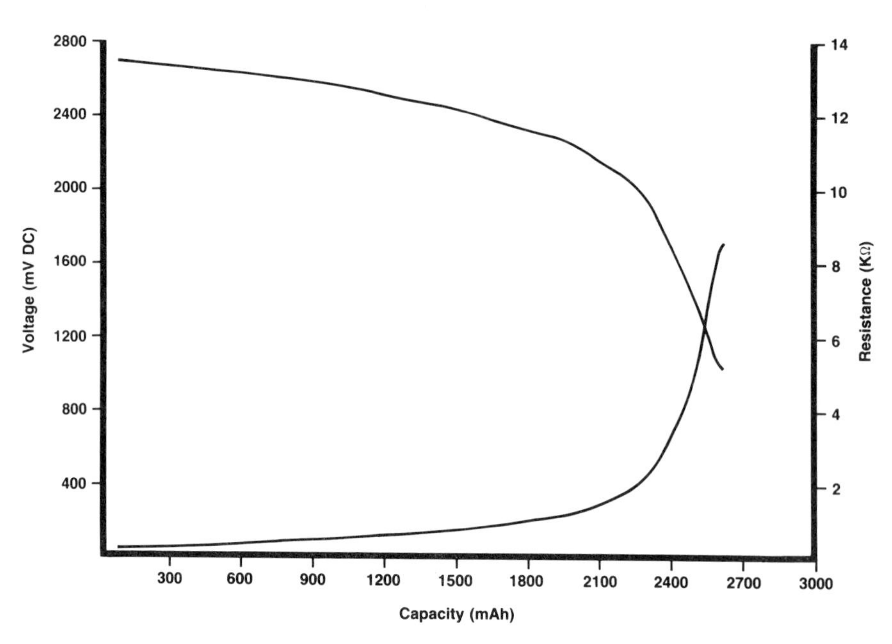

FIGURE 20. Discharge Data for the Model 920 Double-Anode Cell Discharged Under a 5 kΩ Constant Load (Data Supplied Courtesy CRC).

6.3. The Effect of Current Drain on Cell Performance

The current-limiting phenomena present in the Li/I_2–PVP cell have been discussed in Section 2.1 of this chapter. The cell remains basically a low-rate system that fills the needs presented by implantable low current-drain devices such as cardiac pacemakers. With the advent of higher current-drain devices such as implantable drug delivery devices and neurostimulators, the performance of the cell at higher current drains becomes of some interest. Many improvements in current-delivery capability have been made in the 10 years that the system has been in use.[29,30] These have been discussed in previous sections of this chapter. Cells made today are in fact capable of delivering considerably higher current than was possible in previous designs, although higher current drains generally result in lower deliverable capacities because of polarization effects previously discussed. Figure 21 illustrates this phenomenon. It shows the average discharge results from groups of cells on test at constant loads ranging from 110 to 3.74 kΩ. The cell model used in this test is the WG Model 762M cell, rated at 2.5 Ah under a 100 kΩ load. Under the heaviest current drain shown in the curve the cell delivered 1.4 Ah to 1.8 V, or 56% of the capacity at which the cell is rated at 100 kΩ load.

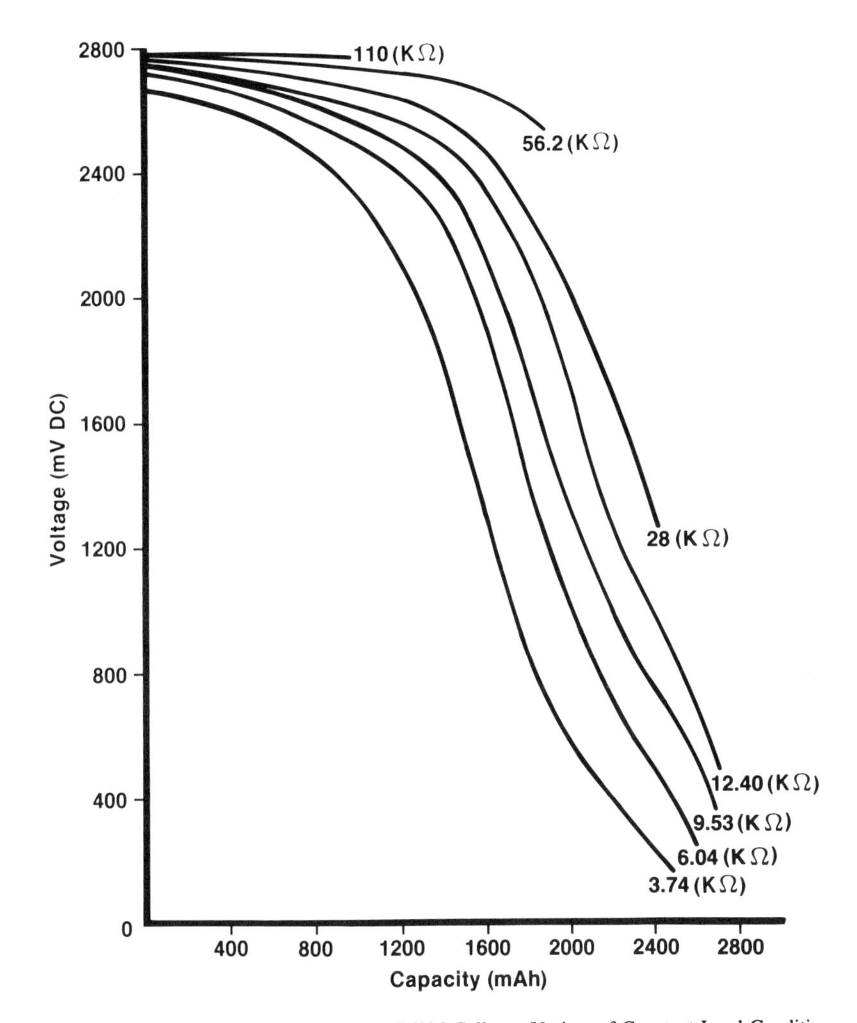

FIGURE 21. Discharge Data for the Model 762M Cell at a Variety of Constant-Load Conditions.

Similar results are seen in other designs of the Li/I_2–PVP system. Figure 22 presents discharge data for the CRC Model 901 pelletized cathode cell at 150, 100, 50, and 20 kΩ constant load. This cell is rated at 2.5 Ah under 100 kΩ load conditions.

6.4. Self-Discharge

The phenomenon of self-discharge is particularly important in low current-drain, long-life applications such as cardiac pacemaker power sources. As a solid

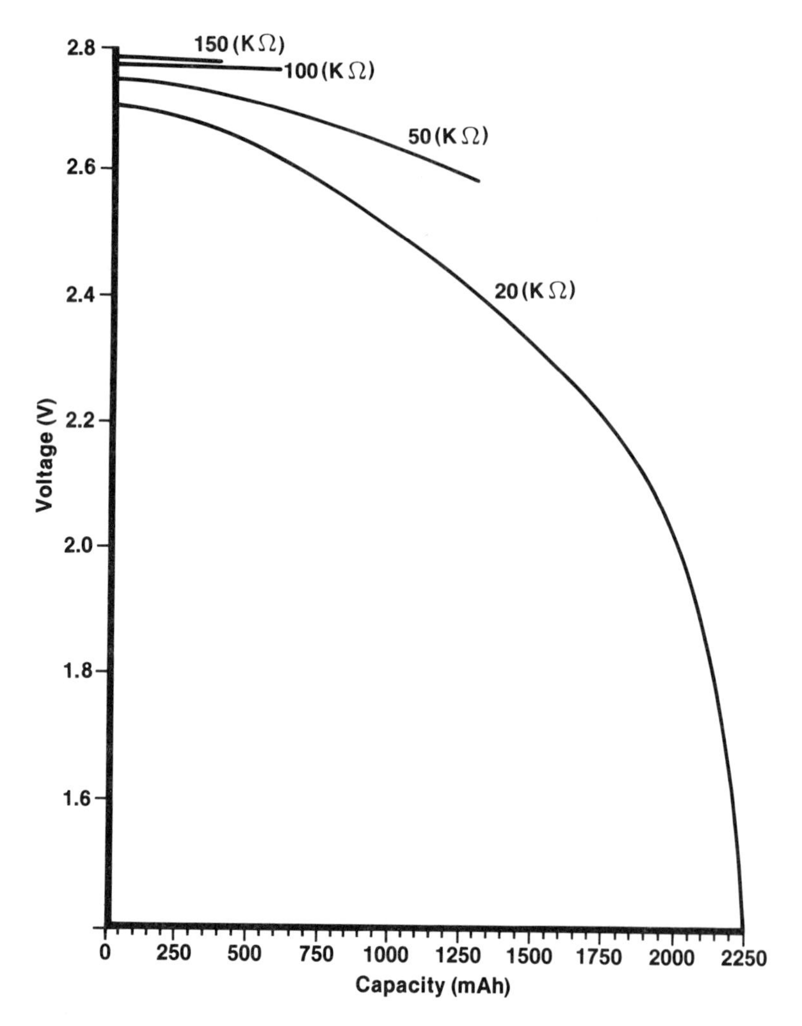

FIGURE 22. Discharge Data for the Model 901 Pelletized Cathode Cell at a Variety of Constant-Load Conditions (Data Supplied Courtesy CRC).

electrolyte system the Li/I_2–PVP battery might be expected to exhibit less self-discharge than some liquid electrolyte systems. Indeed, this has proved to be the case. Real-time data such as that presented in Subsection 6.2 suggest that self-discharge losses amount to only a small percentage of the total capacity of the cell.

The most straightforward mechanism for self-discharge in this system is the migration of iodine molecules through the electrolyte to the anode where direct reaction with lithium can occur. The atomic radius of iodine is 1.33 Å, and the ionic radius of lithium is 0.6 Å,[49] suggesting that iodine will migrate through the electrolyte with difficulty. Since the electrolyte is formed *in situ,* a certain amount of "self-discharge" penalty is paid to form the initial layer of electrolyte. Thereafter, as cell discharge proceeds and the electrolyte layer thickens, one would expect the self-discharge to decrease to negligible amounts after several months of discharge.

The use of microcalorimetry to study self-discharge has become an important research technique since 1976.[50–56] The Li/I_2–PVP system has proved to be well characterizable by this technique. Untereker has presented a detailed study of this system using microcalorimetry.[56] He found that the open-circuit heat dissipation did in fact decrease substantially as discharge proceeded. He estimated the total loss due to self-discharge to be on the order of 10% over the discharge life of the cell. Liang and Holmes noted a similar effect for case-grounded cells and estimated that self-discharge loss was less than 7% in the discharge life of a 3-Ah battery.[57]

Batteries under 100 kΩ load show an initial rise in microcalorimetric heat output followed by a gradual and apparently exponential decrease to negligible values. Figure 23 shows the average microcalorimetric heat output of a group of four case-grounded cells of rated capacity 1.7 Ah. Microcalorimetric measurements were taken under open-circuit conditions at 37°C. After 20 months of discharge the average heat output is under 10 μW. Similar trends are seen for other shapes of cells.

That self-discharge amounts to only a small percentage of total cell capacity appears to be confirmed by the real-time data presented in Subsection 6.2. Batteries on test for over five years under 100 kΩ load yielded capacities slightly greater than those predicted from accelerated discharge data. Naturally, it would be expected that cells discharged under even lower current drains would show larger losses since the time scale is expanded.

Both microcalorimetry and actual data therefore indicate that self-discharge losses constitute a few percent of the total capacity of the cell at application load. The microcalorimetry data further indicate that most of the self-discharge losses occur in the first 20% or less of the cell's discharge. This result is consistent with the model of self-discharge involving the diffusion of iodine molecules through an ever-increasing layer of lithium iodide.

6.5. Modeling and Accelerated Testing

A major challenge facing the producer and user of the lithium/iodine–PVP battery is the prediction of performance several years before real-time data become available. Accelerated testing tends to compress the self-discharge contribution and, as was shown in Subsection 6.3, is limited by the polarization

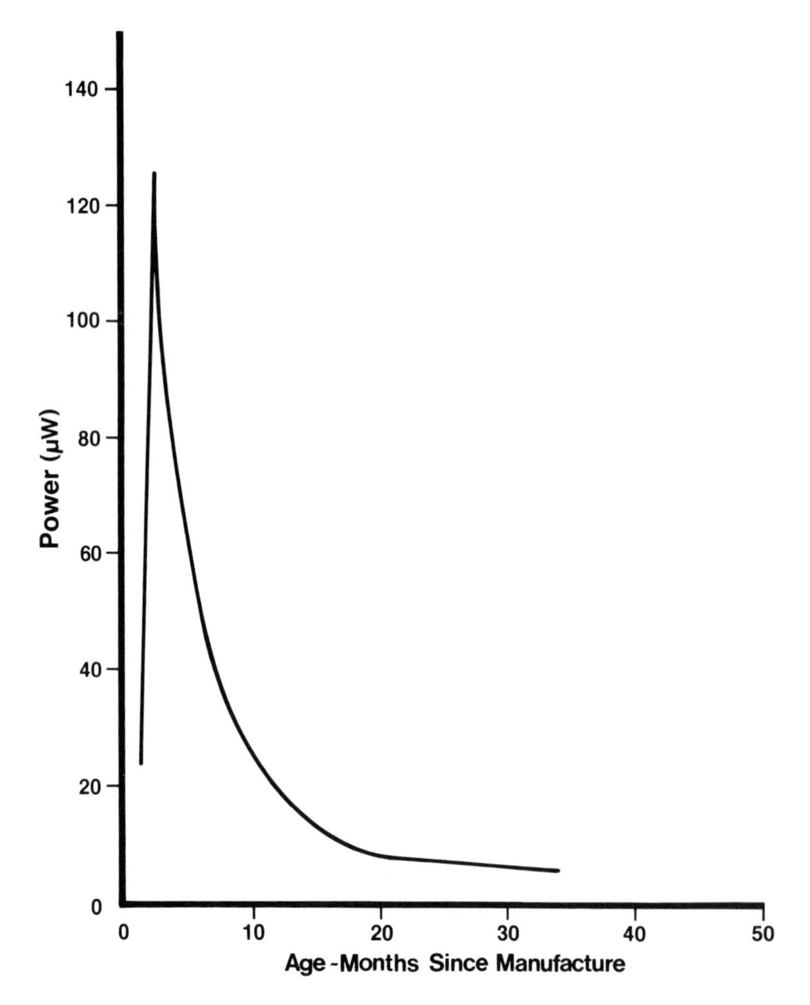

FIGURE 23. Average Net Microcalorimetric Heat Output of a Group of Four Case-Grounded Cells Under 100 kΩ load.

losses that occur at higher current drains. Thus, various techniques have been developed to predict long-term performance from a series of accelerated tests and from modeling.

In 1976 Kraus and Schneider presented a mathematical model of cell polarization based on a linear growth of lithium iodide, measured results of the resistivity of the iodine-PVP cathode material, and a self-discharge term in which self-discharge losses are assumed to be proportional to the square root of time.[26] The form of the equation is

$$\eta(t) = \eta_o + C\,(i)^{1.585} t \exp(8650/RT) + KI\,t^{1/2} + I\Omega\,(t) \qquad (3)$$

where η is the cell polarization at time t, η_o is some initial polarization, C is a constant, i is current density in amperes per unit anode area, R is the gas constant, T is temperature, K is a constant, I is current drain, and Ω (t) is the resistivity of the cathode as a function of time under discharge.

An experimental method for determining cell capacity was developed at WGL. This method involves a cyclic discharge procedure wherein cells are placed under a heavy current drain (e.g., 5 kΩ constant load) for two weeks, held at open circuit for three days, and held at application load (e.g., 100 kΩ or 140 kΩ) for three days. Measurements are taken at each load condition. The cycle is repeated until the voltage under application load attains an arbitrarily-defined end point (1.8 V). In this manner an application load discharge curve can be constructed much faster than a real-time curve can be generated. Results of several such tests are combined and fit to a curve of the form

$$V\,(Q) = 2.800 - A \exp(BQ) \qquad (4)$$

where V (Q) is the voltage at capacity Q in volts and A and B are constants obtained by nonlinear regression analysis.[58] Holmes and coworkers compared results obtained from this procedure with five-year test results under 100 kΩ load.[58,59] They found that the shape of the discharge curve was well predicted by the accelerated test procedure and that the capacity predicted by the accelerated test procedure agreed within 6%, with the average capacity of some 100 cells tested under real-time conditions.

Gerrard *et al.* have presented a computer simulation model for projecting battery longevity.[60] The model used data generated from a variety of constant current accelerated tests to project longevity under application load. The equation predicted by their model was

$$\ln \Delta V = A + B \ln I_L + P_4(Q) \qquad (5)$$

where ΔV is the voltage drop, A and B are constants, I_L is the current under which the cell is being discharged, and P_4 (Q) is a fourth-order polynomial in Q, the charge delivered in mAh.

Later Kim and Brennen improved this model to include self-discharge data generated by microcalorimetry.[61] The equation they obtained for voltage drop as a function of capacity is

$$\ln \Delta V = A + BI_L^{1/2} + P_3\,(Q_T) \qquad (6)$$

where ΔV, A, B, and I_2 are as defined in Eq. (5) and P_3 (Q_T) is a third-order polynomial in Q_T, which is defined as the sum of the charge delivered to the load and the charge lost through self-discharge, estimated by microcalorimetric data.

Combining the results of this modeling with chemical facts obtained by the study of the cathode material allows a convenient graphical summary of the capacity of a Li/I_2–PVP battery. Brennen and Untereker have presented the curve shown in Figure 24 to describe the capacity delivered by a Li/I_2–PVP battery as a function of current drain.[62] Shown on the curve is the stoichiometric capacity, which is simply the total weight of iodine multiplied by 0.211 Ah/g,

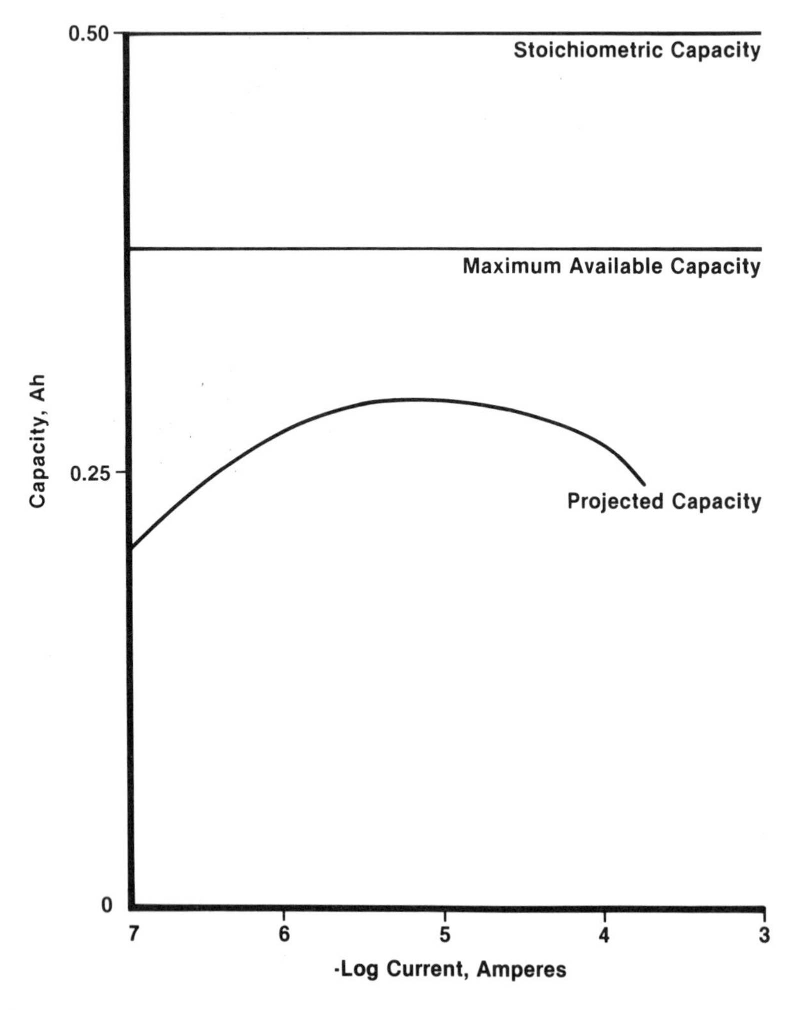

FIGURE 24. Generalized Model for the Deliverable Capacity of a Li/I_2–PVP Battery as a Function of Current Drain.

the "maximum available capacity", a term coined by these workers to denote the stoichiometric capacity minus the three iodine atoms per pyridine ring that are bound to the PVP structure, and the deliverable capacity as calculated by the model represented by Eq. (6). This deliverable capacity curve as shown in Figure 24 is drawn for a particular model of a Li/I_2–PVP cell, but the general form is applicable to any Li/I_2–PVP cell. It is in fact a specific example of the so-called Salem-Bro plot that graphs capacity versus the logarithm of current drain.[63] The general form of the curve, wherein the capacity falls at both very low and high current drains, is characteristic of any primary battery. In the case of the Li/I_2–PVP battery, the capacity falls at very low current drains because of self-discharge. It falls at higher current drains because of polarization losses discussed earlier. It is fortunate that the maximum area of capacity lies in precisely the region of interest to the cardiac pacemaker application.

7. PERFORMANCE OF THE Li/I_2–PVP CELL

7.1. General Remarks

The need for reliable, predictable performance of a cardiac pacemaker power source is obvious. A cell that is depleted much earlier than anticipated can cause extra expense and the trauma of a replacement operation. A sudden depletion or "catastrophic failure" of a cell can lead to more serious consequences, including patient death. Accordingly, considerable effort is taken by manufacturers of Li/I_2–PVP cells to ensure reliable, predictable performance. The number of Li/I_2–PVP cells that have been implanted during the last 11 years is approaching 1 million. Therefore, a very significant data base has been amassed, both in field performance and in manufacturers' in-house testing, that allows an accurate assessment of cell performance. The Li/I_2–PVP system, in fact, has amassed an excellent record of performance, and real-time test results have confirmed the efficacy of early modeling and testing techniques in predicting cell behavior.

The following paragraphs discuss the steps taken by designers and manufacturers to enhance reliability. Results of in-house test programs will be presented, and statistical analyses of field performance of cardiac pacemakers powered by Li/I_2–PVP cells will be discussed.

7.2. The Approach to Cell Reliability

Certain features of the Li/I_2–PVP system impart an inherent reliability to the battery. The most important of these is the self-forming, self-healing nature of the electrolyte/separator. Other features, such as the absence of gas evolution and the inherently stable nature of the interfaces, also contribute to cell reliability.

Of course, cell design is an important aspect of reliability. The first Li/I_2–PVP batteries were designed with great redundancy, and the use of inert material such as polyester and fluorocarbon insulators made the design quite forgiving.

There was, however, an accompanying penalty in energy density. As more experience was gained and more information was amassed on such questions as materials compatibility, the designs became less redundant and more energy dense.[46] The basic design considerations discussed in Section 5 are important in achievement of cell reliability.

Candidate cell designs are typically subjected to a stringent qualification program.[29,39,57] This testing includes such techniques as thermal cycling, shock and vibration testing, humidity and pressure testing, resistance to solvents, feedthrough testing, and high-temperature storage. Accelerated discharge tests are also performed on candidate cell designs.

A stringent program of quality control and quality assurance is followed by all manufacturers of implantable-grade Li/I_2–PVP cells.[29,57,64,65] Incoming inspection of parts and materials is performed. Metallurgical analysis of welds is routinely conducted. Each critical operation is followed by inspection. Complete traceability of all batteries is accomplished by assigning a serial number to each cell from which all manufacturing and inspection data can be traced. Manufacturers typically test each cell under simulated pacemaker conditions for 30–60 days. During that time voltage and impedance measurements are taken at regular intervals. Cells are subjected to X-ray analysis, visual inspection, and leak testing on a 100% basis. Routine destructive analysis is conducted on samples to find potential defects.

The final step in the achievement of cell reliability is the routine maintenance of life test sampling programs by cell manufacturers. A certain percentage of all cells is kept in house on a running sample basis. Such cells are discharged under simulated pacemaker conditions and monitored regularly. Should unexpected performance be noted during such testing, users of the batteries (i.e., pacemaker manufacturers) can be notified so that they can make assessment of such performance on their implanted products.

7.3. Performance of Life Test Batteries

The practice of reserving a percentage of all cells for life testing has made possible the availability of a very large data base from which statistically meaningful conclusions about cell reliability can be made. Several reports on the results of such testing have appeared in the literature.

In 1981 Holmes and Stinebring compared real-time results with accelerated test data for a particular model of the $Li/_2$–PVP battery.[59] A group of 55 of these cells was studied for statistical variability. It was found that the standard deviation was a linear function of the voltage drop and increased more dramatically as cells approached the cutoff voltage. Indeed, for the first three years of the five-year discharge test it was impossible to distinguish the best performer from the worst performer. During the last 18 months or so of discharge the variability became more significant, and the time required to attain cutoff voltage of 1.8 V varied from 60 to 69 months. This represents availability of ±7% of the nominal discharge time. Average performance was within 5% of that predicted by accelerated testing some five years earlier.

Holmes *et al.* performed a statistical analysis of a group of newer cells whose nominal performance is shown in Figure 18.[66] A cumulative survival analysis of 2004 such cells on test for times ranging from 3 to 51 months showed no failures in a total of 57,822 unit-months of testing. A group of 20 of the oldest of these cells showed an average performance some 200 mV above the voltage predicted by accelerated testing. An increase in variability with the length of cell discharge was also noted.

Helgeson has analyzed the life test performance of three different models of Li/I_2–PVP cells.[67] These cell models, all with thermally reacted cathode material and coated anodes, showed random failure rates of 0.0023%, 0.0028%, and 0.0076% per month, respectively. Helgeson pointed out that these failure rates are comparable to those exhibited by electronic parts such as silicon diodes and transistors even though the battery is a considerably more complicated electrochemical device.

Schneider and coworkers have noted that some models of central cathode, uncoated anode cells have shown a higher rate of voltage decline than originally projected while under constant load testing.[68] They also note that these cells perform better under pulse load than under constant load. This observation was also made by Guilleman and Gilet.[69]

Jones has reported that clinical versions of central anode cells with pelletized cathode material have shown no failures in either testing or implanted units as of late 1981.[29]

7.4. Performance of the Li/I_2–PVP Cell in Cardiac Pacemakers

Assessing the performance of a power source by examining data collected on implanted pacemakers presents several difficulties. It is not often possible to tell if an "early depletion" is in fact a poor battery or, for example, the result of a higher than normal current drain due to an electronic component malfunction or a lower than normal pacing lead impedance. Moreover, the inherent variability in the conditions seen by a battery in a particular patient is large. Practically all pacemakers used today are "demand"-type units, which supply a pulse to the heart only if the unit senses that the heart needs stimulation. Some patients are paced 100% of the time; others may need pacing for 40% or less of the time. With the advent of programmable pacemakers whose operation can be changed while the unit is implanted has come an even greater variability in the actual current drain to which a battery is subjected under actual conditions. Owens has pointed out this variability and has estimated that only a small amount of the total variability in pacemaker longevity can be attributed to battery variability.[70] This phenomenon is shown graphically in Figure 25. Nonetheless, it is useful to examine pacemaker survival data in order to obtain a complete picture of battery performance.

In 1979 Luksha reported on the performance of a group of some 50,000 pacemakers powered by a variety of Li/I_2–PVP batteries supplied by two manufacturers.[71] He noted that after 2–6.5 years of service, all units appeared to

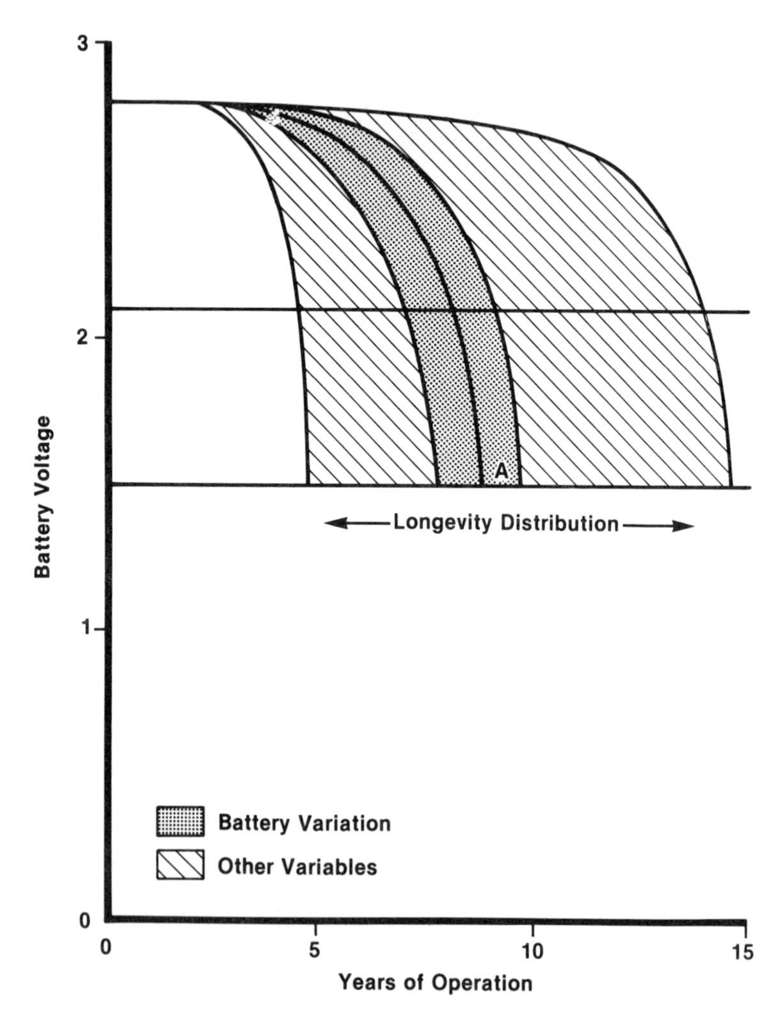

FIGURE 25. Battery Voltage Versus Pacemaker Longevity Comparing Battery Variation to Other Variables.

be functioning as expected and no confirmed battery failures had been seen in any units.

Owens and coworkers have discussed longevity of pacemakers powered by various lithium batteries.[72] Data from several sources were presented in this study. An analysis of data accumulated on three models of a particular manufacturer's Li/I_2–PVP-powered pacemakers showed a random failure rate of 0.0046% per month for the pacemakers. The random failure rate for the batteries was estimated at 0.0004% per month. A second data base, the Lithium Pacemaker Implant Registry of the U.S. Food and Drug Administration, showed a 6-year cumulative survival of 92.7% for some 3000 pacemakers powered by Li/I_2–PVP cells.

Bilitch and coworkers have been tracking pacemaker longevity for several years.[73] They have tracked actuarial survival of pacemakers as a function of battery type and other variables. A total of 5110 Li/I_2–PVP-powered units have been tracked for up to 104 months. Figure 26 shows the overall actuarial survival data for these units. It must be kept in mind that units considered in this study have not been "sorted" according to battery size, and cells with capacities ranging from 1.3 to over 3 Ah are contained in this data base. Thus, many of the depletions seen after 60 months are normal exhaustions of the smaller cells.

8. SUMMARY AND CONCLUSION

The lithium/halogen system with *in situ* electrolyte has been discussed in both general and specific terms. Though some effort has been made toward the development of batteries using halogens above iodine on the periodic chart, no such batteries are being manufactured today for implantable applications.

The lithium/iodine–PVP system has seen extensive use as a pacemaker power source. Since its introduction in 1972 this system has become the battery of choice for approximately 80% of the pacemakers being produced today. A significant data base consisting of in-house test data and field performance records has established that the battery is safe and reliable in this application. By the end of 1984 it is estimated that over 1 million such cells have been implanted.

Considerable progress has been made in enhancement of current-delivery capacity of the cell. In spite of the inherent limitations imposed by the solid electrolyte, cells produced today are capable of delivering considerably greater current densities than earlier versions.

The first commercially successful Li/I_2–PVP battery had a volumetric energy density of 0.28 Wh/cm^3. By elimination of unneeded inert material and enhancement of cathode properties, that number approaches 1.0 Wh/cm^3 for cells produced today.

It is interesting to speculate as to the future of Li/halogen systems in im-

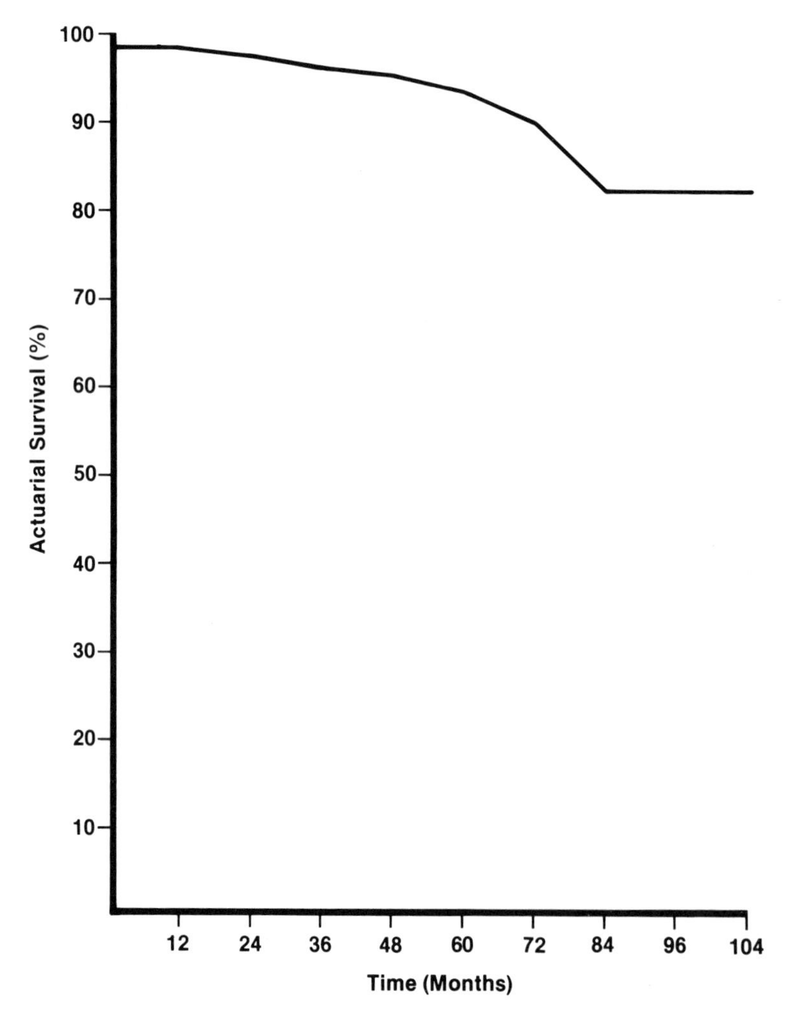

FIGURE 26. Overall Actuarial Survival Statistics for 5110 Li/I_2–PVP-Powered Pacemakers.

plantable applications. The general approach to design and composition of the Li/I_2–PVP system as seen today is likely to be around for the foreseeable future. As both pacemakers and other implantable devices need higher current delivery, further progress is likely to be seen in this area. Alternate iodine-containing cathodes may prove interesting and useful. One alternate system, employing molten onium polyiodide salt-iodine cathodes, was recently discussed by Skarstad and coworkers.[74] It is doubtful that halogens other than iodine will prove useful in implantable power sources, but such an occurrence cannot be ruled out totally.

In conclusion, the Li/I_2–PVP system continues to be a useful, safe, and reliable power source for low-current, long-life implantable devices. It is expected that further improvements and modifications will be made, but the basic system is likely to see considerable use for many years to come.

REFERENCES

1. F. Gutmann, A. M. Hermann, and A. Rembaum, Solid-State Electrochemical Cells Based on Charge Transfer Complexes, *J. Electrochem. Soc. 114,* 323–329 (1967).
2. J. R. Moser, Solid State Lithium–Iodine Primary Battery, U.S. Patent 3,660,163 (1972).
3. A. A. Schneider and J. R. Moser, Primary Cells and Iodine Containing Cathodes Therefor, U.S. Patent 3,674,562 (1972).
4. R. T. Mead, Solid State Battery, U.S. Patent 3,773,557 (1973).
5. W. Greatbatch, J. H. Lee, W. Mathias, M. Eldridge, J. R. Moser, and A. A. Schneider, The Solid-State Lithium Battery: A New Improved Chemical Power Source for Implantable Cardiac Pacemakers, *IEEE Transactions on Bio-Medical Engineering, BME-18,* 317–324 (1971).
6. G. Antonioli, F. Baggioni, F. Consiglio, G. Grassi, R. LeBrun, and F. Zanardi, Stimulatore cardiaco impiantabile con nuova battaria a stato solido al litio, *Minerva Med. 64,* 2298 (1973).
7. W. Greatbatch, R. Mead, R. McLean, F. Rudolph, and N. W. Frenz, Lithium Bromine Cell, U.S. Patent 3,994,747 (1976).
8. R. McLean and W. Greatbatch, Lithium Chlorine Cell, U.K. Patent 1,566,243 (1978).
9. W. Greatbatch, Alkali Metal–Halogen Cell Having Mixed Halogen Cathode, U.S. Patent 4,132,836 (1979).
10. D. R. Stull and H. Prophet, Project Directors, *JANAF Thermochemical Tables, National Standard Reference Data System, NBS-37,* U.S. Government Printing Office, Washington, D.C. (1971).
11. C. C. Liang, in: *Applications of Solid Electrolytes* (T. Takahashi and A. Kozawa, eds.), pp. 60–66, JEC Press, Cleveland.
12. K. Shahi, J. B. Wagner, and B. B. Owens, in: *Lithium Batteries* (J. P. Gabano, ed.), pp. 418, Academic Press, New York (1983).
13. M. Mueller, R. McLean, C. Holmes, and W. Greatbatch, Lithium–Halogen Cell Including Activated Charcoal, U.S. Patent 4,166,887 (1979).
14. W. G. Howard, P. M. Skarstad, T. G. Hayes, and B. B. Owens, in: *Power Sources for Biomedical Implantable Applications and Ambient Temperature Lithium Batteries* (B. B. Owens and N. Margalit, eds.), *PV80-4,* pp. 122–135, The Electrochemical Society, Pennington, New Jersey (1980).
15. A. M. Hermann and E. Luksha, in: *Power Sources for Biomedical Implantable Applications and Ambient Temperature Lithium Batteries* (B. B. Owens and N. Margalit, eds.), *PV80-4,* pp. 110–121, The Electrochemical Society, Pennington, New Jersey (1980).
16. W. Greatbatch, in: *Power Sources for Biomedical Implantable Applications and Ambient Temperature Lithium Batteries* (B. B. Owens and N. Margalit, eds.), *PV80-4,* pp. 63, The Electrochemical Society, Pennington, New Jersey (1980).
17. W. Greatbatch, R. Mead, R. McLean, F. Rudolph, and N. W. Frenz, Lithium–Bromine Cell and Method of Making the Same, U.S. Patent 4,105,833 (1978).
18. W. Greatbatch, R. Mead, F. Rudolph, R. McLean, and N. W. Frenz, Method of Making a Lithium–Bromine Cell, U.S. Patent 4,164,070 (1978).
19. C. F. Holmes and M. Mueller, Lithium–Bromine Cell, U.S. Patent 4,147,842 (1978).
20. A. A. Schneider, G. C. Bowser, and L. H. Foxwell, Lithium Iodine Primary Cells Having Novel Pelletized Depolarizer, U.S. Patent 4,148,975 (1979).

21. W. Greatbatch, C. F. Holmes, and M. Mueller, in: *Extended Abstracts of the Fall 1977 Electrochemical Society Meeting,* Abstract 21, pp. 58–60, The Electrochemical Society, Princeton, New Jersey (1977).
22. K. R. Brennen and D. F. Untereker, in: *Extended Absracts of the Fall 1978 Electrochemical Society Meeting,* Abstract 60, pp. 165–166, The Electrochemical Society, Pennington, New Jersey (1978).
23. K. R. Brennen and D. F. Untereker, in: *Power Sources for Biomedical Implantable Applications and Ambient Temperature Lithium Batteries* (B. B. Owens and N. Margalit, eds.), *PV 80-4,* pp. 161–173, The Electrochemical Society, Pennington, New Jersey (1980).
24. R. L. McLean and J. Bleecher, in: *Power Sources for Biomedical Implantable Applications and Ambient Temperature Lithium Batteries* (B. B. Owens and N. Margalit, eds.), *PV 80-4,* pp. 207–220, The Electrochemical Society, Pennington, New Jersey (1980).
25. G. M. Phillips and D. F. Untereker, in: *Power Sources for Biomedical Implantable Applications and Ambient Temperature Lithium Batteries* (B. B. Owens and N. Margalit, eds.), *PV 80-4,* pp. 195–206, The Electrochemical Society, Pennington, New Jersey (1980).
26. F. E. Kraus and A. A. Schneider, in: *Proc. 27th Power Sources Symposium,* pp. 144–147, PSC Publications Committee, Red Bank, New Jersey (1976).
27. G. M. Phillips and D. F. Untereker, in: *Proceedings of the Symposium on Lithium Batteries* (H. V. Venkatasetty, ed.), *PV 81-4,* pp. 303–322, The Electrochemical Society, Pennington, New Jersey (1981).
28. A. A. Schneider, S. E. Snyder, T. DeVan, M. J. Harney, and D. E. Harney, in: *Power Sources for Biomedical Implantable Applications and Ambient Temperature Lithium Batteries* (B. B. Owens and N. Margalit, eds.), *PV 80-4,* pp. 144–153, The Electrochemical Society, Pennington, New Jersey (1980).
29. K. J. Jones, Lithium Iodine Batteries for Cardiac Pacemakers, *Med. Electron. 72,* 86–91 (1981).
30. C. F. Holmes and J. Greenwood, in: *Proc. 29th Power Sources Symposium,* pp. 62–64, The Electrochemical Society, Pennington, New Jersey (1980).
31. W. Greatbatch, R. Mead, and F. Rudolph, Lithium–Iodine Battery Having Coated Anode, U.S. Patent 3,957,533 (1976).
32. C. F. Holmes and W. R. Brown, in: *Power Sources for Biomedical Implantable Applications and Ambient Temperature Lithium Batteries (B. B. Owens and N. Margalit, eds.), PV 80-4,* pp. 187–194, The Electrochemical Society, Pennington, New Jersey (1980).
33. B. B. Owens and P. M. Skarstad, in: *Fast Ion Transport in Solids* (P. Vashishta, J. N. Mundy, and G. K. Shenoy, eds.), pp. 61–67, Elsevier North-Holland, Amsterdam (1979).
34. P. M. Skarstad and B. B. Owens, in: *Proc. Workshop on Lithium Nonaqueous Battery Electrochemistry* (E. B. Yeager, B. Schumm, Jr., G. Blomgren, D. R. Blankenship, V. Leger, J. Akridge, eds.), *PV 80-7,* pp. 276–292, The Eletrochemical Society, Pennington, New Jersey (1980).
35. W. R. Brown, W. R. Fairchild, H. A. Hornung, and C. F. Holmes, in *Extended Abstracts of the Fall 1984 Electrochemical Society Meeting,* Abstract 174, pp. 257, The Electrochemical Society, Pennington, New Jersey (1984).
36. J. B. Phipps, T. G. Hayes, P. M. Skarstad, and D. F. Untereker, *Extended Abstracts of the Fall 1984 Electrochemical Society Meeting,* Abstract 175, pp. 258, The Electrochemical Society, Pennington, New Jersey (1984).
37. A. A. Schneider, private communication (1982).
38. R. T. Mead, private communication (1977).
39. F. E. Kraus, M. V. Tyler, and A. A. Schneider, in: *Reliability Technology for Cardiac Pacemakers III. A Workshop Report* (Harry A. Schafft, ed.), NBS Special Publication 400-50, pp. 35–37, National Bureau of Standards, Washington, D.C. (1979).
40. W. R. Brown, C. F. Holmes, and R. C. Stinebring, in: *Proceedings of the Symposia on Corrosion in Batteries and Fuel Cells and Corrosion in Solar Energy Systems* (C. J. Johnson and S. L.

Pohlman, eds.), *PV 83-1,* pp. 27–31, The Electrochemical Society, Pennington, New Jersey (1983).
41. A. A. Schneider, J. R. Moser, T. H. E. Webb, J. E. Desmond, in: *Proceedings of the 24th Power Sources Symposium,* pp. 27–30, PSC Publications Committee, Red Bank, New Jersey (1970).
42. R. Mead, N. W. Frenz, and F. Rudolph, Lithium–Iodine Battery, U.S. Patent 3,996,066 (1976).
43. W. Greatbatch, R. Mead, F. Rudolph, and N. W. Frenz, Lithium–Iodine Battery, U.S. Patent 3,969,142 (1976).
44. W. Greatbatch, Lithium–Iodine Battery, U.S. Patent 3,874,929 (1975).
45. R. Mead, W. Greatbatch, F. Rudolph, and N. W. Frenz, Lithium–Iodine Cell, U.S. Patent 4,210,708 (1980).
46. R. T. Mead, C. F. Holmes, and W. Greatbatch, in: *Proceedings of the Symposium on Cell Design and Optimization* (S. Gross, ed.), *PV 79-1,* pp. 327–334, The Electrochemical Society, Pennington, New Jersey (1979).
47. J. D. Jolson and A. A. Schneider, in: *Proc. 30th Power Sources Symposium,* pp. 185–187, The Electrochemical Society, Inc., Pennington, New Jersey (1982).
48. A. A. Schneider and H. D. Goldman, Lithium iodine cells for DDD and antitachycardia pacemakers, *Stimucoeur Med. 11,* 223 (1984).
49. M. C. Day and J. Selbin, *Theoretical Inorganic Chemistry,* Reinhold, New York, 1962, p. 98.
50. W. S. Holmes, in: *Reliability Technology for Cardiac Pacemakers III. A Workshop Report* (Harry A. Schafft, ed.), NBS Special Publication 400-50, pp. 6–9, National Bureau of Standards, Washington, D.C. (1979).
51. L. D. Hansen and R. M. Hart, in: *Reliability Technology for Cardiac Pacemakers III. A Workshop Report* (Harry A. Schafft, ed.), NBS Special Publication 400–50, pp. 10-16, National Bureau of Standards, Washington, D.C. (1979).
52. D. F. Untereker and B. B. Owens, in: *Reliability Technology for Cardiac Pacemakers III. A Workshop Report* (Harry A. Schafft, ed.), NBS Special Publication 400-50, pp. 17–22, National Bureau of Standards, Washington, D.C. (1979).
53. E. J. Prosen and J. C. Colbert, in: *Reliability Technology for Cardiac Pacemakers III. A Workshop Report* (Harry A. Schafft, ed.) NBS Special Publication 400–50, pp. 23-26, National Bureau of Standards, Washington, D.C. (1979).
54. C. F. Holmes, W. S. Holmes, R. L. McLean, and W. Greatbatch, in: *Proc. 28th Power Sources Symposium,* pp. 226–229, PSC Publications Committee, Red Bank, New Jersey (1978).
55. W. Greatbatch, R. McLean, W. Holmes and C. F. Holmes, A microcalorimeter for nondestructive analysis of pacemakers and pacemaker batteries, *IEEE Trans. Biomed. Eng. BME-26,* 306–309 (1979).
56. D. F. Untereker, The use of a microcalorimeter for analysis of load-dependent processes occurring in a primary battery, *J. Electrochem. Soc. 125,* 1907–1912 (1978).
57. C. C. Liang and C. F. Holmes, Performance and reliability of the lithium iodine battery, *J. Power Sources 5,* 3–13 (1980).
58. C. F. Holmes and S. T. Farrell, in: *Cardiac Pacing, Electrophysiology and Pacemaker Technology* (G. A. Feruglio, ed.), pp. 1193–1196, Piccin Medical Books, Padova (1983).
59. C. F. Holmes and R. C. Stinebring, in: *Proceedings of the Symposium on Lithium Batteries,* (H. V. Venkatasetty, ed.), *PV 81-4,* pp. 293–302, The Electrochemical Society, Pennington, New Jersey (1981).
60. D. J. Gerrard, B. B. Owens, and K. Fester, in: *Reliability Technology for Cardiac Pacemakers III. A Workshop Report,* (Harry A. Schafft, ed.), NBS Special Publication 400–50, pp. 30-35, National Bureau of Standards, Washington, D.C. (1979).
61. J. S. Kim and K. R. Brennen, in: *Power Sources for Biomedical Implantable Applications and Ambient Temperature Lithium Batteries* (B. B. Owens and N. Margalit, eds.), *PV 80-4,* pp. 174–186, The Electrochemical Society, Pennington, New Jersey (1980).

62. K. R. Brennen, K. E. Fester, B. B. Owens, and D. F. Untereker, A capacity rating system for cardiac pacemaker batteries, *J. Power Sources 5,* 25–34 (1980).
63. R. Salem and P. Bro, Performance domain analysis of primary batteries, *J. Electrochem. Soc. 118,* 829–831 (1971).
64. C. F. Holmes and D. Taub, Les piles lithium-iode: evaluation, fiabilité, et performances, *Stimucoeur Med. 8,* 318–322 (1980).
65. C. F. Holmes, Powering the pacemaker, *Qual. Prog.* January, 21–22 (1979).
66. C. F. Holmes, W. R. Fairchild, and R. C. Stinebring, Real time performance of lithium/iodine pacemaker batteries, *Rev. Eur. Technol. Biomed. 4,* 386 (1982).
67. W D. Helgeson, in: *Extended Abstracts of the Spring 1983 Electrochemical Society Meeting,* Abstract No. 28, p. 44, The Electrochemical Society, Pennington, New Jersey (1983).
68. A. A. Schneider, S. E. Snyder, and C. P. Bennett, in: *Cardiac Pacing, Electrophysiology and Pacemaker Technology,* (G. A. Feruglio, ed.), pp. 1203–1208, Piccin Medical Books, Padova (1983).
69. M. Guilleman and D. Gilet, in: *Abstracts of the VIIth World Symposium on Cardiac Pacing,* published in *Pace 6,* 366 (1983).
70. B. B. Owens, Pacemaker longevity projections from battery capacity, *Stimucoeur Med. 9,* 161–162 (1981).
71. E. Luksha, in: *Power Sources for Biomedical Implantable Applications and Ambient Temperature Lithium Batteries* (B. B. Owens and N. Margalit, eds.), *PV 80-4,* pp. 71–80, The Electrochemical Society, Pennington, New Jersey (1980).
72. B. B. Owens, K. R. Brennen, and J. Kim, Lithium pacemaker reliability, *Stimucoeur Med. 9,* 371–381 (1981).
73. M. Bilitch, R. G. Hauser, B. S. Goldman, S. Furman, and V. Parsonnet, Performance of cardiac pacemaker pulse generators, *Pace 6,* 670–672 (1983).
74. T. G. Hayes, B. B. Owens, J. B. Phipps, P. M. Skarstad, and D. F. Untereker, in *Extended Abstracts of the Spring 1983 Electrochemical Society Meeting,* Abstract 764, pp. 1126, The Electrochemical Society, Pennington, New Jersey (1983).

7

Lithium Solid Cathode Batteries for Biomedical Implantable Applications

JEAN PAUL GABANO, MICHAEL BROUSSELY, and MICHAEL GRIMM

1. INTRODUCTION

The particular types of systems considered in this chapter are batteries using lithium as the negative electrode, a compatible liquid organic electrolyte or solid inorganic electrolyte, and a transition metal salt or oxide as the positive electrode. The main reasons for reviewing such systems derive from the fact that some have been used as implantable power sources for cardiac pacemakers since the mid-1970s.

Today there are at least five different types of lithium solid cathode batteries that have been proposed as power sources for medical applications. These cells are produced in a variety of shapes, sizes, and operating characteristics.

The chemistry, mechanical construction and main characteristics of these couples are quite different, even though they have a common anode material. They have been selected according to well-defined criteria, such as energy density, shelf life, and the ability to deliver pulses, as required by the various medical applications.

JEAN-PAUL GABANO, MICHAEL BROUSSELY, and MICHAEL GRIMM • Departement Générateurs de Technologies Avanceés, SAFT, Poitiers 86000, France.

2. GENERAL FEATURES OF LITHIUM SOLID CATHODE SYSTEMS

2.1. Thermodynamic Considerations

Among the multitude of conceivable electrochemical systems based on a lithium anode and solid cathode, having enough specific energy for a particular application, the choice inevitably reduces to how much electrical energy can be obtained theoretically from a given weight or volume of reactants. Consequently, it is of prime importance to compare the properties of lithium-solid cathode batteries using a thermodynamic approach in order to determine the upper limit of the specific performances that might be obtained in the most favorable case.

For a given theoretical cell reaction such an approach provides information on the theoretical cell voltage and capacity density from which the theoretical energy density can be computed, such parameters being used very often to select couples for a first evaluation. Summarized in Table 1 are theoretical features of lithium anode cells with some of the most important solid cathodes.

2.2. Some Properties of Electrodes and Electrolytes

The selection of a given couple cannot be made solely from the consideration of theoretical specific performances as indicated in the last table. All electro-

TABLE 1 Theoretical Energy Densities and Reversible Cell Potential for Various Lithium Solid Cathode Battery Systems

Cell Reaction	E/V	Wh/kg	Wh/dm^3
$Li + V_2O_5 \rightarrow V_2O_5Li$	3.40[a]	482	1357
$Li + MnO_2 \rightarrow MnO_2Li$	3.50[a]	1027	3185
$2\,Li + Ag_2CrO_4 \rightarrow Li_2CrO_4 + 2\,Ag$	3.31[a]	513	2088
$Li + CF \rightarrow LiF + C$	2.82–3.3[a]	1992[c]	2976[c]
$Li + TiS_2 \rightarrow TiS_2Li$	2.45[a]	552	1374
$2\,Li + CuO \rightarrow Cu + Li_2O$	2.24	1285	3140
$2\,Li + S \rightarrow Li_2S$	2.18[b]	2250	2826
$2\,Li + 2\,CuS \rightarrow Li_2S + Cu_2S$	2.12[b]	557	1683
$6\,Li + Bi_2O_3 \rightarrow 3\,Li_2O + 2\,Bi$	2.04	646	2478
$10\,Li + Bi_2Pb_2O_5 \rightarrow 5\,Li_2O + 2\,Bi + 5\,Pb$	2	546	2318
$2\,Li + PbI_2 \rightarrow 2LiI + Pb$	1.9	211	997
$4\,Li + FeS_2 \rightarrow 2Li_2S + Fe$	1.75	1273	2474
$2\,Li + Cu_2S \rightarrow Li_2S + 2Cu$	1.74	539	1714
$2Li + PbS \rightarrow Li_2S + Pb$	1.70	360	1574

[a] Based on open-circuit voltage of experimental cells.
[b] Based on ΔG. $Li_2S = -100.8$ kcal proposed by Freeman, Thermochemical Properties of Binary Sulfides, *Okla. States Res. Found.* Report 60 (1962).
[c] Based on $E = 2.82$ V.

chemical cells require that the electrodes be separated by a material known as the electrolyte. This material transfers charge between the electrodes, via the transport of appropriate ionic species; that is, it allows ionic as distinct from electronic conduction. The choice of a suitable electrolyte, its properties, and its interactions with the plate materials are the principal factors that now control the performance of these battery systems.

Lithium batteries using a solid cathode can be classified according to the nature of the electrolyte, with such classifications including organic liquid electrolytes and ionically conductive solids. Because of the lack of sufficiently conductive solid electrolytes most of the lithium solid cathode systems existing today are based on the use of a liquid electrolyte composed of an inorganic lithium solute dissolved in an aprotic organic solvent.

As mentioned earlier, it is required that this electrolyte be compatible with the electrodes, that is, the strong reducing and oxidizing agents that compose the plate materials. Unfortunately, theoretical calculations of solvent stability have generally not been made because of insufficient data on the path of decomposition and on the free energy values for the products and reactants.

Practically, the electrochemical stability of the electrolyte can be determined, by using the three electrode voltametric method, which consists of simply varying linearly the potential of an inert working electrode versus a reference electrode and observing the Faradaic current in the electrochemical cell composed of the working and counter electrodes. In the presence of a nonelectroactive solute, the only processes possible must be due to the solvent decomposition.

In practice, it is difficult to use the results of such a method because the electrochemical voltage range thus determined is strongly dependent on the nature of the working electrode; moreover, the so-called inert solute often undergoes discharge before solvent decomposition occurs. If these problems can be overcome, the effect of impurities can still cause some problems in interpreting the data. For these reasons an empirical approach is generally preferred for testing the compatibility of an electrolyte solution with the active materials and associated electrode structure.

One criterion that must be satisfied is that of lithium stability. While it was thought at one time that some organic electrolytes using aprotic solvents were unreactive with lithium, it is now generally believed that all or nearly all electrolyte solutions react to some extent with the lithium metal surface. Thus, a successful electrolyte solution must show metastability with lithium for extended periods of storage even at high temperature. This apparent stability has been found to be due to the formation of a protective film at the lithium/electrolyte interface, the nature of which is strongly dependent on the composition and purity of the electrolyte. For instance, this stability can be very sensitive to contaminants such as remaining water or metallic impurities. Some evidence for this mechanism has been reported by Dey through electrochemical kinetic measurements undertaken for a lithium electrode in the presence of a lithium perchlorate propylene carbonate solution.[1]

More recently, Peled[2] introduced the "solid electrolyte interphase" (SEI) concept for explaining the properties of such protective films (the thin layer that forms between the electrolyte solvent and lithium being of a conductive nature for lithium ions but not electrons). However, this ideal model can be disturbed if some electronic conductivity appears in the film as a result of an impurity doping effect, leading to a continuous growth of the SEI with time and temperature.

Table 2 gives some physical properties of selected aprotic solvents that have been used in association with various lithium salts for constituting electrolytes having suitable metastability with lithium and sufficient mass transfer properties.

Because of their poor anion solvating abilities, these solvents have been found able to dissolve only lithium salts with low lattice energy. Thus, lithium salts with large monovalent anions such as perchlorate, hexafluoarsenate, and tetrafluoroborate have been used generally. Such solutions are characterized by a greater or lesser degree of ion pairing, a high dielectric constant favoring ionization. However, the latter is not the only criterion for providing electrolytes with high ionic conductivities. The effect of electron donor properties of the solvent for the cation (which means, in other terms, its ability to solvate lithium ions) and the influence of viscosity on ionic mobility are also important factors that have to be considered.

An ideal solvent would combine the properties of low viscosity, high dielectric constant, and good electron donor properties, but such a material has not

TABLE 2 Physical Properties of Various Organic Solvents Commonly Used in Lithium Solid Cathode Battery Systems

Solvent	Formula	Melting Point (0°C)	Boiling Point (0°C)	Dielectric Constant	Viscosity (cP)	Density (g/cm³)
Propylene carbonate (PC)	CH_3—CH——CH_2 (ring: —OCOO—)	−49	241	64.4	2.53	1.19
γ-Butyrolactone (GBL)	CH_2 $(CH_2)_2$ COO (ring)	−43	202	39.1	1.75	1.13
Methyl formate (MF)	H $COOCH_3$	−99	31	8.5	0.33	0.97
1,2-Dimethoxyethane (DME)	$CH_3O(CH_2)_2OCH_3$	−58	84	7.2	0.45	0.86
Tetrahydrofuran (THF)	CH_2 $(CH_2)_3O$ (ring)	−65	65	7.4	0.456	0.88
1,3-Dioxolane (DOL)	$OCH_2OCH_2CH_2$ (ring)	−95	78	7.1	—	1.06

yet been identified. Consequently, typical organic electrolytes have lower conductivities than aqueous ones, in the range of 3×10^{-3} to $2 \times 10^{-2}\ \Omega^{-1}\ cm^{-1}$.

In Table 3 we have reported specific conductivities for some of these electrolytes. As indicated in this table, solvent mixtures demonstrate interesting properties as illustrated by the $LiClO_4$–propylene carbonate/1,2-dimethoxyethane electrolyte, where the adverse effects of viscosity and dielectric constant on conductivity have been minimized.

Another criterion of importance for the electrolyte is its compatibility with the positive plate. Cathode stability with the electrolyte means the absence of chemical interactions between solutes and/or organic solvents and the strongly oxidizing materials such as MnO_2 or V_2O_5 that compose common positives, and it also refers to the insolubility of these materials. The latter is a prime requirement for the stability of the various lithium battery systems. It excludes the use of highly energetic materials, such as transition metal halides ($CuCl_2$, CuF_2, . . .) that have been found to show some direct or indirect solubilization (through complexation with associated discharge products), and limits the choice to nonmetal halogen compounds, such as carbonmonofluoride, or to transition metal oxides, sulfides, and chromates, the solubility of which is low enough to avoid direct chemical reaction with the anode through the electrolyte. Some of the lithium solid cathode battery systems using the aforementioned electrolytes are also indicated in Table 3.

Other criteria, such as material compatibility, separators, current collectors, and cell cases, have to be checked specifically according to the type of system under consideration.

TABLE 3 Specific Conductivities of Some Commonly Used Electrolytes and Associated Lithium Solid Cathode Battery Systems

Organic Electrolytes	Specific Conductivities at 25°C (Ω^{-1} cm^{-1})	Cells
PC + 1 *M* $LiClO_4$	5×10^{-3}	Li/Ag_2CrO_4, Li/MnO_2 Li/CF
PC + 12 DME (50/50) 1 *M* $LiClO_4$	12.7×10^{-3}	Li/MnO_2, Li/CF, Li/CuO
PC + 12 DME (50/50) 1 *M* $LiBF_4$	8×10^{-3}	Li/CF, Li/CuO
GBL + 1 *M* $LiBF_4$	6.2×10^{-3}	Li/CF
GBL/THF (50/50) + 1 *M* $LiBF_4$	7.5×10^{-3}	Li/CF
MF + 2 *M* $LiAsF_6$ + 0.4 *M* $LiBF_4$	44×10^{-3}	Li/V_2O_5
THF + 12 DME (70/30) + 1 *M* $LiClO_4$	3.5×10^{-3}	Li/CuS, Li/CuO
DOL + 2 *M* $LiClO_4$	7.7×10^{-3}	Li/CuO
DOL + 1.2 DME (x) + DMI[a] 1.6 *M* $LiClO_4$	7.2×10^{-3}	Li/CuS

[a] DMI = 3,5-dimethylisoxazole used as a solvent stabilizer.

Further narrowing of the choice of electrolytes and electrodes must be based finally on results of experimental cells designed to provide, for example, information on rate processes for the electrochemical reactions involved at the electrodes, shelf life in various environmental conditions, and the ability to work in a specified range of temperatures.

2.3. Electrode and Cell Configurations

Considerable art is required to construct lithium-solid cathode battery systems with minimal weight and volume and maximum energy and performance.

For the lithium anode an expanded stainless steel grid shaped to the appropriate dimensions required by the cell design and having a connecting tab for external contact is usually pressed into a sheet or piece of lithium, the thickness of which depends on cell capacity.

As far as the cathode is concerned, the construction is strongly dependent on type of cell. High-rate cells based on cylindrical spiral wound or prismatic multiplate electrodes use thin electrodes with an extended surface area. For spiral wound cells, cathodes are made by a slurry process; a mixture consisting of cathode active material conductive additive and a binder usually composed of a Teflon water emulsion is spread into an expanded material grid, then dried to evaporate the water, and finally sintered by a short heat treatment at an adequate temperature. For prismatic cells the positive plates are generally made by a cold-pressed technique, the binder being sometimes eliminated from the mix.

Low-rate cells can be designed as cylindrical-button or special-shaped cells. In cylindrical cells using the inside–outside construction and button cells, the cathode is formed directly on the can from a cathodic dry mix. Suitable techniques have been developed for special-shaped cells.

In all cases, cell sealing is achieved either by a crimped seal using a plastic grommet between can and cover or by a glass to metal feedthrough, depending on the system or specifications required by the applications.

3. SPECIFIC SYSTEMS USED FOR BIOMEDICAL APPLICATIONS

The specific solid cathode lithium batteries discussed in this section involve various metal salts or oxides and have been used or proposed for medical applications, primarily as heart pacer batteries. These compounds have been selected from those materials that demonstrated the best compatibility with the existing electrolytes and provided high-energy density and/or cell voltage. They included in most cases a specific feature constituted by a two-step or multistep discharge profile generally used in such cells as an indication of the approach of end of life (Ag_2CrO_4, V_2O_5).

3.1. The Lithium–Silver Chromate Organic Electrolyte System

The lithium–silver chromate cell was one of the earliest and most promising solid cathode systems for use in high-energy pacemaker cells requiring a 3-V voltage range and a long storage life. The latter characteristic is mainly due to the low solubility of silver chromate (Ag_2CrO_4) in the lithium perchlorate propylene carbonate based electrolyte and the absence of any complication arising from additional cathode solubility via complex formation during discharge. Such features have not been fulfilled in the past with the 3-V solid metal salt lithium systems, mainly based on metal halides. The major problem of these metal halide systems is a high degree of self-discharge associated with an excessive solubility of the material in the electrolyte and an ability to form soluble complexes with the excess lithium halide produced during discharge reactions.[3]

Dey *et al.*[4,5] and Gabano *et al.*[6,7] simultaneously discovered the couple in the early 1970s but the latter developed the system commercially at Saft for medical applications.

3.1.1. *Cell Chemistry and Discharge Mechanism*

As commercial products were not found suitable for direct use in lithium cells, a special process was developed for producing highly pure silver chromate.[8] This process consists of first reacting a solution of silver nitrate with sodium dichromate according to the following:

$$2AgNO_3 + Na_2Cr_2O_7 \rightarrow \underset{\downarrow}{\underline{Ag_2Cr_2O_7}} + 2NaNO_3 \quad (1)$$

Then treating the resulting insoluble silver dichromate with a solution of magnesium chromate,

$$Ag_2Cr_2O_7 + MgCrO_4 \rightarrow Ag_2CrO_4 + MgCr_2O_7 \quad (2)$$

results in a quite pure material (99.9%) free of silver oxide.

Various nonaqueous electrolytes for use in Li/Ag_2CrO_4 batteries have been investigated, but only lithium perchlorate–propylene carbonate solutions were found of interest for cathode insolubility reasons. Typically, 1 M $LiClO_4$ in PC is presently used in practical cells.

The proposed cell reaction has been believed for many years to occur according to the following process[4]:

$$Ag_2CrO_4 + 4Li \rightarrow 2Ag + CrO_2 + 2Li_2O \quad (3)$$

Experimentally, four equivalents per mole of Ag_2CrO_4 were effectively obtained and an initial open-circuit voltage of 3.35 V realized in electrochemical cells. Nevertheless, the theoretical potential calculated from preceding reaction was not found to correspond to the observed value, involving a more complex mechanism.

Recent studies[8] demonstrated the existence of intermediate reaction steps, which were found to be dependent upon the temperature of reaction and the specific discharge rate. Such studies led to the proposal of the general mechanism, which is summarized in Figure 1.

This mechanism shows the importance of the initial reduction process of Ag_2CrO_4 to Ag according to the following:

For high specific discharge rate roughly corresponding to cell potential below 3.1 V, the formation of a metastable intercalation compound $Ag_2CrO_4Li_x$

$$Ag_2CrO_4 + x\mathrm{Li} + x\mathrm{e} \rightarrow Ag_2CrO_4Li_x \quad (4)$$

containing Cr^{VI} and Ag^{I}, leads by fast decomposition to an intermediate compound that should contain Cr^{VI} and atoms of zero valent silver

$$Ag_2CrO_4Li_x \rightarrow \left(1 - \frac{x}{2}\right) Ag_2CrO_4 + \frac{x}{2} Ag_2^0,Li_2Cr^{VI}O_4 \quad (5)$$

$Ag_2^0,Li_2Cr^{VI}O_4$ either evolves to a deactivated mixture of metallic phase Ag and Li_2CrO_4 according to:

$$Ag_2^0,Li_2Cr^{VI}O_4 \rightarrow 2Ag + Li_2CrO_4 \quad (6)$$

or is reduced at 2.6 V with the formation of the corresponding Cr^{V} compound, which may be written $Ag_2^0,Li_3Cr^{V}O_4$.

For low-discharge specific rates the process leads first to the formation of silver and Li_2CrO_4, probably through a dissolution precipitation mechanism as reflected by the rather high cell potential value of 3.2 V.

$$Ag_2CrO_4 + 2\mathrm{e} + 2Li^+ \rightarrow Ag + Li_2CrO_4 \quad (7)$$

This process has also been found to occur to some extent when the temperature of discharge is increased or if the electrolyte composition is changed (addition of ethers to propylene carbonate–lithium perchlorate electrolyte) through the solubility effect.

In these different conditions the second step is observed at 2.4 V instead of 2.6 V as before; it should correspond to the reduction of Li_2CrO_4 to a chromium (V) compound, probably Li_3CrO_4:

$$Li_2CrO_4 + \mathrm{e} + Li^+ \rightarrow Li_3CrO_4 \quad (8)$$

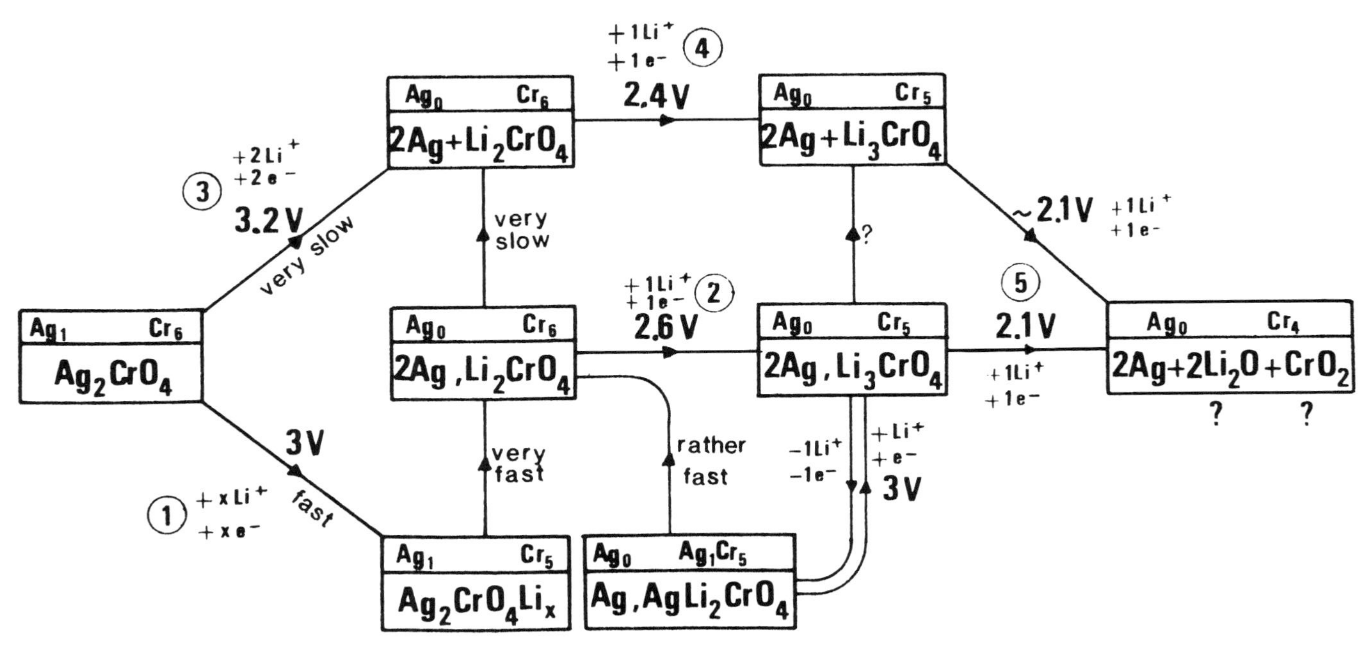

FIGURE 1. Proposed Mechanism for Ag_2CrO_4 Electrochemical Reduction in Organic Electrolyte.

Further reduction of Li_3CrO_4 to Li_2O and CrO_2 has been observed in all cases at 2.1 V, involving one equivalent per mole.

This discharge mechanism is of prime importance for the behavior of lithium/silver chromate cells, especially when a second plateau discharge is needed. For instance, Figure 2 illustrates the effect of temperature on discharge profiles obtained for Li/Ag_2CrO_4 button cells. This figure shows that for a discharge conducted at 55°C, the voltage step at 2.6 V has a tendency to be minimized as compared with a discharge run at 37°C in agreement with the aforementioned process, the overall cell capacity being the same regarding lithium cell balance used in this case.

3.1.2. Cell Construction and Types

Different configurations were proposed for medical applications. Most of the commercial cells manufactured until now have been based on the button-type technology using the crimped seal version. Because of the presence of high-viscosity, low-vapor-pressure 1 *M* $LiClO_4$ propylene carbonate electrolyte, it has been possible to use this type of sealing, which proved to be entirely hermetic. The cathode, composed of silver chromate, was directly cold-pressed in the positive stainless steel bottom can; the anode was lithium as usual, with the

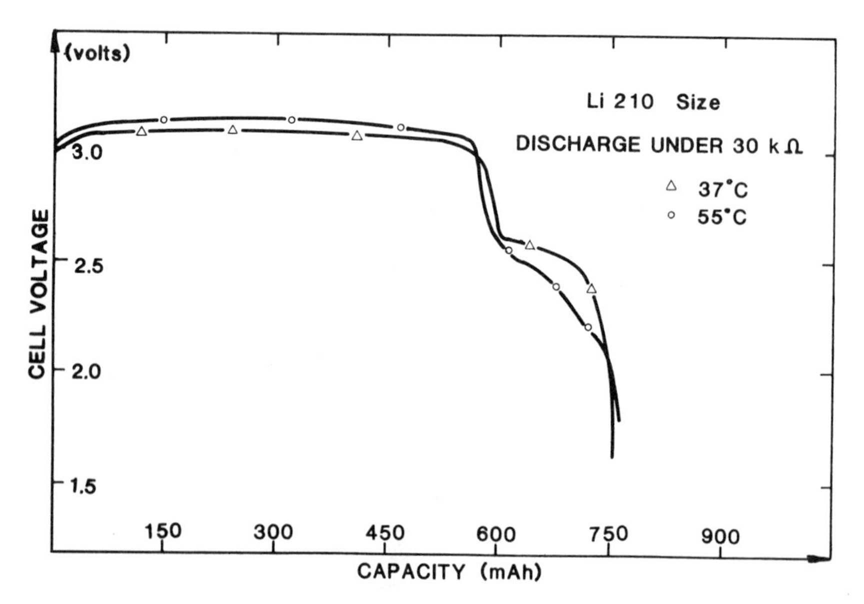

FIGURE 2. Discharge Characteristics for SAFT 210 Li/Ag_2CrO_4 Button Cells (Diameter = 21 mm, Height = 9 mm) under 30 kΩ at Two Different Temperatures (Stoichiometrical Lithium) Capacity = 800 mAh.

anode current collector being welded to the top can and the electrolyte absorbed in a nonwoven cellulosic separator. An additional barrier layer was placed next to the cathode to avoid any short circuit between the two electrodes during discharge. Rectangular and D-shaped cells using glass to metal seal feedthroughs have been designed recently. In all cases the cell balance was anode limited to involve most of the discharge around 3 V and a practical second discharge plateau in the 2.4–2.6-V range according to discharge conditions as seen before representing about 20% of the total cell capacity.

These different designs have demonstrated practical energy densities ranging from 700 to 800 Wh/dm^3 to a 2-V cutoff voltage. An interesting feature of this lithium solid cathode system is the compatibility of the cell chemistry with metals such as titanium, which offers the possibility of selecting this material as a lightweight container for the glass-to-metal seal designs (contrary to the lithium couples based on halogen or oxyalide depolarizers) or to integrate the cell directly as part of the medical device, the latter corresponding to a significant improvement of the packaging efficiency for the energy stored.

The range of Saft lithium silver chromate cells is indicated in Figure 3 and Table 4, which summarize their main characteristics.

FIGURE 3. SAFT Li/Ag_2CrO_4 Cells Designed for Use in Medical Applications.

TABLE 4 Lithium–Silver Chromate Cells for Medical Implantable Devices

Saft Model Number		Thickness (Maximum Swelling) (mm)	Corner Radius (mm)	Weight (g)	Lithium Content (mAh)	Silver Chromate Content (mAh)	Expected Usable Capacity at 2.0 V (mAh)
Li series	Diameter (mm)						
Li 114	11.4	5.4 (0.7)	–	1.7	150	95	135
Li 210	21.0	9.0 (1.0)	–	8.9	810	600	730
Li 273	27.3	7.9 (0.7)	–	13.1	1220	900	1100
Li 273/9	27.3	8.9 (0.7)	–	14.5	1220	1100	
Li 355/6	35.5	6.1 (0.5)	–	19.3	1750	1310	1575
Li 355/10	35.5	10.2 (0.7)	–	29.4	3100	2300	2790
Li R series	Dimensions (mm)						
Li R 123	45.,3 × 23	8.0	8.9	23.0	1900	1500	1710
Li R 132	45.3 × 32.2	8.0	36.0	2450	2120	2205	
Li R 223	45.0 × 23	9.0	11.0	26.0	2150	1720	1940
Li D series							
Li D 122	45.3 × 22.5	8.0	22.3	19.0	1480	1150	1330
Li D 128	45.3 × 28.0	8.0	22.3	24.0	2100	1600	1890
Li D 224	45.3 × 24.0	9.0	22.3	24.0	1750	1480	1570

3.1.3. *Cell Performance*

Typical discharge characteristics for LiD 128 size cells (taken as an example) at 37°C are reported in Figure 4 for different continuous loads. Such discharge curves show the ability of this system to be efficiently discharged over a wide range of currents, with good voltage regulation for the moderate ones. This result is in agreement with the relatively low impedance of the cell, the values of which remain in the 10-Ω range during the discharge for the aforementioned discharge load.

The Li/Ag_2CrO_4 cell can also deliver heavy pulses, as is shown in Figure 5, where we have illustrated the performance of LiD 128 cells for a typical duty corresponding to a drug releaser application (insulin dispenser). According to this requirement, the cell is discharged on a continuous 150-kΩ background current drain with a repeated 2-sec current pulse per minute under 100 Ω (≃28 mA). This figure shows that in these conditions the cell delivers 95% of its stoichiometric capacity.

The Li/Ag_2CrO_4 cell is also noted for its excellent storage capability, even at elevated temperature, as illustrated in Figure 6, where we have reported typical shelf-life data obtained with the Li 210 button crimped seal design. These results are consistent with the absence of parasitic heat flow determined through cell microcalorimetry measurements, both on open circuit and under discharge.

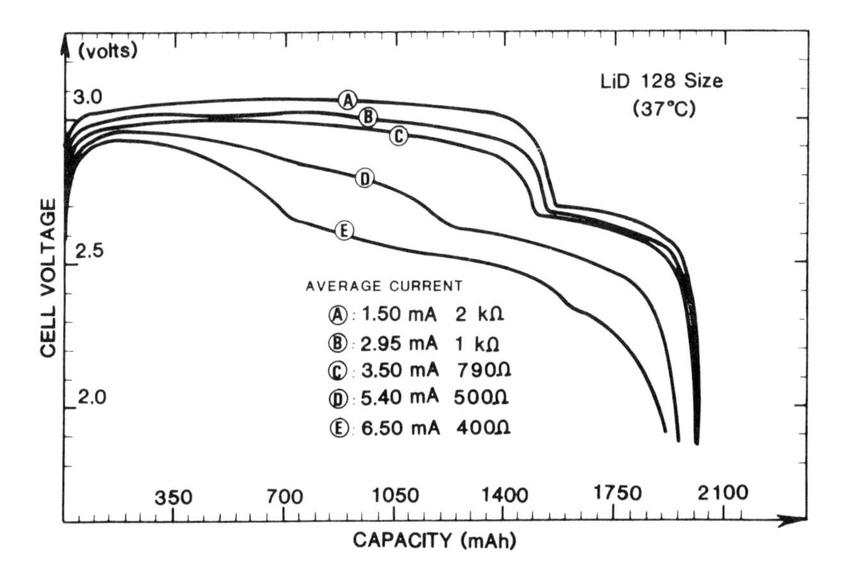

FIGURE 4. Discharge Characteristics for SAFT Li/Ag_2CrO_4 D-Shaped Cells (LiD 128) at 37°C Through Different Loads.

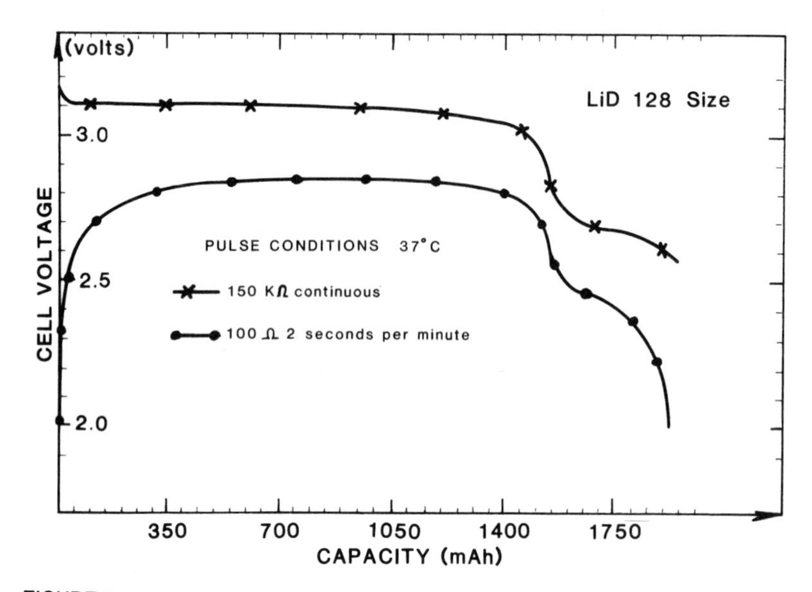

FIGURE 5. Pulse Discharge Ability for SAFT Li/Ag_2CrO_4 D-Shaped Cells (LiD 128) at 37°C.

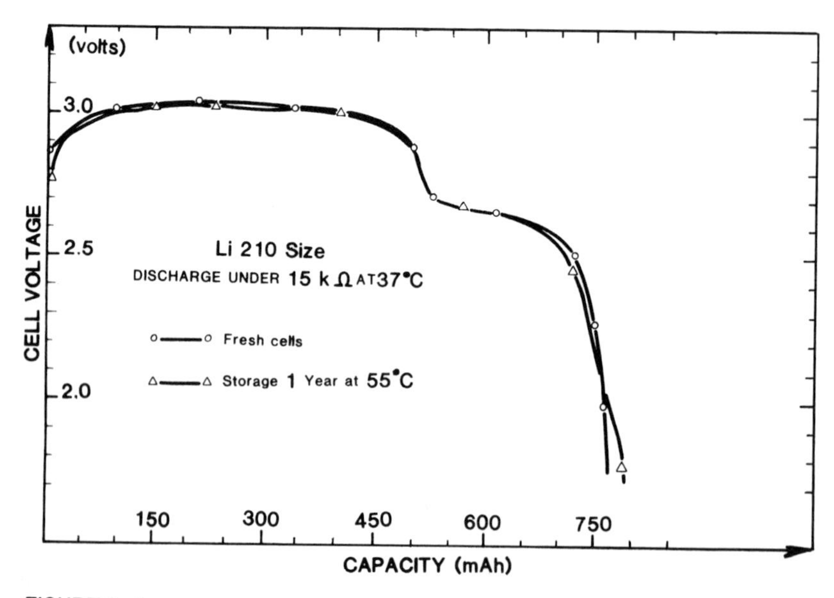

FIGURE 6. Storage Ability for SAFT 210 Li/Ag_2CrO_4 Button Cells at Elevated Temperature.

3.1.4. Silver Bismuth Chromate: An Alternative Material

Results obtained with silver chromate have led to the investigation of other metallic chromates.[9] Special attention has been devoted to silver bismuth chromate, $AgBi(CrO_4)_2$, since the presence of silver and chromate ions should lead to an electrochemical behavior similar to Ag_2CrO_4.

Figure 7 shows typical discharge characteristics run at 37°C under various loads ranging from 15 to 300 kΩ for a crimp sealed Li 210 button cell design using this compound as cathodic material.

The overall discharge process involves five equivalents per mole of $AgBi(CrO_4)_2$, and specific reactions attributed to each step have been studied. Contrary to silver chromate, the first plateau at about 3.1 V did not correspond to the reduction of silver only, but included a simultaneous reduction of silver and chromium ions, as demonstrated by the two equivalents per mole obtained in these condtions. The sloping part was found to correspond to a more complex mechanism, but in any case bismuth ions were not found to be electroactive for the voltage range considered.

As observed with the Li/Ag_2CrO_4 cell, this couple also exhibited excellent storage characteristics. In addition, it was found able to operate at temperatures as high as 70°C during extended periods of time, as reflected by results of discharges conducted at the three-year rate, where no deterioration of cell capacity was recorded when compared with those run at 37°C.

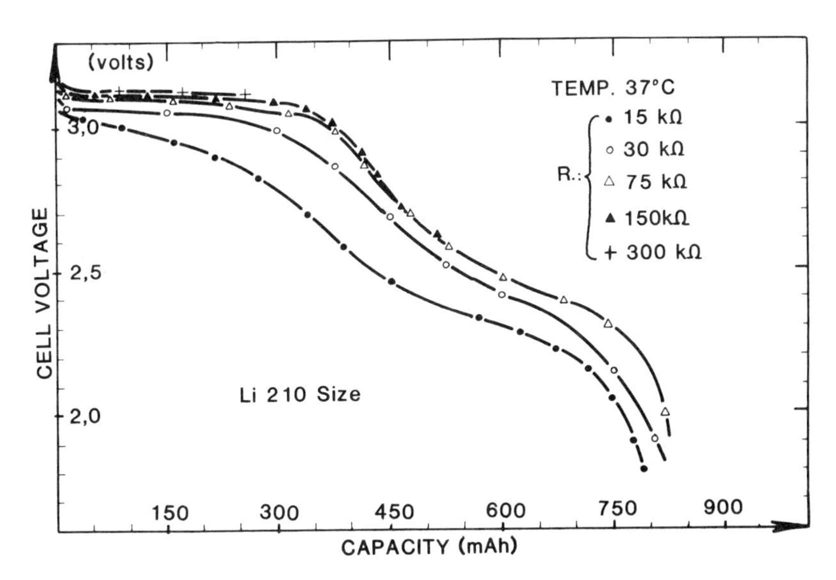

FIGURE 7. Discharge Characteristics for SAFT 210 $Li/AgBi(CrO_4)_2$ Button Cells at 37°C Through Different Loads.

3.2. The Lithium–Cupric Sulfide Organic Electrolyte Battery*

3.2.1. *Electrolytes for Li/CuS Cell*

The use of metallic sulfides as cathodic materials in lithium cells using a lithium isopropylamine electrolyte was first mentioned as a patent claim by Herbert and Ulam in the early 1960s. Long-term compatibility of lithium with the aliphatic amine limited the development of these systems until Saft introduced ether-based electrolytes,[11–13] which were found to be entirely compatible with these battery systems. Large-capacity 30-Ah Li/CuS cells using a mixed $LiClO_4$ tetrahydrofuran/ – 2 dimethoxyethane (THF/DME) electrolyte demonstrated less than 10% capacity loss after 13 years of storage at ambient temperature.

Later specific references[14–16] in the patent literature described the use of 1-3 dioxolane or its derivatives as the primary solvent and their mixtures with straight-chain ethers as cosolvents, $LiClO_4$ being the preferred solute in all cases. A serious gassing problem associated with lithium was noted with dioxolane electrolytes, but this effect was somewhat reduced by the addition of the ether cosolvent.[16] Moreover, the addition of tertiary nitrogen bases as stabilizers was found to reduce gassing problems to a satisfactory level.[17] Of this group of electrolytes, solutions of lithium perchlorate in mixtures of 1,3-dioxolane (DOL) and 1,2-dimethoxyethane (DME) including 3,5-dimethylisoxazole (DMI) as a stabilizer were finally selected for use in present Li/CuS cells. Such a specific choice was mainly justified on the basis of the observed high electrochemical utilization of the cupric sulfide cathodic material over a wide range of discharge rates.

3.2.2. *Cell Chemistry and Cathode Preparation*

The cell chemistry involves two different reactions, which are:

$$2Li + 2CuS \rightarrow Li_2S + Cu_2S \text{ (OCV = 2.12 V)} \tag{9}$$

$$2Li + Cu_2S \rightarrow Li_2S + 2Cu \text{ (OCV = 1.75 V)} \tag{10}$$

No material solubility was encountered with very pure cupric or cuprous sulfides in the aforementioned electrolytes, indicating the absence of any cathode/electrolyte-related interactions, which generally leads to a poor activated shelf life. As with silver chromate, the lack of high-purity cupric sulfide had led the majority of investigators who dealt with the battery system to propose a method for purifying commercial products or to prepare their own cupric sulfide

* The authors acknowledge the valuable review and additional information provided by Drs. K. Jones and A. DeHaan, Cordis Corp. (1984).

from high-purity copper and sulfur powders under carefully controlled conditions, the latter being the preferred choice. A preparation method favored by DuPont[18] involved the direct reaction between finely divided copper and sulfur. After mixing, the powder is spread in a thin layer and aged for several days under carefully controlled temperature and humidity conditions. The highly exothermic reaction to form CuS proceeds slowly until 50–60% of the mass is reacted. Cathodes are molded from this powder at room temperature, and the reaction is completed by firing at about 240°C for several minutes. The resultant wafers have a high porosity (up to 50%), are electronically conductive, and are mechanically strong. However, the process is difficult to control and the product may contain undesirable amounts of unreacted copper and sulfur. Recently, it has been found at Cordis that any elemental sulfur in the cathode reacts with Li_2S to form polysulfides, which act to solubilize the CuS.

Cordis has developed a more controllable preparation method for CuS that yields high-purity powder with less than 0.1% sulfur content. Sublimed sulfur and copper shot are the starting materials; the reaction is carried out in aqueous media acidified with HCl. A small amount of Cu_2O is added as a catalyst, and with controlled heating the reaction goes to completion within several hours. The reaction mechanism is not completely understood, but the main elements seem clear. Cuprous ions, either from contamination on the original copper or from added Cu_2O, are definitely required to promote the reaction. The cuprous ions react with sulfur according to the following equation:

$$2Cu^+ + S^0 \rightarrow CuS + Cu^{2+} \tag{11}$$

The free cupric ion then reacts with the elemental copper to regenerate the supply of cuprous ions:

$$Cu^0 + Cu^{2+} \rightarrow 2Cu^+ \tag{12}$$

The product CuS is washed until free of Cl^- and then vacuum-dried to give a fine (22-μm) powder that is compacted to the desired cathode shape. The final step is to fire at about 240°C to remove the last vestiges of sulfur. Cathodes made by this process are dense and conductive and have adequate mechanical strength.

3.2.3. *Cell Construction and Performance*

Various configurations of parallelepipedic, cylindrical, and flat cells have been designed in the past, both by Saft[18,19] and DuPont/Ray-O-Vac[17,20,21] using their own technology. The best performance reported at low drain corresponds to energy densities ranging between 400 and 500 Wh/dm^3.

For use in heart pacemakers Cordis first developed flat D-shaped cells based on the DuPont electrolyte. The design of the Cordis D cell is shown in Figure 8. This cell is about 5.1 cm long, 3.2 cm wide, and 0.51 cm thick and is of crimp sealed type. The cell is balanced so that about 90% of its total capacity is delivered through the first reaction step; the voltage drop between the two discharge steps provides, as in the case of Li/Ag_2CrO_4 cells, a useful indication of end of life.

The rated capacity of this D cell is 1.8 Ah under pacemaker loads. The discharge behavior of such cells at 10 times pacemaker drain (12.3 kΩ) and higher drains at 37°C is shown in Figure 9. Storage data indicate no loss of capacity after storage for 60 days at 54°C and 18 months storage at room temperature, such data being confirmed through microcalorimetry.

Cordis also produces a line of hermetically sealed cells that employ a welded header and glass-insulated feedthrough construction. Since two cells in series must be used for pacemaker applications, these cells have a form of one-half of a D shape (see Figure 10). These cells employ a dual-anode construction so that the electrode surface area is nearly equivalent to the older D cell. The central anode structure is completely enclosed in a heat-sealed bag composed of two layers of nonwoven polypropylene (Pellon) or microporous polypropylene (Celgard). These cells are made in two sizes; one has a capacity of 0.9 Ah on the first plateau and the other has a 1.9-Ah capacity.

The volumetric energy density of current Li/CuS cells is 750 Wh/dm^3. This has been achieved by (1) optimization of the external to internal volume ratio, (2) reduction of inert components, (3) reduction of tolerances on the active components so that minimum performances are closer to theoretical, and (4) effective utilization of up to 97% of the available lithium in the cell.

The overall efficiency of the Li/CuS cell is quite high. About 2% of the

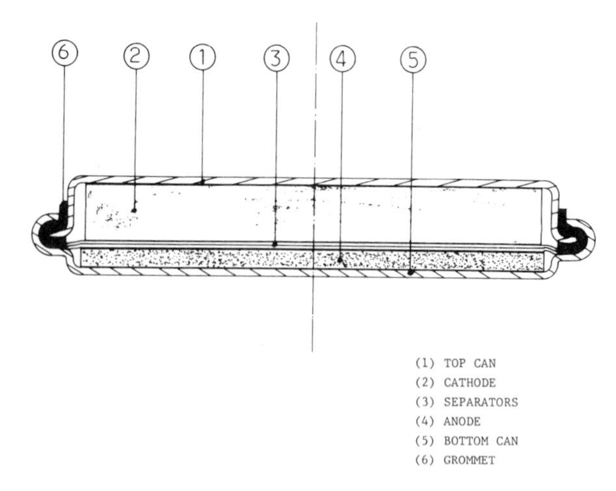

FIGURE 8. A Schematic Cross Section of the Crimped Seal Li/CuS Cordis D Cell.

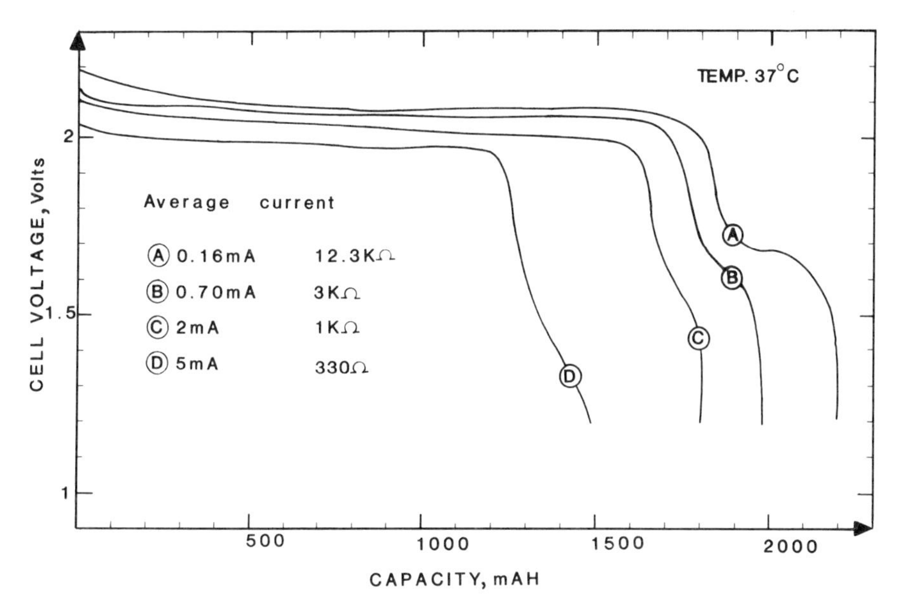

FIGURE 9. Discharge Characteristics for the Li/CuS Cordis D Cell at 37°C Through Different Loads (After Ref. 22).

stoichiometric capacity of the cell is used up in the burn-in process, a technique that removes high-voltage impurities and stabilizes the performance. Allowing for 3% unavailable lithium yields an available capacity that is 95% of the stoichiometric. Because of the relatively low impedance of the system (10–50 Ω), virtually all this capacity is available at drain rates up to 100 μA (upper limit for dual-chamber physiological pacers).

3.3. The Lithium–Vanadium Pentoxide Organic Electrolyte System

The electrochemistry of this battery system was initially studied by Honeywell for a NASA application in 1968.[24] Some work in this area has been also reported by P. R. Mallory[25] and Eagle Picher.[26] It has been developed mainly for various military applications as a reserve cell. However, considerable progress made in the field of electrolyte stability has allowed the system to be used as an active primary cell, with a reasonably good shelf life.

Based on the latter and taking into account its good current handling capabilities, high voltage, and good energy density, the Li/V_2O_5 cell was selected specifically for some medical applications requiring high specific power and a

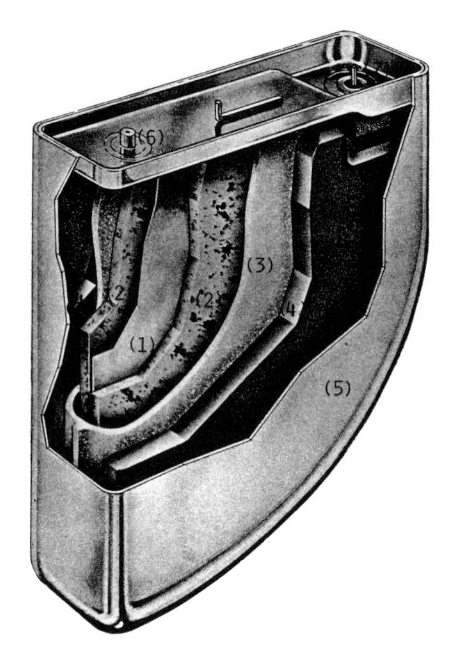

(1) Current Collector
(2) Lithium Anode
(3) Polypropylene Separator
(4) Cupric Sulfide Cathode
(5) Stainless Steel Case
(6) Backfill Tube
(7) Negative Electrode

FIGURE 10. A Schematic Cross Section of the Glass-to-Metal Sealed Cordis Li/CuS Gamma Cell.

suitable level of reliability established through its development maturity during the last 10 years.

3.3.1. Electrolytes for Li/V_2O_5 Cell

Various organic electrolytes have been proposed for Li/V_2O_5 experimental cells, but only methyl-formate-based electrolytes[27] received considerable attention, as they were found to offer good stability to oxidation by V_2O_5 and demonstrated the best ionic conductivities and cell performance of the solutions.

Lithium hexafluoroarsenate dissolved in methyl formate was first selected as the electrolyte for Li/V_2O_5 primary cells, but was found to be thermally unstable in the cell environment. The major decomposition reaction of this solution was the hydrolysis of methyl formate followed by the dehydration of formic acid to produce carbon monoxide and dimethyl ether according to the following series of reactions[28]:

$$H(CO)OCH_3 + H_2O \rightarrow H(CO)OH + CH_3OH \tag{13}$$

$$H(CO)OH \rightarrow CO + H_2O \tag{14}$$

$$2CH_3OH \rightarrow CH_3OCH_3 + H_2O \tag{15}$$

The hydrolysis reaction was found to occur in either acidic or basic solution, and both formic acid and methanol reacted with the lithium anode to form lithium formate (HCO_2Li) or lithium methoxide ($LiOCH_3$) and hydrogen until all the water impurity present in the cell was consumed. However, in the presence of acidic impurities carried by the solute ($LiAsF_6$), formic acid and methanol were found to dehydrate, producing additional free water that can hydrolyze more methyl formate until the hydrolysis/dehydration process is completed.

More pure grades of $LiAsF_6$ containing less free acid impurities were found to be effective in electrolyte stabilization, but the problem was really solved when it was discovered that lithium tetrafluoroborate ($LiBF_4$) inhibited the dehydration reactions. A 2 M $LiAsF_6$ + 0.4 M $LiBF_4$ solution in methyl formate was finally selected as the optimum electrolyte.[28]

3.3.2. *Cell Discharge Reactions*

The Li/V_2O_5 battery system exhibits an open-circuit voltage (OCV) of 3.4 *V* and can deliver a coulombic capacity equivalent to about three electrons per V_2O_5 unit above 2 *V* through a multistep sloping discharge curve. The discharge reaction for the first step occurring in the 3–3.4-V range is now generally accepted as going through an intercalation compound,[30]

$$Li + V_2O_5 \rightarrow LiV_2O_5 \tag{16}$$

In fact, this mechanism is more complex,[29] than this overall process as indicated by a slight OCV change from 3.4 to 3.2 V when 0.5 equivalent per mole of V_2O_5 has been obtained. This phenomenon has been interpreted as the formation of an intermediate phase according to

$$0.5Li + V_2O_5 \rightarrow 0.5LiV_2O_5{\cdot}V_2O_5 \tag{17}$$

followed by

$$0.5Li + 0.5LiV_2O_5.V_2O_5 \rightarrow LiV_2O_5 \tag{18}$$

The two additional voltage steps occurring, respectively, at 2.4 and 2.1 V are not presently known in detail and need further investigation.

3.3.3. Cell Construction and Performances

Much of the present work appears to be in the development of reserve cells, but Honeywell also realized some primary active designs using the improved electrolyte (2 *M* $LiAsF_6$ + 0.4 *M* $LiBF_4$ in methyl formate).

The most popular cell is the DD 30-Ah cell,[29] which is the same diameter as a D cell but twice the length. Taking into account the high initial discharge voltage of this system, the DD cell is intended to be equivalent to two aqueous electrolyte D cells in series.

The DD cell uses multiple flat electrodes of the appropriate dimension that are connected in parallel. The lithium anode current is collected by a central rod; pressed powder V_2O_5-based cathodes are isolated from the central rod by insulating rings and are press-fit to the cell case wall. Energy densities of 550–660 Wh/dm^3 have been obtained for this design at low rates to a cut-off voltage of 2 V. Thick button cells have also been made, but the most recent cell specifically designed for high-rate medical applications is of prismatic configuration.[29–31]

This cell has the following external dimensions: $4.1 \times 2.2 \times 1.1$ cm. The anode consists of two pieces of lithium laminated on either side of a nickel screen current collector. The cathode is made by pressing a mixture of V_2O_5 and graphite in a ratio of 90/10 on both sides of a stainless steel current collector. A few percent of binder, such as microthene powder, may also be pressed into the cathode. Forming pressure is usually around 700 kg/cm^2.

The electrodes are assembled in a stainless steel case using a microporous polypropylene separator. Multiple cathode plates and a continuous interleafed anode provide the required electrode area to meet the power requirements of the intended applications. The cover assembly is equipped with a glass-to-metal seal feedthrough and is laser welded to the stainless steel case. A small hole in the cover assembly is used for filling the cell with the electrolyte, which is introduced under vacuum before closure by a laser weld. The terminal pin is negative and the case is positive in polarity.

As a high stable voltage is required in these applications, the cell is balanced on the first equivalent with a stoichiometric capacity of 930 mAh (operation above 3 V), but a part of the second plateau occurring at 2.4 V may be used as a warning that the battery should be replaced if needed. In addition, the small voltage drop observed in the first step at 50% of its duration can also be used for the indication of cell mid-life. Available volumetric energy densities are thus reduced in this case to 250–260 Wh/dm^3.

Figure 11 shows a typical discharge curve obtained under pulsing duty through 2-A, 20-sec constant current pulses corresponding to the automatic defibrillator application. Typical characteristics of the cell are:

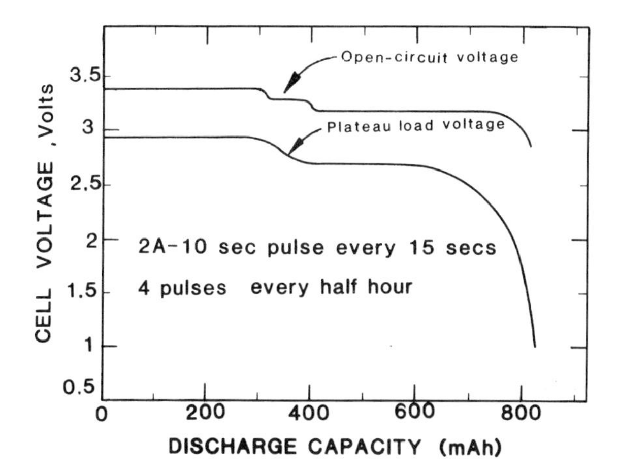

FIGURE 11. Pulse Discharge Characteristics for Honeywell Li/V_2O_5 G 3061 B Prismatic Cell at 37°C (After Ref. 31).

Capacity recovered: 0.805 Ah
Coulombic efficiency: 87%
Current density: 45 mA/cm^2
Power density: 200 W/kg (610 Wh/dm^3)
Energy density: 80 Wh/kg (240 Wh/dm^3)

About 150 pulses have been delivered in these conditions.

No storage experiments after extended periods of time have been reported, but based on results obtained with thick button cells, which use basically the same electrochemistry, little cell capacity deterioration is expected after two year storage at ambient temperature.

No danger with these high-rate cells has been indicated when they were subjected to shorting: None vented or leaked, in spite of peak current and temperature reaching values of 12 A ($\simeq$270 mA/cm^2) and 93°C, respectively, in nonadiabatic conditions.

3.4. The Lithium–Manganese Dioxide Cell

A lot of work has been done with this battery system over the last 10 years.[32–36] The results, however, were not very attractive because of a reported capacity equivalence significantly less than 1 faraday per mole of MnO_2[32,34] and loss of capacity and gassing on storage[33] until an improvement was achieved by Ikeda *et al.*[35] through a suitable heat treatment of synthetic active forms of MnO_2. The best combination and storage stability were obtained when MnO_2 was heated at about 330–400°C and transformed from the gamma phase to a mixed beta phase containing some gamma phase.

The cell discharge process has been studied by Ikeda[35] and co-workers, who proposed the following overall mechanism:

$$Li + MnO_2 \rightarrow MnO_2\ (Li) \quad (19)$$

indicating that lithium cations diffuse into the MnO_2 lattice upon discharge. The cells exhibit an OCV of 3.6 V and provide most of their service life in the 2.7–3-V range, depending on the discharge rate. Various electrolytes have been tested, but the preferred one is 1 *M* LiClO∈4 using the mixture of PC–DME.

Commercial cells in flat button cell types and cylindrical configurations are presently available. For medical applications some modifications of the present system have been brought by Gerbier and Lehmann[36] for adapting it to pacing in terms of improving energy density and cell reliability. These cells used a slightly different electrochemistry. The design details such as exact cathode material and electrolyte composition have not been clearly indicated.

Since these authors reported the utilization of a highly dense pure MnO_2 form, it is believed that the cathode material is prepared by thermal decomposition of manganese nitrate at temperatures ranging between 150 and 200°C, as recently mentioned.[37] Moreover, a 1 *M* $LiClO_4$ solution in pure propylene carbonate is probably the electrolyte.

A rectangular glass-to-metal sealed cell having the dimensions $4.5 \times 2.3 \times 0.9$ cm, for a total volume of 8.1 cm^3, has been designed with a nominal capacity of about 2.1 Ah. About 630 Wh/dm^3 is obtained at the three-month rate but 750–770 Wh/dm^3 is projected at the pacemaker rate. Typical discharge curves for this cell at various discharge rates and 37°C are reported in Figure 12. No long-term shelf-life experiments have been reported with this system.

3.5. Solid Electrolyte Lithium Cells

The search for highly Li^+-ion-conducting materials at ambient temperature with a high decomposition voltage has not been successful up to now, limiting the use of Li/metal salts solid electrolyte battery systems to applications that do not require high or moderate discharge rate, but only low rate and reasonably good specific energy. One such battery was the heart pacemaker battery developed and manufactured at P. R. Mallory (Duracell), which used lithium iodide in the solid electrolyte.

3.5.1. *Lithium-Iodide-Based Electrolytes*

Among the various existing materials, lithium halides possess high decomposition voltages (2.8–4 V), but only lithium iodide, which provides the highest conductivity for this group (about $4 \times 10^{-7}\ \Omega^{-1}\ cm^{-1}$ at 28°C), has been found

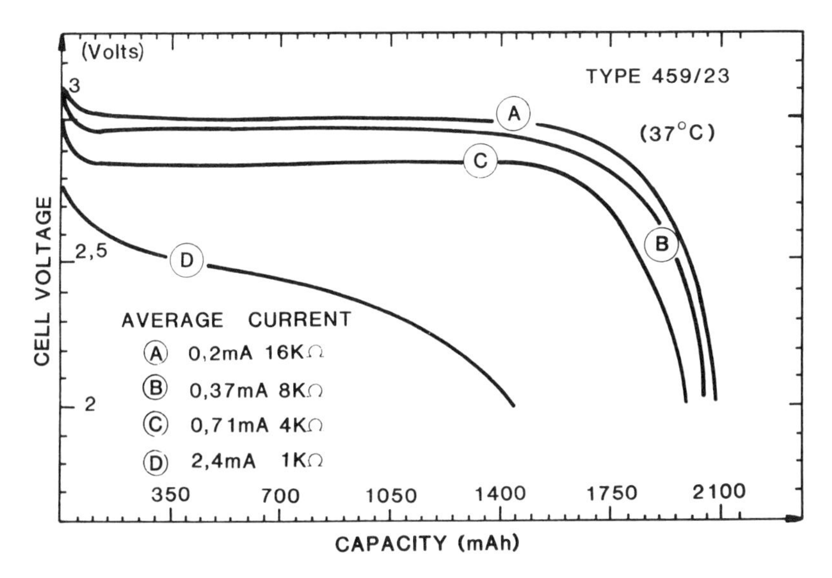

FIGURE 12. Discharge Characteristics for Celsa Li/MnO_2 D-Shaped Cells (459/23) at 37°C Through Different Loads (From Celsa Janvier, 1983).

of interest in battery systems. Some attempts for increasing this conductivity by doping with CaI_2 have been undertaken by Schlaikjer and Liang[38]; the maximum conductivity at 28°C that they obtained was $1.7 \times 10^{-5}\ \Omega^{-1}\ cm^{-1}$ for LiI containing 2 mol % of CaI_2. Nevertheless, this enhanced conductivity was not sustained on aging, and this was attributed to a slow precipitation of CaI_2 from the metastable solid solution to a stable solution containing only 0.23 mol % of CaI_2. Subsequent studies by Liang[39,40] on the conductivity of LiI containing a dispersion of insulating submicron-sized Al_2O_3 particles, a solid-state two-phase mechanical mixture, demonstrated significant improvements. Liang reported an enhancement in the conductivity of LiI by about a factor of 500 at 25°C by the dispersion of 40–50% Al_2O_3 corresponding to a stable ionic conductivity greater than $10^{-5}\ \Omega^{-1}$ cm⁻;1.

Owens and Hanson[41] extended this work to SiO_2 in LiI and observed similar behavior. They reported that the presence of a small amount of water introduced by the dispersing phase could be partially responsible for the conductivity enhancement, but Jow and Wagner[42] suggested that fine Al_2O_3 particles may possess a surface charge that when dispersed in another matrix, such as LiI, provides excess compensating defects in the immediate neighborhood of the dispersoid and thereby enhances the surface ionic conductivity of the halide.

Another important feature of Al_2O_3 and other dispersoids is that they do not appreciably change the electronic conductivity of the halide, which remains negligible, as reported by Liang *et al.*,[43] with the LiI (Al_2O_3) dispersed phase solid electrolytes, thus justifying the choice of this combination made some years ago by Mallory (Duracell) for building all solid state lithium battery systems.

3.5.2. *Lithium–Metal Salt Cells*

The utilization of different transition metal salts such as PbI_2, PbS, and TiS_2 as cathode materials in all solid-state lithium batteries and the use of $LiI(Al_2O_3)$-based electrolytes, was reported by Liang and co-authors in various papers.[44–46] In these systems, the three essential components of the battery—namely, the anode, the solid cathode, and the solid electrolyte that separate them were made separately and then brought together by co-pressing them generally at a somewhat higher pressure ($\approx$100K psi) than that used for forming the individual wafer components. The cells were hermetically packaged in steel battery housings with a glass-to-metal seal current feedthrough. Moreover, the contact between the different battery components during discharge was maintained by spring-loading. The first system to be developed was the Li/PbI_2 couple.[43] The cathode of the cell consisted of a mixture of 40% PbI_2, 30% LiI (Al_2O_3), and 30% Pb, by weight. Pb and LiI (Al_2O_3) were added to facilitate, respectively, the electronic and ionic conductivities in the cathode. The reported cell reaction is:

$$2Li + PbI_2 \rightarrow 2LiI + Pb \tag{20}$$

which, according to the thermodynamics, is expected to generate an OCV of 1.9 V (Table I). This prediction is in excellent agreement with the observed values. Practical batteries of this system were found to deliver the relatively low energy density of about 200 Wh/dm^3 at room temperature and low rate.

In an effort to increase the energy density of the original Li/PbI_2 cells, a part of the Pb and electrolyte in the cathode were replaced by PbS. Such a modification was found to improve substantially the volumetric energy density up to 500 Wh/dm^3 without changing the open-circuit voltage.

The most successful cathode composition to date was a mixture of 40% PbI_2, 40% PbS, 15% Pb, and 15% electrolyte, as described by Liang *et al.*[44] The reported cell reactions are:

$$2Li + PbI_2 \rightarrow 2LiI + Pb \ (\text{OCV} = 1.9 \text{ V}) \tag{21}$$

$$2Li + PbS \rightarrow Li_2S + Pb \ (\text{OCV} = 1.8 \text{ V}) \tag{22}$$

Single cells having a stoichiometric capacity of 0.136 Ah have been designed for pacemaker applications. Figure 13 shows typical discharge curves obtained for such cells at 37°C under various constant currents ranging from 10 to 1000 times the pacemaker rate. Because of the lack of rate capability the total cell

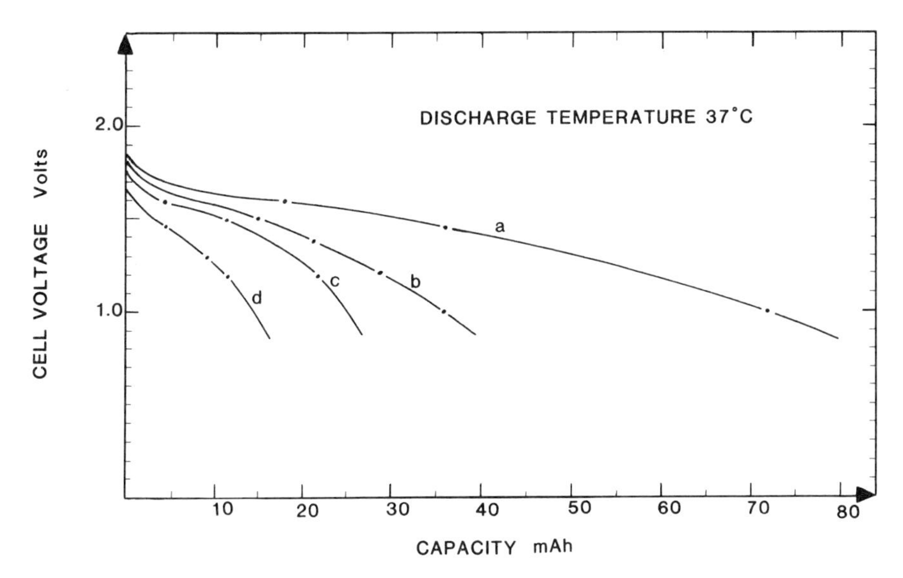

FIGURE 13. Discharge Characteristics for the 0.136 Ah Li/PbI_2, PbS Solid Electrolyte Mallory Cell at 37°C Under Various Constant Currents. Curve a: 18 μA; Curve b: 36 μA; Curve c: 36 μA; Curve d: 90 μA. (After Ref. 45, Reprinted by Permission of the Publisher, The Electrochemical Society, Inc.)

capacity was not recovered under these conditions. Thus, as an alternative, accelerated discharge tests were conducted according to duty cycles run at temperatures (≃100°C) and rates (≃200 μA) higher than 37°C and pacemaker rates to determine whether or not the stoichiometric capacity of the 0.136-Ah cells could be realized and to determine the load voltage at various depths of discharge at the pacemaker rate. Results obtained under these conditions demonstrated that after a recovered capacity of 0.13 Ah the load voltage under the pacemaker rate was still in excess of 1.3 V at 37°C, strongly suggesting the possibility of achieving such a capacity under pacemaker discharge conditions.

The storage capability of this system has been studied using an experimental 33-mAh test cell. No capacity loss has been reported after storage periods of up to two years at room temperature, or one year at 45 and 60°C, demonstrating that the Li/LiI(AL_2O_3)/PbI_2, PbS, Pb batteries are stable and have a potentially long shelf life.

In units used clinically, the battery contained seven parallel-stacks of three cells in series, having a total stochiometric capacity of 0.95 Ah and an open-circuit voltage of about 6 V. The dimensions of the packaged battery were approximately 20 mm OD × 50 mm long.

A second system for the pacemaker application developed at P. R. Mallory and Co. was based on titanium disulfide as the solid cathode.[46] However, as the practical volumetric energy density of batteries using TiS_2 would not be higher than the energy density of the previously reported Li/LiI(AL_2O_3)/PbI_2, PbS, Pb solid cells, a mixture of TiS_2 and sulfur was used to increase the specific capacity of the cathode. Although sulfur by itself cannot be discharged in a solid cell because of its poor conducting properties, it was found that it could be discharged at low rates when mixed with a sufficient amount of TiS_2, which has some electronic conductivity.

The reported cell reactions are in this case:

$$Li + TiS_2 \rightarrow LiTiS_2 \quad (23)$$

$$2Li + S \rightarrow Li_2S \quad (24)$$

The cell exhibited a measured open-circuit voltage of 2.45 V.

The construction of the cell was similar to PbI_2, PbS cell. Typically, the cathode formula has consisted of a mixture of 74% TiS_2, 21% S, and 5% electrolyte by weight. Energy densities of 900–1000 Wh/dm^3 were projected at the 2-μA pacemaker rate. Practical 0.38-Ah cells discharged at twice the pacemaker rate were reported to deliver approximately 48% of their capacity with a working voltage in excess of 2 V.

Limited versions of this pacemaker battery consisting of two parallel stacks of three cells in series were designed for Coratomic.[47]

4. USE OF LITHIUM SOLID CATHODE SYSTEMS IN IMPLANTED MEDICAL DEVICES

Table 5 summarizes the power needs for implantable devices. Five solid cathode systems have been used as implantable-device batteries. These cathode systems are Ag_2CrO_4, CuS, V_2O_5, MnO_2, and the mixture of PbI_2 with PbS.

TABLE 5 Power Needs for Implantable Devices

Devices	Voltage (V)	Stand-by Drain (μA)	Pulse Drain	Desired Longevities
Pacemakers	2–10	≃5	10 to 100 μA	4–10 years
Nerve stimulators	1–15	≃5	Several mA	Several years
Releasers (pumps)	>3	Several tens of μA	Several tens of mA	2–5 years
Defibrillators	3	≃20	≃2 A	2–5 years
Bone healers	1.5/3	20	—	A few weeks

No safety problems have been reported. However, there have been some design-related distinctions in the area of predictability of performance.

4.1. Lithium–Silver Chromate

Lithium–silver chromate was initially the most widely used lithium solid cathode system for implantable devices. The Li/Ag_2CrO_4 cell interested many implantable device manufacturers for several reasons. The Li/Ag_2CrO_4 cells show a two-step voltage discharge profile at 3.1 V and 2.6–2.4 V. This offers a convenient end-of-life indication.

The cells exhibit high energy densities (in excess of 700 Wh/dm^3 on the basis of volume). This is in part due to the very low self-discharge characteristic, below 0.5% per year, according to microcalorimetric studies. Because of the low internal impedance of this system on a high current drain scale, the cells can deliver up to several milliamperes of current.

The Li/Ag_2CrO_4 cell chemistry is compatible with titanium. This means that cells can be made with titanium cases, which induces a weight saving in comparison with other couples such as lithium-iodine which are not compatible with titanium. Furthermore, since implantable devices use titanium as the biocompatible container, it was possible to eliminate the separate battery case to allow pacemakers to weigh up to 10 g less or to increase longevity by 20–30% while having the same volume.

Two parameters of special interest to describe the behavior of power sources for implanted heart pacers are reliability (i.e., the absence of any abrupt voltage decrease) and longevity (i.e., the absence of progressive cell performance degradation).

From December 1974 to December 1982, over 8000 Li/Ag_2CrO_4 cells were placed under long-term observation (tested over several years) at body temperature. So far, not a single catastrophic failure has occurred, which means no leakage, no internal connection break, no spotweld failure, no separator breakdown, and no corrosion process. Over 470,000 Li/Ag_2CrO_4 cells were manufactured by the end of 1982. None of these cells exhibited any catastrophic failure due to poor quality. Calculations made six years after the beginning of the production of these cells showed 4.084 million implanted pacemaker months with no reported failures.

Confidence Level	Failure Rate per Month (%)
90%	0.00006
99.9%	0.00017

Cell longevity performances depend entirely on the longevities of the devices that contain them. For instance, over 50% of the implanted pacemakers do not work under the same conditions. Each device manufacturer adopts his own cutoff voltage, thus taking advantage (or not) of the second voltage plateau capacity. Modern medical devices are programmable, which means the drain delivered by the cells can be very different. The pacemaker lead impedance can vary from one patient to another, which again has the same consequence. A premature pacer depletion does not necessarily mean a bad cell discharge efficiency.

The cell manufacturers cannot discharge their control cells according to the discharge characteristics of all the circuits of their customers. Consequently, they have all decided to discharge their cells on fixed loads, which for some cells, such as Li/Ag_2CrO_4, leads to efficiency under evaluations. Today, with the same cell model, longevities can vary from three years to seven years or more.

4.2. Lithium–Cupric Sulfide

The American pacemaker manufacturer Cordis manufactures Li/CuS cells for their own use. Over 300,000 Li/CuS cells have been manufactured for pacemakers.

The Li/CuS cells present potential interest for several reasons, similar to the Li/Ag_2CrO_4 cells. The Li/CuS cells show a two-step voltage discharge profile at 2.2 V and then at 1.7 V, which allows a convenient end-of-life indication. The cells have a low self-discharge according to microcalorimetric tests and therefore offer good energy densities (750 Wh/dm^3). These cells also can deliver currents of up to several milliamps because of the low internal impedance, and there have been no safety problems.

Again, no catastrophic failures have been reported in cardiology magazines or reliability survey reports, which illustrates the very good reliability of the Li/CuS cells. All lithium-cell-powered pacemaker types have experienced premature depletions, but unsatisfactory battery discharge efficiencies have not necessarily been responsible for these premature voltage drops. The Li/CuS cell reliability records are good, as indicated in publications of the journal *PACE* reporting the data of eight North American hospitals and in longevity data published by a transtelephone service company that monitors over 10,000 pacemakers in the United States.

4.3. Lithium–Vanadium Pentoxide

The main significance of lithium–vanadium pentoxide cells is their very high drain capability. Their main drawback is their inability to work for a long period of time. These two facts explain why Li/V_2O_5 cells, manufactured by Honeywell, have only been used to manufacture a small number of automatic implantable defibrillators, which require cells able to deliver current drains as high as 2 A.

4.4. Lithium–Manganese Dioxide

This system is the most popular lithium system in consumer products such as watches, calculators, or photographic devices. It is often used for long-term applications such as memory back-up in telephone sets. Li/MnO_2 cells have also been offered by one French manufacturer, Celsa, to pacemakers manufacturers, and Li/MnO_2 cells have been implanted since 1982 in a few pacemaker models.

The Li/MnO_2 system has been widely studied and is in commercial use. It is a 3 V system with flat voltage profile and gradual voltage drop at the end. Similarly, it has a relatively low internal impedance and can deliver currents up to several milliamperes. The cell has a low self-discharge and energy densities near 700 Wh/dm^3.

4.5. Lithium–Lead Iodide, Lead Sulfide

Lithium–lead iodide, lead sulfide solid electrolyte cells for pacemakers were manufactured by P. R. Mallory (Duracell) for a few years. As seen before, cylindrical batteries containing three series-connected groups of seven electrochemical stages were used for this application (5.7 V – 0.91 Ah), each pacemaker containing two redundant hermetically sealed batteries.

Elemental cells were found to demonstrate slight internal volume change during discharge and consequently required an internal spring construction.[49] The batteries, composed of 21 spring-loaded cells in parallel sets of three groups in series, were difficult to produce with a high degree of reliability when compared with the other systems.

5. SUMMARY AND CONCLUSIONS

The relevant properties of available lithium solid cathode batteries using either organic or solid Li^+-based electrolytes, were reviewed and examined in view of their practical use in implantable medical devices. The intensive research activities carried out in the field of new materials, for developing high-energy lithium solid cathode battery systems, have led to several different chemistries for these particular applications.

There exists a whole spectrum for possible cathode materials and associated electrolytes, as illustrated by the number of combinations investigated during the past 10–15 years; however, the requirements needed for clinical use have limited their number to some specific Li solid cathode systems. Further R&D in the search for new solid cathodes with improved capacity and energy densities are presently continuing. Some recently reported materials, such as copper oxyphosphate,[50] that have been studied for high-temperature organic electrolyte lithium batteries could find applications in implantable medical devices.

REFERENCES

1. A. N. Dey, *Thin Solid Film 43,* 131 (1977).
2. E. Peled, *Lithium Batteries* (J. P. Gabano, ed.), pp. 43–73 Academic Press, London (1983).
3. G. Eichinger and J. P. Besenhard, *J. Electroanal. Chem. 72,* 1 (1976).
4. A. N. Dey and M. B. L. Rao, *Extended Abstract* No. 53, Electrochemical Society Fall Meeting, Boston (1973).
5. A. N. Dey, U.S. Patents 3,658,592 (1972), 3,947,289 (1976).
6. G. Lehmann, T. Rassinoux, G. Gerbier, and J. P. Gabano, in: *Power Sources,* Vol. 4 (D. H. Collins, ed.) pp. 493, Oriel Press, Newcastle Upon-Tyne (1973).
7. G. Lehmann and J. P. Gabano, U.S. Patent 3,853,627 (1974).
8. J. P. Rivault, M. Broussely, *Lithium Batteries* (J. P. Gabano, ed.), pp. 245–261, Academic Press, London (1983); and R. Messina, J. Perichon and M. Broussely, *J. Electro Anal. Chem. 133* 115–123 (1982).
9. R. Messina, A. Tranchant, J. Perichon, A. Lecerf, and M. Broussely, *J. Power Sources 8,* 277–288 (1982); and M. Broussely, J. P. Rivault, and J. P. Gabano, International Meeting on Lithium Batteries, Rome, 1982, *J. Power sources 9,* 339–344 (1983).
10. D. Herbert and J. Ulam, U.S. Patent 3,043,896 (1962).
11. J. P. Gabano, U.S. Patent, 3,542,601 (1970).
12. J. P. Gabano and G. Gerbier, U.S. Patent 3,511,716 (1970).
13. J. P. Gabano, G. Gerbier, and J. F. Laurent, *Proceedings of the 23rd Power Sources Symposium, 80,* PSC, Publications Committee, Red Bank, New Jersey (1969).
14. K. G. Wuttke, U.S. Patent 3,884,723 (1975).
15. B. H. Garth, U.S. Patent 3,788,310 (1973).
16. B. H. Garth, U.S. Patent 4,071,665 (1978).
17. A. M. Bredland, T. G. Messing, and J. W. Paulsen, *Proceedings of the 29th Power Sources Symposium, 82,* The Electrochemical Society, Pennington, New Jersey (1980).
18. M. R. Kegelman, U.S. Patent 3,847,647 (1974).
19. J. P. Gabano, V. Dechenaux, G. Gerbier, and J. Jammet, *J. Electrochem. Soc. 119,* 459 (1972).
20. "DuPont Electrochemical Systems" data sheet issued by E. I. DuPont de Nemours and Co., Wilmington, Delaware.
21. B. C. Bergum, A. M. Bredland, and T. J. Messing, *Progress in Batteries and Solar Cells, 3,* 90 (1980).
22. A. J. Cuesta and D. O. Bump, *Proceedings of the Symposium on Power Sources for Biomedical Implantable Applications and Ambient Temperature Lithium Batteries, 95,* (B. B. Owens and N. Margalit, ed.), The Electrochemical Society, Princeton, New Jersey (1980).
23. Cordis Corporation, The Cordis Gamma Lithium Battery, Miami, Florida (1980).
24. R. J. Horning and F. W. Rhoback, *Progress in Batteries and Solar Cells,* 4, 97 (1982).
25. A. N. Dey, in Extended Abstract, No. 54, Electrochemical Society Fall Meeting, Boston (1973).
26. J. L. Russel, U.S. Patent 3,985,577 (1976).
27. S. G. Abens, U.S. Patent 3,918,988 (1975).
28. W. B. Ebner and C. R. Walk, *Proceedings of the 27th Power Sources Symposium,* Publications Committee, Red Bank, New Jersey (1976).
29. C. R. Walk, *Lithium Batteries* (J. P. Gabano, ed.) pp. 265–279, Academic Press, London (1983).

30. M. S. Whittingham, *J. Electrochem. Soc. 123,* 315 (1976).
31. R. J. Horning and S. Viswanathan, *Proceedings of 29th Power Sources Symposium, 64,* The Electrochemical Society, Pennington, New Jersey (1980).
32. H. F. Hunger and G. H. Heymach, *J. Electrochem. Soc. 120,* 1161 (1973).
33. F. W. Dampier, *J. Electrochem. Soc. 121,* 656 (1974).
34. A. N. Dey, Extended Abstract No. 54, Electrochemical Society Fall Meeting, Boston (1973).
35. H. Ikeda, T. Saito, and H. Tamura, *Manganese Dioxide Symposium,* Vol. I, pp. 384, IC Sample Office, Cleveland (1975).
36. G. Gerbier and G. Lehmann, *Proceedings of the Symposium on Power Sources for Biomedical Implantable Application and Ambient Temperature Lithium Batteries, 95* (B. B. Owens and N. Margalit, ed.), The Electrochemical Society, Princeton, New Jersey (1980).
37. G. Gerbier, BF 24.66.872 (1979).
38. C. R. Schlaikjer and C. C. Liang, *J. Electrochem. Soc. 118,* 1147 (1971).
39. C. C. Liang, *J. Electrochem. Soc. 120,* 1289 (1973).
40. C. C. Liang, U.S. Patent 3,713,897 (1973).
41. B. B. Owens and H. J. Hansen, U.S. Patent 4,007,122 (1977).
42. T. Jow and J. B. Wagner, Jr., *J. Electrochem. Soc. 126,* 1963 (1979).
43. C. C. Liang, A. V. Joshi, and N. E. Hamilton, *J. Applied Electrochem. 8,* 445 (1978).
44. C. C. Liang, A. V. Joshi, and L. H. Barnette, *Proceedings of the 27th Power Sources Symposium, 141,* PSC Publications Committee, Red Bank, New Jersey (1976).
45. C. C. Liang, and L. H. Barnette, *J. Electrochem. Soc. 123,* 453 (1976).
46. J. R. Rea, L. H. Barnette, C. C. Liang, and A. V. Joshi, *Proceedings of the Symposium on Power Sources for Biomedical Implantable Applications and Ambient Temperature Lithium Batteries, 245,* (B. B. Owens and N. Margalit, ed.) The Electrochemical Society, Princeton, New Jersey (1980).
47. Ovalith 13 Pacemaker Coratomic 1979, product literature.
48. D. S. Schechter, *J. Med. 72* 1166 (1972).
49. Solid State Battery Systems, Mallory (1975).
50. M. Broussely, A. Lecerf, and J. P. Gabano, *13th Power Sources International Symposium,* Brighton, England, September (1982).

8

Lithium–Liquid Oxidant Batteries

PAUL M. SKARSTAD

1. INTRODUCTION

A battery used to power an implanted medical device must be safe and reliable. The deliverable capacity and energy at the intended application rate must be known or predicted with a suitable degree of confidence. As an added precaution, because of the possibility of unpredicted time-dependent losses and because the load may vary, depending on the needs of the individual patient, there should be a positive warning of approaching depletion in order to allow timely replacement. Finally, the capacity and energy delivered under application conditions should be the highest possible, consistent with the requirements of safety, reliability, predictability, and depletion warning.

At this time the cardiac pacemaker has accounted for the vast majority of battery-powered implanted devices. The power requirements of the cardiac pacemaker have been met by at least eight lithium battery chemistries.(1) Of these the most widely and frequently used at the present time is the cell system $Li/LiI/I_2$.(2) Batteries of this type have demonstrated reliabilities comparable to those reported for silicon transistors and diodes.(3) In addition, for this system, predictions of deliverable capacity based on empirical modeling have been found to agree with the results of discharge experiments at simulated application discharge rates to within a few months during tests of as long as eight years.(3) In optimized designs delivered capacities are expected to be as much as 0.9 Wh/cm^3. However, discharge at higher rates leads to a significant decline in capacity and energy density.

A variety of implantable medical devices is under development, with sig-

PAUL M. SKARSTAD • Medtronic, Inc., Minneapolis, Minnesota 55440.

nificantly higher mean and peak power requirements than pacemakers. Examples are shown in Table 1. An application discharge for these devices typically consists of a series of higher power pulses superimposed upon a background level of a few microwatts. Pulse requirements range from about 10^{-2} W to a few watts and last up to a few seconds.

The lithium/thionylchloride cell[4] has been investigated for such applications because it offers high power capability with high energy density. This system was introduced as a pacemaker power source in 1974 by ARCO. It exhibited excellent performance in accelerated tests and initially in clinical application.[5] However, the deliverable capacity at application discharge rate turned out to be significantly lower than predicted.[6] This, coupled with the lack of a positive depletion warning, led to sudden, unexpected depletion of the batteries and occasioned the recommendation of replacement of many implanted units.[6]

Since that unfortunate occurrence, questions concerning the predictability of performance, plus the availability of highly satisfactory alternatives for the lower rates of discharge have limited the application of lithium/thionylchloride batteries in implantable devices.

It should be emphasized that the reliability reported for the lithium/thionylchloride cells used by ARCO prior to depletion was excellent.[5,6] Thus, the failure of these batteries to perform as expected came about not because of an inherent instability in the system, but because of inadequate characterization of rate-dependent losses.

In the decade since these lithium/thionylchloride batteries were introduced as pacemaker power sources, considerable progress has been made in the characterization of the system. In particular, the wide availability of calorimeters with sensitivities on the order of a microwatt has revolutionized the assessment of parasitic losses. Moreover, 10 additional years of discharge data have been obtained over a range of discharge rates on cells of varying designs by several manufacturers. Thus, it is now possible to estimate the performance of a lithium/thionylchloride battery with a given set of design parameters with a high degree of confidence. Because of this it is reasonable to consider once again the use of thionylchloride cells for implantable applications.

TABLE 1 Device Power Requirements

Device	Peak Power Requirement	Device Lifetime (years)
Pacemaker	50 μW	5–10
Drug pump	10 mW	3–5
Neurological stimulator	10 mW	3–5
Tachyrhythmia control	0.5 W	3
Defibrillation	10 W	3

Several excellent recent reviews of the lithium/thionylchloride system and its chemistry exist.[7–9] This article will focus upon issues that are of particular importance in the development of the lithium/thionylchloride system for implantable devices. These include volumetric energy density, depletion warning, voltage delay, and safety.

High energy density is important for implantable medical applications because of the need to minimize the volume and maximize the lifetime of the implantable device. Small medical lithium/thionylchloride batteries are reported to deliver as much as 1.2 Wh/cm^3, based on the external volume of the battery case. This high energy density arises from a number of factors. The chemical energy density for the balanced lithium/thionylchloride reactants is 2.0 Wh/cm^3. In addition, the system lends itself to the design of batteries that convert chemical energy to electrical energy efficiently. Overall volumetric packaging efficiencies of carefully designed cells are about 65%. Electrochemical efficiencies under optimum discharge conditions are as high as 90%, leading to an overall volumetric coulombic efficiency of about 60%. We shall examine in some detail how these efficiency factors come about in Sections 3.4 and 3.5.

Particular attention is given to the calorimetric estimation of efficiency at the lower rates characteristic of background power levels of devices with intermittent high-power loads. The intermittent nature of the high rate discharge superposed on a low but nonzero background discharge rate raises unusual and formidable problems in the assessment of voltage delay as well as in the estimation of electrochemical efficiency. The history of lithium/thionylchloride use in pacemakers demonstrates the importance of a positive warning of impending battery depletion. Finally, several design issues converge in ensuring the safety of small lithium/thionylchloride batteries in implantable medical applications.

2. DESCRIPTION OF THE SYSTEM

2.1. Liquid Oxidant Systems

The lithium/thionylchloride cell belongs to a family of liquid oxidant cell systems in which the reductant component (lithium) and the oxidant component (thionylchloride solution) are placed initially in direct contact with one another. No separator element needs to be included between the cell reactants when the cell is constructed. Other members of this family that have been developed include lithium/sulfur dioxide, lithium/thionylchloride with bromine chloride dissolved in the catholyte, lithium/sulfurylcholoride, and lithium/sulfurylchloride with chlorine dissolved in the catholyte. Of these only the thionylchloride-based systems have been developed for medical applications. Lithium/iodine cells, often used in low-rate medical applications, in which the conductive cathode phase is a polyiodide liquid,[10] can also be considered members of this family.

This system is treated in detail in the lithium/halogen chapter of this book. Fundamental differences in the mode of charge transport within the cathodic reactant medium account for the vast difference in the performance characteristics of thionylchloride-type and iodine-type cells.

A chemical reaction initiates on the direct contact of the two reactive components. This reaction results in the formation of a layer of insoluble material on the surface of the anode. Growth of this layer slows the rate of chemical reaction and allows the reactants to exist together in a battery. In addition to separating the reactants physically, this layer must allow the transport of at least one ion involved in the cell reaction, and it must insulate the reactants from each other electronically. In other words, the layer of insoluble chemical reaction product must act as a solid electrolyte.[11] For this reason the layer has been termed the *solid electrolyte interphase,* or SEI.[12]

In the case of the lithium/iodine cell, the discharge reaction also contributes to the growth of this layer and is the main source of increasing cell impedance. For the liquid oxyhalide and oxide cell systems noted previously, including lithium/thionylchloride, however, the SEI does not grow as the cell is discharged; in fact, the SEI may diminish as the cell begins to discharge and then grow again if discharge is interrupted. This dynamic character of the SEI allows these systems to have high rate capability and yet a high degree of kinetic stability. The chemical constitution of the SEI may vary from system to system.[9] For a given system such as lithium/thionylchloride, the morphology and kinetic parameters depend on the electrolyte salt used, the concentration of the salt, impurities present, and anode treatments such as polymeric coatings.

2.2. Cell Reaction

Considerable disagreement has surrounded the identification of the cell reaction in the lithium/thionylchloride cell.[13–16] The overall reaction that has been identified as the most probable under ordinary discharge conditions is the following:[13]

$$2Li + SOCl_2 = 2LiCl + \tfrac{1}{2}S + \tfrac{1}{2}SO_2 \qquad (1)$$

The mechanism and rate of formation of SO_2 have likewise been matters of disagreement. The formation of reaction intermediates has been proposed.[17,18] Chemical analysis,[19,20] cyclic voltammetry, and various spectroscopic measurements[21–23] have suggested the existence of, but have not definitively identified, such reaction intermediates.

In order for thionylchloride to be reduced electrochemically on a continuing basis in a discharging cell, the thionylchloride phase must transport electrical charge. Pure thionylchloride is a poor conductor; however, it is sufficiently polar to act as an electrolytic solvent for a number of lithium salts, notably those with complex anions, yielding solutions with conductivities greater than 10^{-2} s/cm.[24] Although many salts have been studied, lithium tetrachloroaluminate in the concentration range 1–2 *M* is most often used to make thionylchloride conductive for use in a cell. Thus, the active oxidizing component of the lithium/thionylchloride cell is also the solvent of an electrolytic solution. For this reason the thionylchloride solution in a battery is often termed the electrolyte; because it is also the cathode-active component of the cell, it is sometimes termed the *catholyte*.

Lithium/thionylchloride cells in which interhalogens have been dissolved have also been described.[25–27] One such system containing BrCl is available commercially in a medical-grade battery from Wilson Greatbatch Ltd. under the name BCX. Preferential discharge of chlorine is consistent with the initial high voltage observed in the discharge curves of these cells.[27] It is not reported whether the bromine co-discharges to any appreciable extent with the chloride to form a solid-solution discharge product. The bromine chloride is reported to have beneficial effects both on voltage delay and on cell safety.[27] Experimental cells with the interhalogen ICl_3 dissolved in thionylchloride electrolyte yield LiCl,S, and I_2 as solid discharge products.[28]

2.3. Principles of Operation

The discharge products of the lithium/thionylchloride cell reaction consist in part of the highly resistive solids LiCl and S; yet the salient operating characteristic of the lithium/thionylchloride system is its high-rate capability throughout the discharge life of a battery. This is in marked contrast to the lithium/iodine cell system, widely used in low-rate medical applications. In this system the discharge product, lithium iodide, while still a poor conductor, is a significantly better conductor than the thionylchloride discharge products. Yet the impedance of a lithium/iodine cell may rise above that of a lithium/thionylchloride cell by a factor of 1000 during the course of typical discharge. It is important to understand how this difference in operating characteristics between these two similarly constructed cell types arises.

The critical difference lies in the disposition of the discharge product. In the case of lithium/iodine cells, the discharge product is deposited over the limited surface area of the lithium anode. The discharge product forms a more or less dense, increasing layer in series with the other components of the cell. In lithium/thionylchloride cells the discharge product is deposited within a high-surface-area porous carbon cathode element. Because of the high surface area over which the high-resistivity discharge product forms, the overall resistance remains low.

The difference in the site of formation of the discharge product is related to the difference in the way in which charge is passed through the conductive cathode medium. In lithium/thionylchloride cells the steady-state discharge current is passed through the thionylchloride solution via motion of lithium ions dissolved in the solution. Reduction occurs near the surface of the porous carbon element, and the lithium chloride discharge product is deposited at that point. In the lithium/iodine system there is no evidence for any significant level of lithium ion conduction or solubility of lithium ions in the conductive liquid polyiodide medium. As a result, the lithium iodide discharge product forms near the surface of the anode, increasing the thickness of the SEI and the overall cell impedance.

If the cathodic surface area in a thionylchloride cell were as limited as the anode surface area in a lithium/iodine cell, the result would be rapid passivation of the cathodic surface, and a cell capable of only very low rates of discharge. However, lithium ion transport within the thionylchloride solution and the use of the high-surface-area porous conductor for the cathode current collector allow the thionylchloride cell to maintain low impedance in spite of the high resistivity of the discharge products.

Experimentally, different carbon blacks are found to give different levels of performance.[15,29] Dibutylphthalate absorption has been found to correlate well with the capacity of a carbon.[30] The most commonly used carbon is Shawinigan acetylene black bonded with Teflon, although other carbons such as Ketjenblack EC have been identified as having excellent performance characteristics.[30]

In addition to accommodating the precipitation of the solid products of the cell reaction, the carbon element of a thionylchloride cell may catalyze the reduction of thionylchloride. Various carbons show various degrees of catalytic activity, as evidenced by rate capability or by the cathode efficiencies determined at high discharge rate.[31] Additives such as phthalocyanines[31] and platinum[32] have been shown to enhance the catalytic activity of carbons toward thionylchloride reduction, increasing the rate capability.

3. CAPACITY AND ENERGY DENSITY

3.1. Classification of Losses

In order to meet the capacity levels required by implantable devices, within the constraints of limited available volume and sufficient discharge time to justify implantation, careful attention must be given to factors that affect the volumetric efficiency of a particular battery chemistry or design. Losses in efficiency may be classified into packaging and electrochemical losses.

Packaging losses are those that arise because components other than the balanced cell reactants must be included to make a working cell. Examples include the battery case, current collectors including the carbon element of a thionylchloride cell, internal separators, conductivity additives such as $LiAlCl_4$, and incomplete filling of the available volume. In general, packaging losses can be determined on a volumetric or gravimetric basis. For any battery design these loss factors can be completely determined as part of the design process.

Electrochemical losses arise from incomplete electrochemical utilization of the limiting electrode component. In thionylchloride cells the most important source of electrochemical inefficiency is self-discharge by direct chemical combination of the cell reactants. Other sources of loss may be premature choking of the carbon element at high discharge rates or anode passivation, particularly if utilization is defined with respect to arbitrary load voltage levels. Electrochemical loss factors are determined as part of the evaluation process.

3.2. Stoichiometric Energy and Capacity Density

The stoichiometric capacity and free energy densities of the balanced reactants represent upper bounds of capacity and energy density for any battery couple. As such they present opportunities for comparison of different battery chemistries and serve as bench marks in the evaluation of practical battery designs. The limiting volumetric capacity density for a couple can be determined by simply adding the volumes per unit charge (according to the governing reaction) and inverting the sum. For reference later in this discussion, however, it is useful to consider the quantity as arising from two fundamental constraints, charge balance and volume additivity:

$$\text{Charge balance: } C_+ - C_- = 0 \tag{2}$$

$$\text{Volume additivity: } V_+ + V_- = V_{+-} \tag{3}$$

or, with $v_i = V_i/V_{+-}$, and $Q_i = C_i/V_{+-}$

$$Q_+ - Q_- = 0 \tag{4}$$

$$v_+ + v_- = 1 \tag{5}$$

where C_i and Q_i represent capacity(Ah) and capacity density (Ah/cm^3) elements of the ith component, respectively, and V_i and v_i represent the corresponding volume (cm^3) and volume fraction.

Let k_i be the specific capacity (Ah/g) of the ith component, m_i the corresponding mass per unit volume V_{+-}, and d_i the corresponding density. Substituting $Q_i = k_i\, m_i$ and $v_i = m_i/d_i$, and solving we get the capacity density and energy density of the balanced couple shown in Eqs. (6) and (7).

$$Q = \frac{d_+ d_- k_+ k_-}{d_+k_+ + d_-k_-} \quad (\text{Ah/cm}^3) \tag{6}$$

and

$$W = \int_0^1 E^0\,(x)\,(dQ/dx)\,dx \qquad (\text{Wh/cm}^3) \tag{7}$$

where the densities and specific capacities are for the pure reactants and represent the known or supposed stoichiometry of the reaction.

If either reactant is contained in a solution phase, the preceding equations hold, provided the density and specific capacity of the solution are used. If either electrode is a mixed-phase composite, such as a composite anode for depletion warning, or a cathode composed of solid interhalogen plus liquid oxyhalide (*e.g.*, ICl_3 and thionylchloride), then additional constraints must be added.

The extra constraints may be obtained by specifying the proportion of each phase. For example, for a composite anode consisting of lithium (L) and an intermetallic compound (I) in a Faradaic ratio of *n*, the third constraint is

$$Q_L = nQ_I \tag{8}$$

and thus

$$Q_- = Q_I(1 + n) \tag{9}$$

It may be useful to solve the linear system of Eqs. (10) for the mass of each component per unit of electrode volume V_{+-} (the m_1 values):

$$\begin{aligned} k_+m_+ - k_Lm_L - k_Im_I &= 0 \\ m_+/d_+ + m_L/d_L + m_I/d_I &= 1 \\ k_Lm_L - nk_Im_I &= 0 \end{aligned} \tag{10}$$

Specific capacities and other pertinent data for a number of anode and cathode materials are listed in Tables 2 and 3.[33] Balanced couple capacity and free energy densities and associated data obtained from these equations for lithium/-thionylchloride and a number of other important couples are listed in Table 4.

TABLE 2 Properties of Selected Anode Metals

Metal	Atomic Mass (g/mol)	*n* (eq/mol)	Density (g/cm^3)	Specific Capacity (Ah/g)	Capacity Density (Ah/cm^3)
Li	6.939	1	0.534	3.862	2.062
Na	22.99	1	0.97	1.166	1.131
Ag	107.87	1	10.5	0.248	2.609
Cu	63.55	1	8.92	0.422	3.762
Mg	24.31	2	1.745	2.205	3.848
Ca	40.08	2	1.54	1.337	2.060
Zn	65.38	2	7.14	0.820	5.854
Cd	112.41	2	8.642	0.477	4.121
Pb	207.19	2	11.343	0.259	2.935
Al	26.98	3	2.702	2.908	8.053

Source: From Weast.[33]

TABLE 3 Properties of Selected Cathode Materials[33]

Oxidant	Formula Weight (g/mol)	*n* (eq/mol)	Density (g/cm^3)	Specific Capacity (Ah/g)	Capacity Density (Ah/cm^3)
Oxyhalides					
$SOCl_2$	118.97	2	1.655	0.451	0.746
SO_2Cl_2	134.97	2	1.667	0.397	0.662
$POCl_3$	153.33	3	1.675	0.524	0.878
$SeOCl_2$	165.87	2	2.42	0.323	0.782
BCX[a]			1.75	0.407	0.712
CSC[b]			1.73	0.434	0.751
Halogens, Interhalogens					
$Br_2(l)$	159.81	2	3.12	0.335	1.047
$I_2(s)$	253.81	2	4.93	0.211	1.041
ICl(s)	162.36	1	3.24	0.165	0.535
$ICl_3(s)$	233.26	3	3.117	0.345	1.074
BrCl	115.36	2	–	0.465	–
γ-MnO_2	86.94	1	4.81	0.308	1.483
$SO_2(l)$	64.06	2	1.434	0.418	0.600
CuS	95.6	1	4.6	0.280	1.288
CuS	95.6	2	4.6	0.560	2.576

[a] BCX is a solution of 2.23 *M* (14 mol %) BrCl in $SOCl_2$ used in batteries made by Wilson Greatbatch Ltd. of Clarence, New York.

[b] CSC is a solution of 0.45 *M* Cl_2 in SO_2Cl_2 used in batteries made by Wilson Greatbatch Ltd. of Clarence, New York.

TABLE 4 Stoichiometric Capacity and Energy Densities of Selected Balanced Couples

Couple	Capacity Density (Ah/cm^3)	OCV (V)	Free Energy Density (Wh/cm^3)
$Li/SOCl_2$	0.548	3.65	2.00
Li/BCX[a]	0.529	3.65–3.9	1.9–2.1
$LiSO_2Cl_2$	0.501	3.92	1.96
Li/CSC[b]	0.550	>3.92	>2.15
$Li/POCL_3$	0.616	3.4	2.1
Li/ICl_3	0.706	3.65–3.9	2.56–2.75
Li/I_2	0.692	2.805	1.94
Li/Br_2	0.693	3.50	2.43
$Li/\gamma\text{-}MnO_2$	0.863	3.0	2.6
Li/CuS (1e-)	0.793	2.12,1.74	2.22

[a] BCX is a solution of 2.23 *M* (14 mol %) BrCl in $SOCl_2$ used in batteries made by Wilson Greatbatch Ltd, of Clarence, New York.
[b] CSC is a solution of 0.45 *M* Cl_2 in SO_2Cl_2 used in batteries made by Wilson Greatbatch Ltd, of Clarence, New York.

3.3. Capacity Density of Practical Electrodes

The calculation of capacity and energy density can be refined to include homogeneous conductivity additives by using the appropriate densities and specific capacities of the respective phases in the preceding systems of equations. Further refinement can allow inclusion of heterogeneous additives such as the carbon element of a thionylchloride cell or the electrolyte in the case of cells with a nonreacting electrolyte phase. Results of such calculations allow comparison of different battery chemistries on a more realistic basis than the simple balanced couple values shown in Table 5. For a given system such calculations allow the quantification of losses that may be governed by design or materials selection.

Many levels of sophistication are possible, including dynamic properties of the porous electrode in a liquid cathode cell such as thionylchloride.[34–36] Two simple approaches to the calculation of cell capacity density for a cell couple including a porous carbon cathode element are outlined here.

The first employs a pseudo-Faradaic equivalency for the carbon of the carbon element[37]:

$$q_c = k_c m_c \tag{11}$$

where the specific capacity of the carbon used varies with the porosity, the specific carbon used, mechanical constraints on the ability of the element to

TABLE 5 Stoichiometric Capacity Densities of Selected Couples of Practical Electrodes

Couple	Capacity Density (Ah/cm^3)	OCV (V)	Energy Density (Wh/cm^3)	Comments
$Li/SOCl_2$	0.52	3.65	1.9	80% porous
Li/BCX[a]	0.49	3.65–2.9	1.8–1.9	80% porous
Li/ICl_3 $SOCl_2$	0.68	3.65–3.9	2.5–2.65	80% porous
Li/I_2	0.62	2.80	1.7	5 wt % PVP
Li/CuS(1e-)	0.64	2.12	1.6	20% porous 5 wt % inert
Li/CuS(2e-)	0.98	2.12–1.74	2.0	20% porous 5 wt % inert
$Li/\gamma\text{-}MnO_2$	0.56	3.0	1.7	35% porous 10 wt % inert

[a] BCX is a solution of 2.23 *M* (14 mol %) BrCl in $SOCl_2$ used in batteries made by Wilson Greatbatch Ltd., of Clarence, New York.

swell to accommodate solid reaction product, and miscellaneous variables such as the discharge rate, temperature, and the nature and proportion of binder used. It is important to note that the carbon does not participate in the Faradaic reaction and that the "specific capacity" of the carbon is determined by the pore volume and the ability of the carbon element as fabricated and discharged to accommodate the solid discharge product. Adding more carbon to the porous carbon element at constant volume (i.e., decreasing the porosity) in general decreases rather than increases the capacity. The specific capacity of the carbon material can be determined from the discharge of carbon-limited cells.[38] Theoretical descriptions of the carbon capacity, giving a functional form with the variation of design or discharge parameters e.g., discharge rate[34] can be included conveniently with this approach.

For thionylchloride cells with Teflon-bound Shawinigan black elements of typical porosity (~85%) and capacity (~4 Ah/g), the volume of the thionylchloride catholyte required to balance the cell will be greater than the pore volume within the pores of the carbon element. Thus, for a balanced cell a reservoir volume of catholyte is needed outside the carbon element. The external reservoir also allows the carbon to remain completely immersed as the cell discharges if the cell is appropriately oriented.

It is convenient to include the reservoir volume explicitly as a variable, with the pore volume plus the reservoir volume v_R set equal to the volume of thionylchloride electrolyte. The system of equations then becomes in matrix form:

$$\begin{bmatrix} k_c & 0 & -k_L & 0 \\ k_c & -k_s & 0 & 0 \\ 1/d_c & 1/d_s & 1/d_L & 0 \\ P/[(1 - P)d_c] & -1/d_s & 0 & 1 \end{bmatrix} \begin{bmatrix} m_c \\ m_s \\ m_L \\ v_R \end{bmatrix} = \begin{bmatrix} 0 \\ 0 \\ 1 \\ 0 \end{bmatrix} \tag{12}$$

The masses m_I and the reservoir volume v_R are all expressed in units per unit volume of balanced electrode components, including in this case the carbon of the carbon element.

The second approach assumes implicitly that the maximum capacity of the cathode element is determined by the pore volume and the ability of the element to swell but does not express a carbon capacity explicitly. This approach is more generally applicable than the first but does not lend itself so conveniently to the detailed description of the behavior of the carbon element. As applied to the thionylchloride case, we assume that we have a carbon element of porosity P filled initially to the extent F with cathode-active material, where F is the fraction of pore volume filled. In this case we have the system of equations (13):

$$\begin{bmatrix} k_s & 0 & -k_L \\ 1/d_s & 1/d_c & 1/d_L \\ -1/d_s & PF/[(1 - P)d_c] & 0 \end{bmatrix} \begin{bmatrix} m_s \\ m_c \\ m_L \end{bmatrix} = \begin{bmatrix} 0 \\ 1 \\ 0 \end{bmatrix} \tag{13}$$

Figure 1 shows the stoichiometric thionylchloride electrolyte specific capacity as a function of $LiAlCl_4$ concentration, prepared from published density data.[24] Solving Eq. (12) for a Shawinigan black carbon element of 80.5% porosity with k_c taken to be 3.85 Ah/g, we calculate the stoichiometric capacity density of practical electrodes shown in Figure 2. For comparison the stoichiometric capacity density of the balanced reactants and the mean delivered capacities per unit of total electrode volume (V_{+-}) determined for cylindrical case-negative cells as described by Patel *et al.*[39] by discharge under 2-kΩ loads.[28]

Table 5 gives a comparison of capacity densities calculated using Eqs. (12) and (13) for a number of important lithium battery systems with reasonable practical values of the parameters F and P.

The systems of Eqs. (12) and (13) allow the exploration of the volumetric capacity density of lithium/thionylchloride and other battery systems as a function of various design parameters in a systematic way. In either system weighting factors w_i may be included in the charge balance equation to account for a desired electrode imbalance:

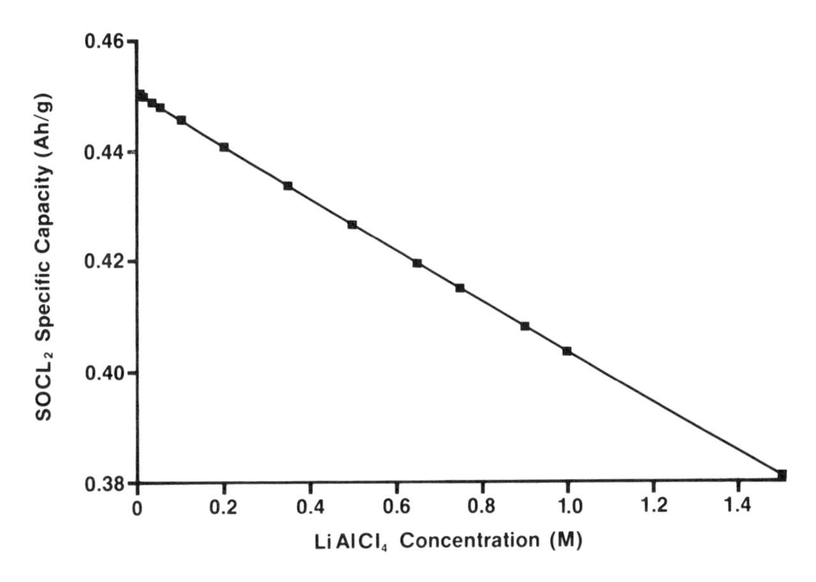

FIGURE 1. Variation in Stoichiometric Specific Capacity of Thionylchloride Electrolyte Solutions with $LiAlCl_4$ Concentration (Data from Ref. 24).

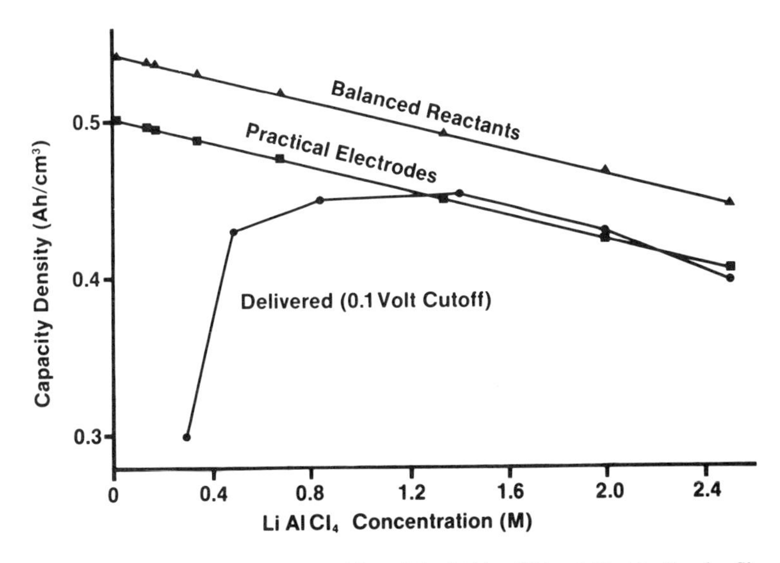

FIGURE 2. Stoichiometric Capacity Densities of the Lithium/Thionylchloride Couple, Showing Balanced Reactants and Practical Electrodes. Deliverable Capacity Densities of Balanced Lithium/Thionylchloride Electrodes Determined in Small (Anode Area, 7 cm^2) Case-Negative Cylindrical Cells (2 kohm Load to 0.5 V).

$$k_+ m_+ / w_+ \; - \; k_- m_- / w_- \; = \; 0 \tag{14}$$

The matrix form lends itself conveniently to computer solution with systematic variation of the parameters. This, in turn, should allow the identification and evaluation of packaging loss or efficiency factors leading to the achievement of less than the stoichiometric capacity density of the balanced couple.

These equations provide a framework for the empirical characterization of electrochemical performance. For any defined experimental situation the matrix elements of 12 or 13 determine the capacity of the cell, deliverable or theoretical. Thus, evaluating battery performance becomes in part a problem of devising experiments to determine these matrix elements, particularly the specific capacities for ranges of conditions of interest. The specific capacities in Eqs. (12) and (13) can in each case be replaced by the product of coulombic efficiency and the stoichiometric or maximum available capacity density. The connection of these systems of equations with cell discharge data is thus made. Analogous systems of constraining equations can be written for gravimetric capacity density. Because limited volume is the more important constraint under which implantable batteries are designed, only this approach is developed here.

The systems of equations developed in this section can play a unifying role in the evaluation of packaging efficiency factors in design and in the practical utilization of electrochemical efficiency factors. In addition, they may provide a framework of fundamental constraints on which to build more sophisticated models of the dynamic behavior of the system.

3.4. Packaging Efficiency

In any battery some components are included in addition to the balanced chemical reactants. Each of these additional components occupies some space and can be considered a source of volumetric or packaging loss. Examples include any excess electrode material and the battery container itself, plus current collectors, connectors, and separators.

In thionylchloride batteries in addition a soluble salt such as $LiAlCl_4$ is needed to make the thionylchloride conductive; carbon and binder are needed for the carbon element; anode coatings may be included in the design. While each of these components may be necessary or may enhance the electrochemical efficiency, each one occupies some volume and therefore contributes to a loss in the packaging efficiency of the battery.

As examples, Figures 1 and 2 show effects of the addition of the electrolyte salt on the stoichiometric specific capacity of $SOCl_2$-$LiAlCl_4$ solutions and on the capacity density of balanced electrodes. The effect of addition of carbon at 80.5% porosity and 3.85 Ah/g is indicated by the difference between the two stoichiometric lines in Figure 2.

The overall packaging efficiency can be factored into a product of individual efficiencies. These efficiency factors can be determined as part of the design process. As an example of a breakdown of the packaging efficiency for a medical lithium/thionylchloride cell, volumetric packaging factors obtained for cells described by Howard *et al.*[40] are shown in Table 6. The battery described by these authors has a D-shaped case 8 mm deep and 36 mm high with a semicircular end 27 mm in diameter. The electrolyte is 1.9 *M* $LiAlCl_4$ in thionylchloride. The electrodes are nearly balanced at 2.45 Ah.[40]

The losses in Table 6 are represented as factors. They are not in general orthogonal: A change in one may require recalculation of the entire set. They were obtained in various ways. The packaging hardware factor was obtained from helium pycnometer measurements of the external volumes of completed batteries and of the case, header, current collectors, and internal connectors. The filling factor was calculated from the available internal volume and the volumes determined by the masses and densities of the electroactive components: lithium and polymeric coating, carbon and binder, thionylchloride solution used to build the cells. The volume of glass fiber paper separators plus any electrode imbalance are included in this factor in addition to any void space. The losses associated with the carbon element and the anode coating were calculated directly from the volumes of the components.

Results are consistent with calculations made according to Eq. (12). The loss associated with electrolyte concentration was calculated using results of Eq. (6) and (12) to solve for the capacity densities of the balanced combinations of lithium, electrolyte solution, and carbon for pure $SOCl_2$ and 1.9 *M* $LiAlCl_4$ in $SOCl_2$. Largest losses in the stoichiometric capacity of the cell come from the salt in the electrolyte and from the container. Although none of the losses are

TABLE 6 Packaging Efficiency Factors for a Medical Lithium/Thionylchloride Battery

Source of Loss	Efficiency (%)
Hardward	84.1
Filling	98.8
Carbon element 80% porous 0.27 g/Ah	94.3
Anode coating	99.5
1.9 *M* $LiAlCl_4$ catholyte concentration	84.8
Overall	66.1

by themselves very large, the product gives a net packaging efficiency of 66% for this design; that is, about two-thirds of the external volume of the battery is occupied by the balanced reactive couple.

Applying the net packaging efficiency factor 0.661 to the capacity density of the balanced Li/$SOCl_2$ couple from Table 4, 0.548 Ah/cm^3, we get the stoichiometric capacity density 0.365 Ah/cm^3. The stoichiometric capacity based on this capacity density and the external cell volume reported by Howard *et al.*, 6.57 cm^3, is 2.45 Ah. The deliverable capacity and capacity density under given discharge conditions are the products of these stoichiometric quantities and the electrochemical efficiency under the conditions of discharge.

Total utilization of the capacity of this cell design at an average of 3.6 V would give an energy density of 1.34 Wh/cm^3. Discharge experiments (*vide infra*) yield a deliverable energy density of about 1.2 Wh/cm^3 when the batteries described by Howard *et al.* are discharged at optimum rates.[40] Analyses of this sort allow the identification of major loss modes in a design and indicate where improvements might be made.

3.5. Electrochemical Efficiency

3.5.1. *Principles*

Any battery transforms the energy of a chemical reaction into electrical energy. The appropriate free energy of the chemical reaction is the upper bound on the electrical energy available, as discussed in Section 3.2. Losses in the transformation, leading to less electrical energy than the free energy of the chemical reaction, are of two types: capacity and polarization. Losses in capacity arise from incomplete electrochemical utilization of the chemical reactants. This may result from a competitive parasitic reaction draining the capacity of one or both electrodes, for example. Polarization losses occur when the cell is discharged under thermodynamically irreversible conditions, that is, at a voltage less than the reversible thermodynamic potential. Because in most practical situations a battery is discharged to a minimum voltage level greater than zero, the two concepts are not entirely independent: A high degree of polarization can lead to a practical capacity loss. Nevertheless, for any completely defined set of discharge conditions the two are unambiguously separable.

The capacity-based or Coulombic efficiency can be defined as

$$e_Q = Q_D/Q_{\max} \tag{15}$$

where Q_D is the total capacity delivered under the specified conditions and $Q_{\max}$ is the stoichiometric capacity or maximum available capacity[41] if this is known. The energy-based efficiency e_E is defined as

$$e_E = W/N\ \Delta A \tag{16}$$

where W is the energy delivered under the specified discharge conditions, ΔA is the appropriate molar free energy of the electrochemical reaction, and N is the number of moles of the reference reactant.

With many battery applications the electrochemical efficiency can be determined directly by simply discharging groups of representative batteries to completion under the conditions of application. However, the long-term nature of a human implant implies a battery life long enough to preclude application-rate testing prior to implantation, at least initially. This has certainly been the case with pacemaker batteries, having expected application times approaching 10 years. In higher-rate devices, two or three years may still be a reasonable minimum period to justify implantation, although in cases of acute need, shorter device lifetimes may be possible.

Estimation of energy and capacity deliverable at application rate remains one of the central problems in implantable batteries. Typically, the estimate is made by a combination of cell discharge experiments over a manifold of discharge rates and calorimetric assessment of parasitic losses as a function of discharge rate and depth. Evaluation techniques are described in Chapter 4 of this book.

3.5.2. *Direct Determination of Discharge Efficiency*

Howard *et al.* described the discharge of flat D-shaped case-positive lithium/thionylchloride cells with anode area 12.5 cm^2 and stoichiometric capacity 2.45 Ah at constant current over the range 0.1–2.0 mA.[40] Their family of discharge curves is shown in Figure 3. They reported capacity and energy delivered above 3.0 V. These are shown in Table 7. The discharge efficiency decreases as the rate is decreased, falling off significantly from 87% to 79% between 0.25 and 0.10 mA (20 and $\mu A/cm^2$). This is thought to result from a greater relative importance of self-discharge at lower rates.

Results of a dual-rate discharge experiment with a 36-μA drain for one year followed by discharge to completion at 2mA are puzzling in that losses in the year at 36 μA were deduced to be insignificant. This is not consistent with other estimates of efficiency at low rate.

Also presented was a comparison of delivered capacity and energy density of lithium/iodine and lithium/thionylchloride cells of similar design and dimensions. These are shown in Figures 4 and 5. The thionylchloride-lithium aluminum chloride solution was 1.9 *M* in $LiAlCl_4$; the iodine cathode was 95% by weight iodine with the balance poly-2-vinylpyridine. The thionylchloride cell is clearly superior to lithium iodide in delivered capacity at current densities above about 10 $\mu A/cm^2$. Recalling that the stoichiometric energy densities of the balanced couples with pure reactants in Table 4 are nearly equal for lithium/iodine and

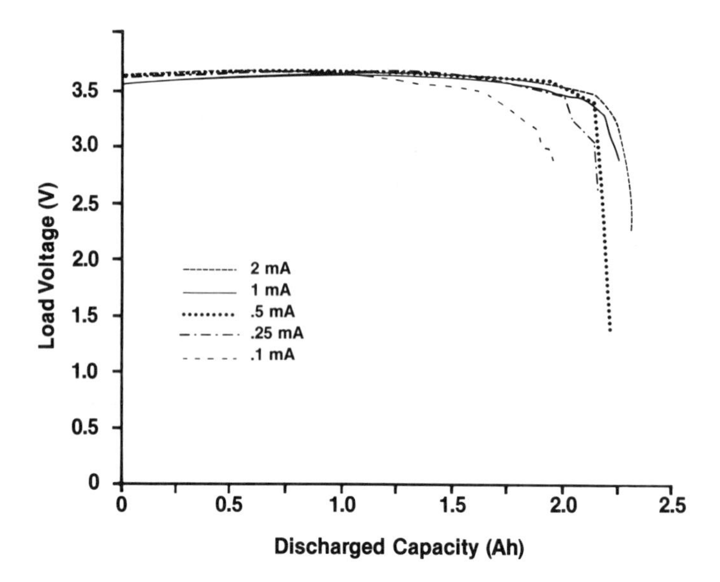

FIGURE 3. Constant-Current Discharge of Case-Positive Lithium/Thionylchloride Cells with Anode Area 12.5 cm^2 Over the Range of Rates 0.1–2.0 mA.[40]

lithium/thionylchloride, it is clear that the latter has a significantly higher volumetric efficiency.

Figure 2 shows the variation with $LiAlCl_4$ concentration of the deliverable capacity density of balanced practical electrodes determined in small (7 cm^2 anode area, ~1 Ah capacity) cylindrical case-negative cells with 1.4 *M* $LiAlCl_4$ electrolyte under a 2-kΩ load to a 0.1-V cutoff. Although the deliverable capacity density at high concentrations is close to stoichiometric for two-electron reduction

TABLE 7 Deliverable Capacity and Energy Density for Case-Positive Lithium/Thionylchloride Batteries[a]

Discharge Current (mA)	Discharge Time (years)	Delivered Capacity (Ah)	Utilization (%)	Energy Density (Wh/cm^3)
2.0	0.13	2.25(5)	91.8	1.21(7)
1.0	0.26	2.24(2)	91.4	1.24(1)
0.5	0.50	2.18(4)	89.0	1.21(2)
0.25	0.97	2.13(4)	86.9	1.19(2)
0.10	2.21	1.94(5)	79.2	1.07(3)
0.036[b]	1.1	2.27(2)	92.6	1.24(1)

[a] Standard deviations in parentheses apply to the final digit; 3.0-V cutoff.
[b] Discharged at 36 μA for one year, followed by 2 mA to 3.0 V

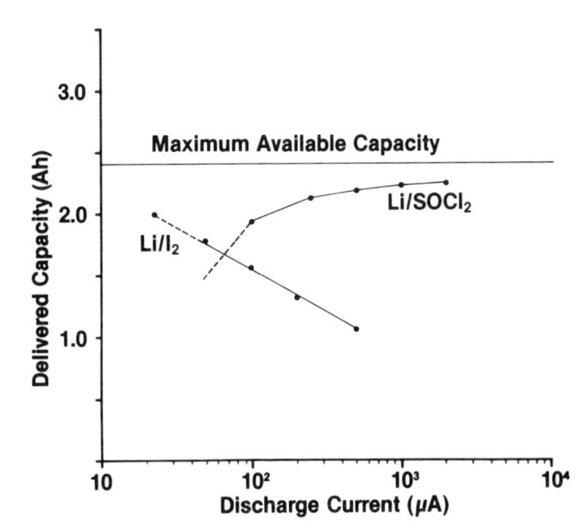

FIGURE 4. Deliverable Capacities of Lithium/Thionylchloride and Lithium/Iodine Cells of Similar External Dimensions.[40]

of the thionylchloride, there is a significant falloff at concentrations below 1 *M*. This may be related to nonuniform distribution of the discharge product in the carbon element with the lower conductivity electrolyte or to relatively higher rates of dynamic self-discharge. What appears to be almost exactly two-electron reduction of thionylchloride at higher compositions may in fact represent more than two-electron reduction, that is, additional reduction of products of reaction (1), because of the low final voltage value used. Analysis of the heat evolved during discharge indicates that thionylchloride cells do not discharge with 100% efficiency, as discussed in Section 3.5.3.

Buchman *et al.* presented data that demonstrate the effects of case polarity,

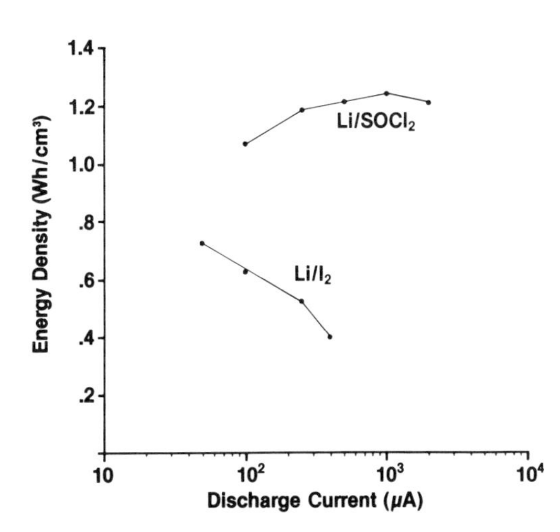

FIGURE 5. Deliverable Energy Densities of Lithium/Thionylchloride and Lithium/Iodine Cells of Similar External Dimensions.[40]

polymeric anode coating and discharge rate on electrochemical efficiency. The cells on which their data are based are similar in dimensions and construction to those of Howard *et al.* All electrolytes were 1.9 *M* $LiAlCl_4$ in thionylchloride. These results are summarized in Table 8 and are discussed in more detail in connection with calorimeter data.

3.5.3. Thermal Assessment of Electrochemical Efficiency

In any battery, chemical reactions may occur that consume active electrode components in competition with the Faradaic reaction. Such reactions are termed *parasitic reactions*. These include reactions of the electroactive components with case and separator materials as well as direct combination of the cell reactants (e.g., by diffusion). In thionylchloride cells the most important parasitic reaction is usually the direct combination of the cell reactants.

In this situation the electrochemical and parasitic reactions are competitive: At high rates of electrochemical discharge the electrochemical reaction predominates; at low rates the parasitic reactions may predominate. As noted previously, the overall electrochemical efficiency at high rates can be measured directly by discharging cells containing known amounts of reactants to completion.

While this may be a practical approach at high discharge rates, an estimate of discharge efficiency at low rates may be needed before the cells can be discharged completely. The availability of modern calorimeters with sensitivities on the order of a microwatt allows such an assessment to be made by analysis of the heat flow from a small battery under discharge. Heats evolved under load have been reported for several types of batteries, including lithium/thionylchloride.[42,43] The analysis of calorimeter data on batteries has been discussed.[44,45]

In order to understand the calorimeter experiment and the analysis of the measured heat, consider the following four experimental situations:

TABLE 8 Factors Affecting Discharge Efficiency

Case Polarity	Anode Coating	Discharge Rate (μA)	Direct Discharge (%)	Calorimetric Estimates[a]	
				ΔS1 %	ΔS2 %
+	Yes	500	89	92	90
−	Yes	500	83	86	84
−	No	500	73	77	74
−	No	250	68	74	72

[a] ΔS1: −90 J/mol-K; ΔS2: +40 J/mol-K.
Source: From Buchman *et al.* [51]

Case 1. Direct chemical combination of the stoichiometrically equivalent electroactive components of a thionylchloride cell in a closed rigid container at constant temperature within a calorimeter. The total heat q_c absorbed* by the system during the course of the chemical reaction is given by Eq. (17):

$$q_c = N\,\Delta U \tag{17}$$

where N is the number of moles of thionylchloride and ΔU† is the internal energy change per mole of thionylchloride for the complete reaction (1). The heat of a parasitic reaction is dissipated internally; thus, if self-discharge in the thionylchloride cell proceeds according to Eq. (1), then (17) gives the heat absorbed by the self-discharge reaction. The instantaneous rate of heat absorption measured by the calorimeter is given by Eq. (18):

$$\dot{q}_c = \dot{N}\,\Delta u = (I_c/nF)\,\Delta u \tag{18}$$

where Δu is the partial molal internal energy change for the reaction, $\dot{N}$ is the rate of reaction of the thionylchloride, I_c is the chemical reaction rate expressed as a current, n is the number of electrons transferred in the reaction (two as reaction (1) is written), and F is Faraday's constant.

Case 2. Discharge of a battery made of the same amounts of reactants through a resistor within the calorimeter chamber. The results are exactly the same as for case 1; in this case we use the subscript F to denote Faradaic reaction:

$$q_F = N\,\Delta U \tag{19}$$

for the overall reaction and

$$\dot{q}_F = \dot{N}\,\Delta u = (I_F/nF)\,\Delta u \tag{20}$$

for the instantaneous heat absorbed. In this case I_F is the current in the circuit composed of the battery and resistor.

Case 3. Discharge of the same battery as in case 2, reversibly, but with the load outside the calorimeter. In this case the internal energy change ΔU is

* The sign convention used throughout is that heat *absorbed* by the system (i.e., the battery) is positive. For an exothermic reaction the heat absorbed is negative.

† The internal energy and Helmholz free energy functions U and A are used in this discussion, rather than the enthalpy and Gibbs free energy functions H and G, because we have assumed reaction in a rigid (i.e., constant-volume) container. In practice the system may not be strictly constant volume, nor will it be constant pressure. However, if the reaction is electrochemically reversible, analysis of the open-circuit potential will give appropriate values of the free energy and entropy.

divided between external work and entropic changes within the cell. The external work is limited by the free energy $N\ \Delta A$. The heat absorbed by the cell during the course of the Faradaic reaction is given by the difference between the internal energy and the free energy:

$$q_F = N\ \Delta U - N\ \Delta A = NT\ \Delta S \tag{21}$$

where ΔS is the entropy change for the complete reaction and T is the temperature.

The instantaneous heat absorption rate in this case is given by

$$\dot{q}_F = \dot{N}T\ \Delta s = (I_F/nF)T\ \Delta s \tag{22}$$

where Δs is the partial molal entropy of the reaction at that depth of discharge.

Case 4. Irreversible discharge of the same battery with the load outside the calorimeter. Under real discharge conditions, some of the free energy of the reaction will be dissipated within the cell through Joule heating (irreversible discharge). This is reflected in the instantaneous heat absorption rate:

$$\dot{q}_F = (I_F/nF)T\ \Delta s + (E_L - E^o)I_F \tag{23}$$

where E_L and E^o represent load voltage and open-circuit voltage, respectively. The total heat of the Faradaic reaction absorbed by the system in the calorimeter is given by

$$q_F = NT\ \Delta S + \langle E_L - E^o \rangle Q_F \tag{24}$$

where the quantity within brackets represents the average polarization, and Q_F, the total capacity discharged.

The Faradaic and parasitic processes occur competitively during the discharge of a thionylchloride cell, so that the total rate of heat absorption by the cell under discharge is given by the sum of the respective rates for each process:

$$\begin{aligned} \dot{q}_{\text{tot}} &= \dot{q}_F + \dot{q}_c \\ &= \{(I_F/nF)T\ \Delta s + (E_L - E^o)I_F \\ &\quad + (I_c/nF)\ \Delta u\} \end{aligned} \tag{25}$$

The partial molal quantities, including the open-circuit voltage, which is proportional to the partial molal free energy of the reaction, are in the strict sense distinct from the overall molar quantities that describe the complete reaction. This is because the thionylchloride solution electrode changes composition as the cell discharges: Thionylchloride is consumed, and SO_2 and sulfur are formed, with both dissolving in the remaining thionylchloride solution to some extent. Because of these changes in composition of the solution electrode as the cell discharges and because of pressure changes within the cell, the thermodynamic properties are not constant throughout the course of reaction. Analysis of the instantaneous rate of heat absorption is further complicated by the possibility of the formation of long-lived reaction intermediates.

The analysis of calorimetric data requires some knowledge of the basic thermodynamic properties of the cell reaction: the heat of reaction, free energy of reaction, and entropy of reaction. In principle these quantities are available from the open-circuit voltage and its temperature variation:[46]

$$(\partial E/\partial T)_v = -(1/nF)\,(\partial\, \Delta a/\partial T)_v = (1/nF)\,\Delta s \qquad (26)$$

In practice there is some disagreement about how the open-circuit potential and its temperature variation should be determined. Godshall and Driscoll have recently reviewed the literature on the thermodynamics of the lithium/thionylchloride.[47] Of six independent determinations of the open-circuit potential at 25°C, five place the value between 3.645 V and 3.67 V. The open-circuit voltage can be determined by direct measurement on equilibrated cells or by extrapolation of a polarization curve to zero current. The first method may seem straightforward enough, but low-level impurities can affect the value, whereas at nonzero currents reactions involving low-level impurities might be expected to polarize.

The main point of disagreement is in the sign of the entropy change. One source using the dynamic method obtained a positive temperature coefficient of +1.026 mV/K.[48] Other workers using either the static technique or analysis of calorimeter data in a current density range where the coulombic efficiency is high[42,43,47,49] report negative values for the temperature coefficient of the OCV in the range −0.20 to −0.90 mV/K. Exact quantitative interpretation of the calorimeter data depends on the entropy value used. The wide range of values reported for the entropy does not preclude useful interpretation of the calorimeter data, however, because the entropic heat is small compared with other energy changes in the system.

Patel[50] obtained a value of +0.21 mV/K on medical cells of the type described by Howard *et al.*[40] by extrapolation of the polarization curve. Static measurements on this same type of cell yield values close to that reported by Gibbard,[43] −0.46 mV/K. Howard *et al.* show plots of calculated instantaneous Coulombic efficiency using +0.21 and −0.6 mV/K as bracketing values for the entropy change. The two curves differ by no more than 10% over the range of current densities 0.1 μA/cm^2 to 0.5 mA/cm^2. Unless otherwise noted, the entropy value used in the subsequent discussion and calculations corresponds to −0.46 mV/K. The open-circuit voltage used is based on the static OCV of the individual cell in question, normally 3.627 V.

The partial molal quantities can be obtained in principle from the open-circuit potential and its temperature variation as a function of discharge depth. However, the available data are too limited to make an accurate assessment of the variation in these quantities with discharge depth, beyond observing that such variations must be small until reaction has proceeded to near completion of two-electron reduction of thionylchloride.

The object of the analysis of the heat evolved by a thionylchloride cell is to obtain an estimate of the discharge efficiency of the cell. To proceed we simplify the analysis with the following assumptions:

1. All heat absorbed or evolved by the cell comes from Faradaic discharge or self-discharge according to reaction (1).
2. Variations with discharge depth in the partial molal thermodynamic quantities for reaction (1) are neglected.
3. Thermal consequences of any long-lived intermediates are neglected.

Given the free energy of reaction from the open-circuit potential and the entropy from the temperature coefficient of the OCV, for a battery discharging in a calorimeter at current I_F and load voltage E_L, we can solve Eq. (25) for the equivalent self-discharge current I_c. Given this we can calculate the instantaneous Coulombic efficiency

$$e_Q^{\,i} = I_F/(I_F + I_c) \tag{28}$$

and the instantaneous energy efficiency

$$e_E^{\,i} = e_Q\, E_L/E^o \tag{28}$$

It has been reported that the total rate of heat absorption for lithium/thionylchloride cells varies widely with depth of discharge and discharge rate.[(51)] If we assume that this variation comes about from a variation in self-discharge rate, then this implies that the instantaneous discharge efficiencies Eq. (27) and (28) are not constant but instead vary with rate and depth of discharge. The cumulative Coulombic and energy efficiencies defined, respectively, by Eqs. (29) and (30) are better estimators of the overall discharge efficiencies:

$$e_Q^{\,c} = \int_0^t I_F\, dt' / \int_0^t (I_F + I_c)\, dt' \tag{29}$$

and

$$e_E^{\,c} = \int_0^t I_F\, E_L\, dt' / \int_0^t (I_F + I_c) E^o\, dt' \tag{30}$$

Estimation of the cumulative efficiency requires knowledge of the variation in rate of heat absorption as a function of discharge depth. Such knowledge implies complete discharge of the battery on which the measurements were made at the rate in question. However, complete discharge of the cell would seem to obviate the entire calorimetric assessment, which has as its object the prediction of discharge efficiency without complete discharge at low discharge rates. Two important things can be determined from the experiment, however: (1) comparison of the estimated discharge efficiency $e_Q{}^c$ with the real discharge efficiency e_Q to establish the validity of the analysis; (2) establishment of the rapidity with which the cumulative efficiency reaches its limiting value at the end of discharge.

3.5.4. *Calorimetric Estimates of Discharge Efficiency*

Two types of calorimeter experiments have been described to obtain an estimate of the electrochemical efficiency of a battery under discharge.[39,51] One involves measuring the rate of heat evolution as a function of current density over a narrow range of discharge depth. This gives a rate profile of the instantaneous discharge efficiency of the cell at that depth of discharge. The other involves discharging a cell at fixed rate with the load outside the calorimeter chamber with periodic monitoring of the rate of heat evolution. This experiment yields both instantaneous and cumulative efficiencies of the cell as functions of discharge depth. If discharge is carried to completion, the experiment also gives a direct comparison of directly determined discharge efficiency and calorimetrically determined electrochemical efficiency. This has been important for giving confidence in the interpretation of calorimetric results.

Figure 6 shows plots of instantaneous coulombic efficiency from the rate of heat evolution observed for a fresh case-positive cell constructed as described by Howard *et al.*[40] as a function of current density. The instantaneous coulombic efficiency was calculated by partitioning the rate of heat evolution according to Eq. (25). The heat and load voltage measurements were made in a Tronac differential heat conduction calorimeter using procedures described by Patel *et al.*[39] The entropy value used, -0.46 mV/K, is based on the temperature variation observed in the OCV on a similar cell.

The three curves show the hysteresis on repetition of the current scan. The hysteresis may be related to depassivation and repassivation of the lithium surface or to slow continuing reaction of intermediate products.

The instantaneous coulombic efficiency curve is sigmoid in shape with a high-efficiency region ($e_Q{}^i > 0.8$) at higher current densities (>10 μA/cm^2) and a low-efficiency region ($e_Q{}^i < 0.20$) at lower current densities (less than about 0.1 μA/cm^2). The important observation here is that the discharge efficiency of a thionylchloride cell has three distinct regions, one of high efficiency, one of low efficiency, and a third transition region between the two, spanning one to two orders of magnitude in current density.

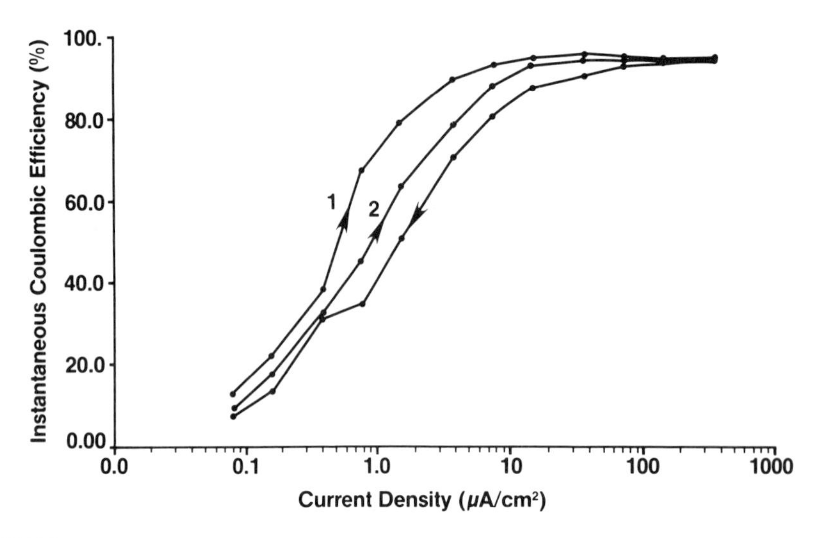

FIGURE 6: Hysteresis in Thermally Determined Instantaneous Coulombic Efficiency of a Fresh Case-Positive Lithium/Thionylchloride Cell on Cycling Current Density, Calculated from Calorimeter Data. The First Cycle and the Second Upward Scan are Shown.

In spite of the rapid drop in efficiency of discharge with decreasing current density, the duration of discharge calculated from the instantaneous efficiency function [e.g., as $(e_Q{}^i/I_F)$] is found to decrease monotonically as the discharge rate is decreased. However, as the discharge rate is decreased, other battery chemistries come to surpass lithium/thionylchloride in deliverable capacity. It should be noted that this interpretation is not totally consistent with the dual-rate discharge data of Howard *et al.* mentioned earlier.

Figure 7 shows the progression of the first-scan instantaneous discharge efficiency as a function of discharge rate for the same cell at three discharge depths. The three curves have similar sigmoid shapes and differ mainly in the current density of the inflection point. It is apparent that the instantaneous discharge efficiency varies with discharge depth.

Buchmann *et al.*[(52)] have recorded the variation in heat absorption rate as a function of discharge depth over a manifold of discharge rates as shown in Figure 8 for the case-positive cell design described by Howard *et al.*[(40)] The heat evolution varies widely with discharge depth, reaching a peak at each discharge rate at about 0.8 Ah. Although the highest discharge rate has the highest rate of heat loss, it also has the most efficient discharge, the Faradaic current rising faster than the self-discharge current. Figure 9 shows the instantaneous Coulombic efficiency calculated from the data in Figure 8. The minimum in the instantaneous efficiency at about 0.8 Ah corresponds to the peak in the rate of heat evolution in Figure 8.

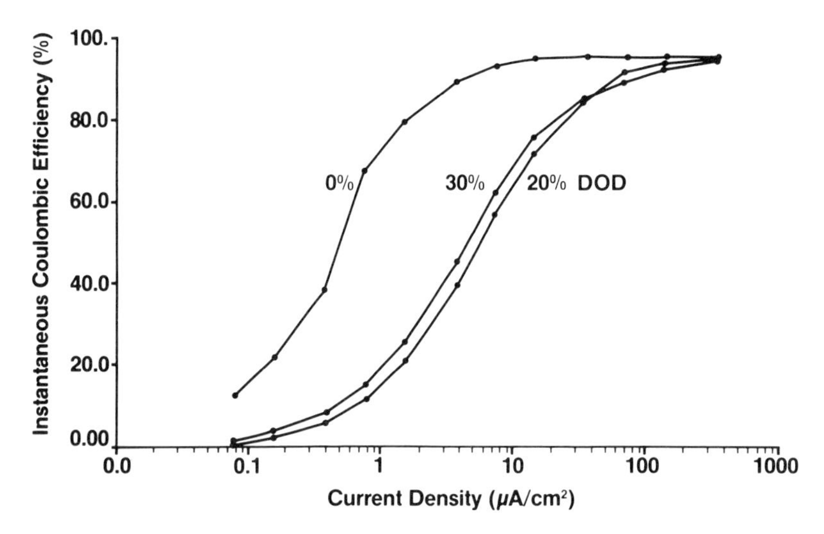

FIGURE 7. Variation of Thermally Determined Instantaneous Coulombic Efficiency of a Lithium/Thionylchloride Cell with Current Density and Depth of Discharge (DOD) Calculated from Calorimeter Data Taken on the First Upward Scan in Current Density.

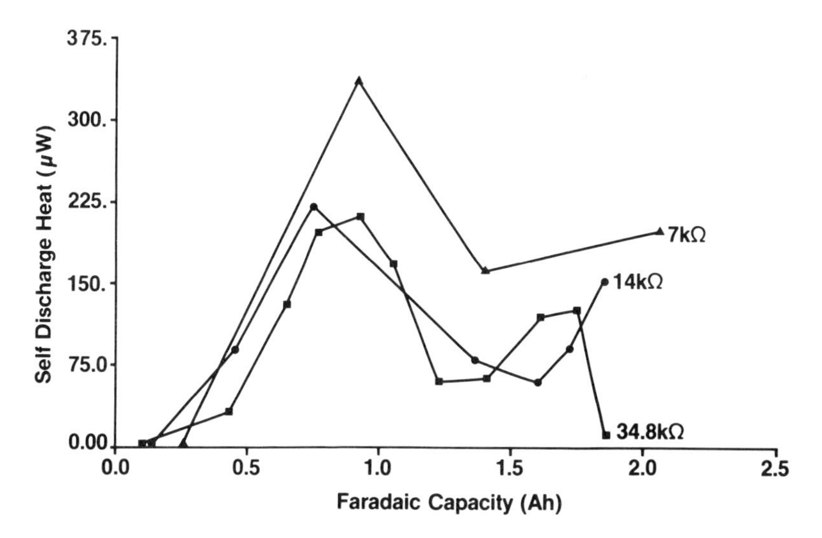

FIGURE 8. Variation in Rate of Parasitic Heat Evolution by Case-Positive Lithium/Thionylchloride Cells Discharged under Constant Loads with Discharged Capacity and Load.

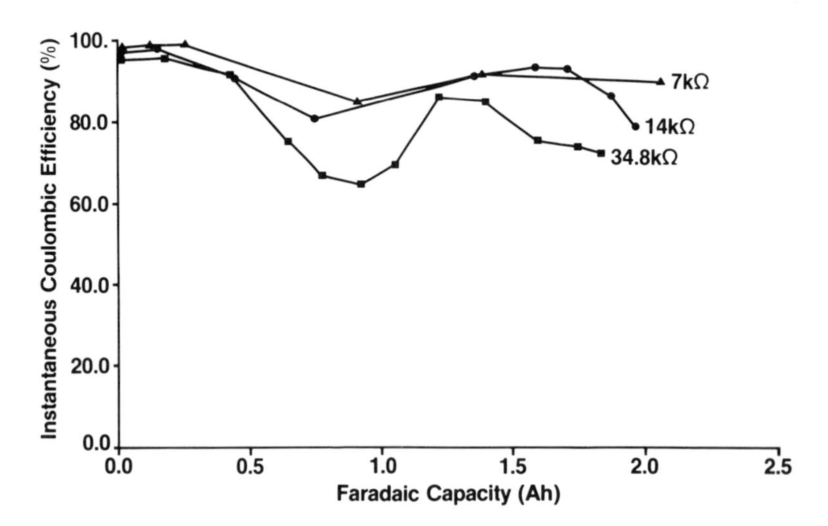

FIGURE 9. Variation in Thermally Determined Instantaneous Coulombic Efficiency of Case-Positive Lithium/Thionylchloride Cells Discharged at Constant Loads with Discharged Capacity and Load; Calculated from Data in Figure 8.

Figures 10 and 11 show the cumulative efficiency calculated from the data of Figure 8 as functions of time and Faradaic capacity, respectively. The cumulative efficiency function shown in Figures 10 and 11 achieves its final value within a percent or two at about 60% discharge depth. Thus, in spite of the wide variation in rate of heat evolution with discharge depth seen in Figure 8, the cumulative efficiency is a well-behaved function that can be used as an early estimator of discharge efficiency.

Figures 12 and 13 show the effect of case polarity and discharge rate on the heat evolved by the cells corresponding to the data in Table 8. Corresponding cumulative efficiencies are shown in Figure 14. The cumulative efficiency function for a case-negative cell goes through a broad minimum before achieving the final value, rising less than 5% in all the examples shown. Agreement between the discharge efficiency and the calorimetric cumulative efficiency in Table 8 again is good. Table 8 also shows the small effect the entropy value used in the calculation has on the calculated efficiency.

It can be seen from this discussion that the discharge efficiency of lithium/thionylchloride cells can be estimated from calorimeter data with a fair degree of agreement at moderate to low rates of discharge. Moreover, the early convergence or near convergence of the cumulative efficiency function allows accurate predictions to be made significantly before discharge is complete. Determination of the functional form of the cumulative efficiency for even earlier prediction of discharge efficiency is likely to be a fruitful area for further work.

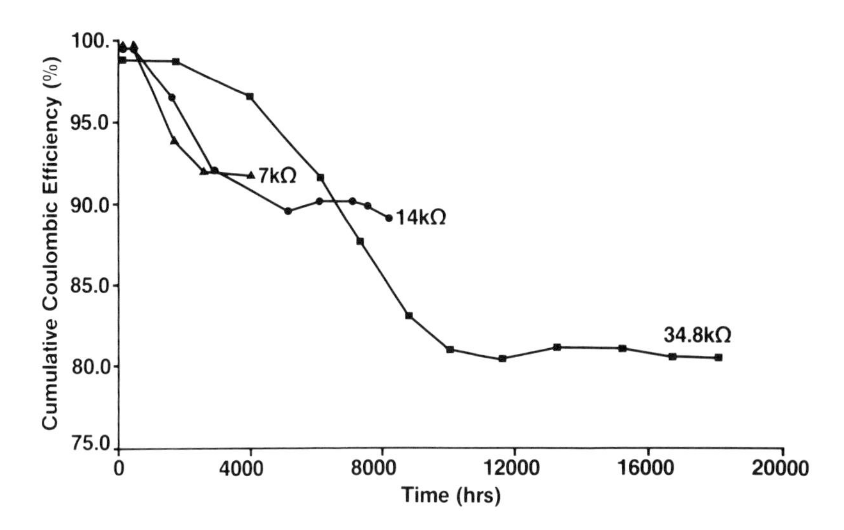

FIGURE 10. Variation in Thermally Determined Cumulative Coulombic Efficiency of Case-Positive Lithium/Thionylchloride Cells Discharged under Constant Loads with Time and Load; Based on Calorimeter Data Shown in Figure 8.

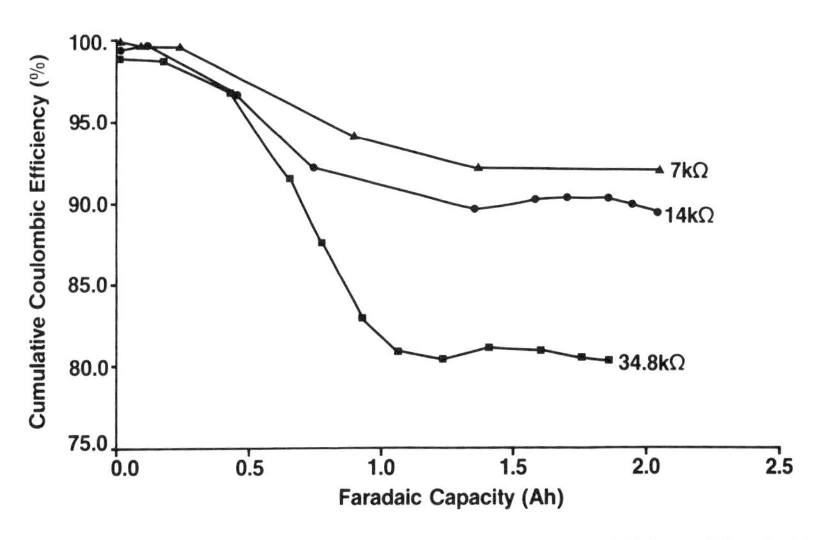

FIGURE 11. Variation in Thermally Determined Cumulative Coulombic Efficiency of Case-Positive Lithium/Thionylchloride Cells with Discharged Capacity and Load; Based on the Data Shown in Figure 8.

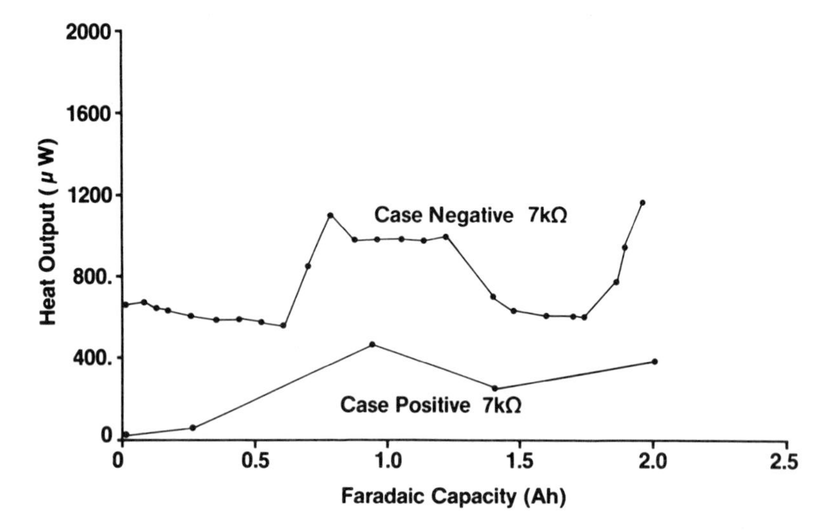

FIGURE 12. Effect of Case Polarity on Rate of Heat Output of Similar Lithium/Thionylchloride Cells with Coated Anodes Discharged under 7 kohm Loads (Data from Ref. 51).

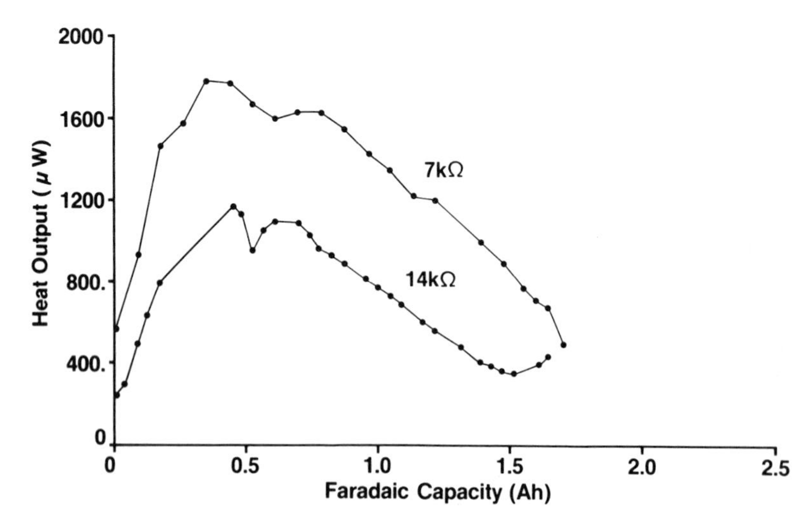

FIGURE 13. Effect of Discharge Rate on the Rate of Heat Output of Case-Negative Lithium/Thionylchloride Cells with Uncoated Anodes (Data from Ref. 51).

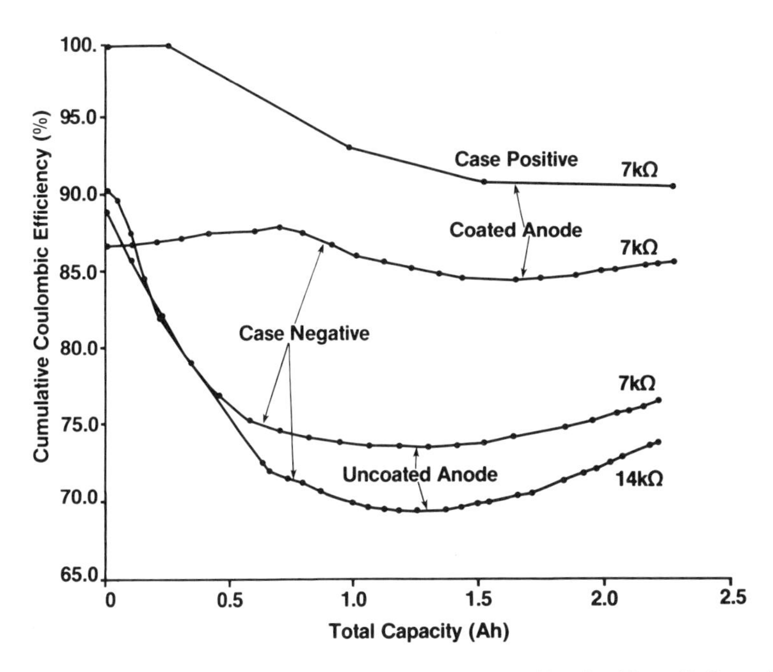

FIGURE 14. Effects of Case Polarity, Polymeric Anode Coating and Load on Thermally Determined Cumulative Coulombic Efficiency of Lithium/Thionylchloride Batteries (Data from Ref. 51).

All the calorimetrically estimated efficiencies are larger than those determined electrochemically. This small discrepancy may reflect a difference between stoichiometric capacity and maximum available capacity above 3.0 V or errors of the analysis. Many assumptions are required for this analysis that ultimately limit its accuracy. For example, there is a need for accurate and detailed measurement and analysis of the open-circuit potential behavior of lithium/thionylchloride batteries and of the partial molal thermodynamic properties of the system $SOCl_2$-SO_2-$LiAlCl_4$. All the assumptions made in the analysis should be reexamined and refined as necessary to ensure the validity of this analysis and to improve its accuracy.

The constant-load or constant-current discharge conditions of the data reviewed here are much simpler than application loads. The power requirements of most implantable medical devices are not constant, but consist of high-power pulses superimposed on a low-power background. It is possible to estimate the discharge efficiency of cells under such load regimes only very crudely based on the data presented here. However, the calorimetry techniques reviewed here can be applied to any periodic load profile.

3.5.5. Storage Losses

Losses in capacity on storage at open circuit can be determined in two ways. First, batteries can be stored for various periods and then discharged to completion. Alternatively, the loss rate can be estimated from calorimeter data.

Howard *et al.* have presented the data shown in Table 9 on small medical batteries, in which they are unable to detect any difference in deliverable capacity between fresh cells and cells stored up to 18 months prior to discharge at 2 mA.[40]

Hansen and Frank[53] have shown that the rate of heat evolution by thionyl-chloride cells progresses in time according to the inverse power law shown in Eq. (31):

$$\dot{q} = K\,t^{-x} \tag{31}$$

where $\dot{q}$ is the time rate of heat evolution and t is time and K is proportional to the heat of reaction and the reaction rate and where x should be 0.5 for strict diffusion limitation of the reaction (9). In their analysis of open-circuit calorimeter data for small case-positive medical cells, Howard *et al.* found reasonable conformity to the inverse power law, as shown in Figure 15. Assuming that the reaction generating this heat is self-discharge according to reaction (1), these authors generated the projected losses shown in Figure 16 for 2.4-Ah cells with 12.6 cm^2 anode area. While it may be somewhat unrealistic to project losses out to five years from three months of data, it is clear that losses on open-circuit storage for these cells are small and, if anything, overestimated by the calorimeter data. The effectiveness of the polymeric anode coating in controlling self-discharge losses is clear from Figures 15 and 16. Losses in cells with coated anodes are estimated to be only 0.1 Ah after five years, in contrast to 0.4 Ah in cells with no anode coating. It should be noted that other factors, such as case polarity,

TABLE 9 Capacity Retention on Open-Circuit Storage

Storage Time (mo)	Number of Cells	Delivered Capacity[a,b] (Ah)	Energy Density[a,b] (Wh/cm^3)
0	10	2.25(5)	1.21(3)
1	2	2.23(3)	1.23(2)
6	2	2.27(2)	1.25(1)
12	2	2.23(7)	1.22(4)
18	10	2.25(8)	1.22(5)

[a] Standard deviations in parentheses apply to the final digit.
[b] Discharged at 2 mA and 37°C to 3.0 V.
Source: From Howard *et al.*[40]

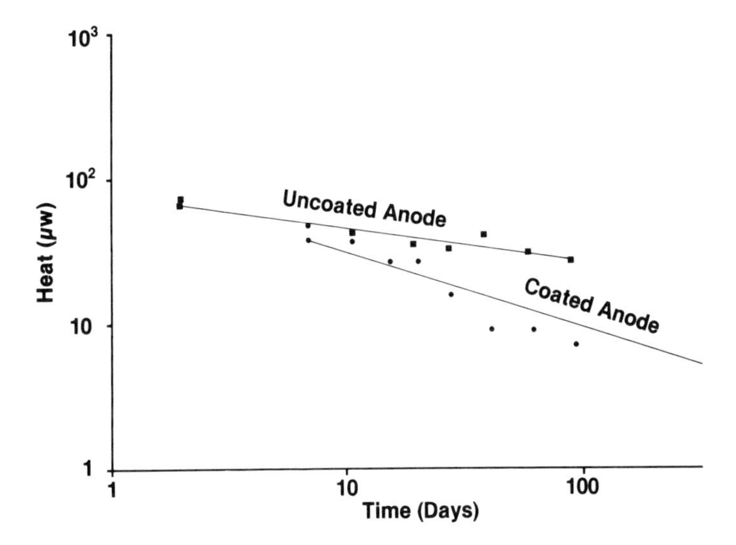

FIGURE 15. Effect of Polymeric Anode Coating on the Rate of Heat Output of Case-Positive Lithium/Thionylchloride Cells on Open-Circuit Storage.[40]

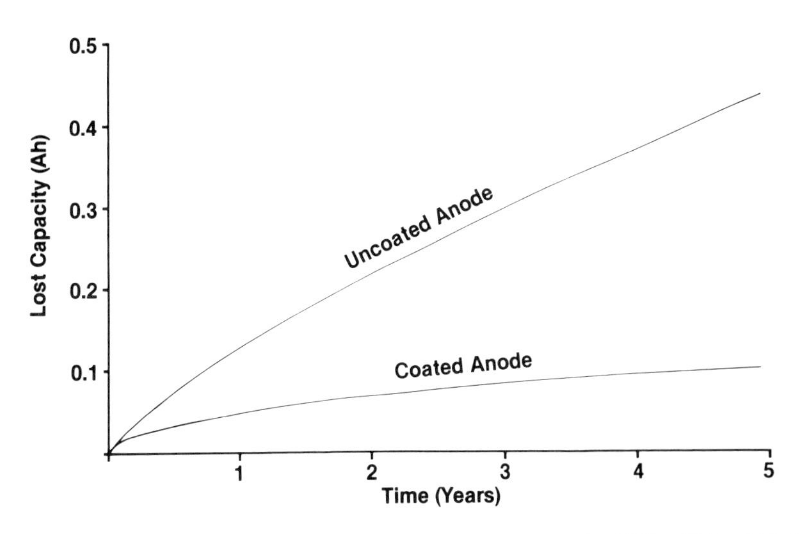

FIGURE 16. Projection of Parasitic Losses on Open-Circuit Storage for Lithium/Thionylchloride Cells with and without Polymeric Anode Coatings; Based on Data of Figure 15.[40]

electrolyte concentration, and impurities in the thionylchloride solution, also affect the rate of static self-discharge.

4. STATE-OF-DISCHARGE INDICATION

One characteristic of most thionylchloride cells that are not thionylchloride limited is a flat discharge curve terminated by a sudden loss in voltage when the reactants are depleted. In a critical application such as an implantable medical device, loss of voltage without warning is, in general, not acceptable. Thus, some means of anticipating the depletion of a thionylchloride cell is needed for it to be a desirable power source for implantable devices. Several approaches to the problem of end-of-life indication in medical lithium/thionylchloride batteries have been reported.

Marincic has described thionylchloride-limited cells in which there is a step in potential down to about 3 V near the end of discharge.[54] The 3-V plateau is of limited duration, but apparently sufficient to provide a depletion warning. Data regarding the extent to which the relative lengths of the two discharge plateaus can be varied are not given. Marincic suggests that the depressed potential may arise from the discharge of a complex formed between residual $SOCl_2$ and SO_2. It should be noted that 3 V is close to the discharge potential of Li/SO_2 cells. Detailed study of the mechanism of voltage depression has not been reported.

Composite anodes of lithium foil and calcium have been described for end-of-life indication in oyxyhalide cell systems in a number of cell configurations.[55,56] When used with thionylchloride, such composite anodes lead to a step in potential down to the 3-V level of the calcium/thionylchloride couple when the lithium is consumed.

One potential problem with this approach is the thermodynamic instability of elemental lithium and calcium in contact with one another. The phase diagram reported for the system lithium–calcium shows an intermediate compound, the hexagonal Laves phase Li_2Ca.[57] Neither Goebel *et al.*[55] nor Kiester *et al.*[56] report data on the rate of reaction of lithium and calcium. Before a composite of this type could be seriously considered for use, some information about the rate or reaction and the consequences of the formation of a zone of intermetallic compound would need to be investigated. Alternatively, the two metals might be kept physically separated within the cell.

Many workers have investigated the discharge behavior of lithium alloys in thionylchloride cells. Kiester *et al.*[56] and Howard *et al.*[40] have specifically reported the investigation for end-of-life indication of composite alloy foils of lithium and calcium in a composition range corresponding to mixtures of Li and Li_2Ca. Figure 17 shows a discharge curve obtained with such an anode.[40] Both groups of workers report that the approach provides flexibility in the duration

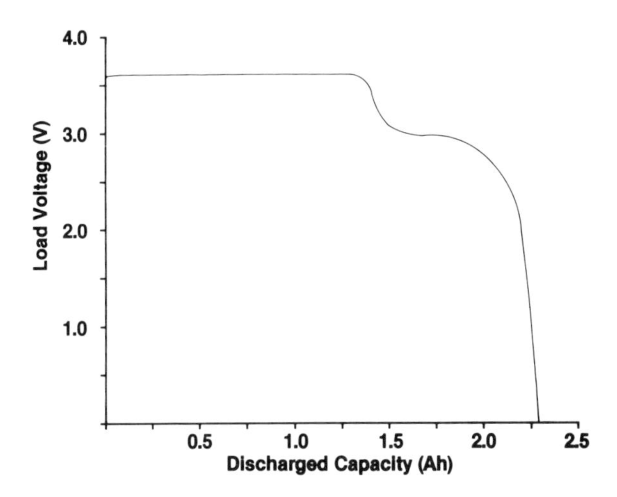

FIGURE 17. Discharge Curve of Lithium/Thionylchloride Cell with 85 at. % Lithium-15 at. % Calcium Alloy Anode.[40]

of the lower potential step. Although the alloy foils are somewhat less ductile the pure lithium, compositions up to about 15 at. % calcium can be handled and formed in much the same way as lithium.

One potential problem with this approach is that the discharge properties of the composite foils are likely to be strongly dependent on the microstructure of the composites, which in turn may be influenced by small amounts of impurities and by fabrication procedures. Ideally, the two phases should form two interpenetrating continuous networks. Discharge of the lithium phase should leave an electrically continuous residual sponge of the intermetallic compound phase. Clearly, the possibility exists of mixed-potential phenomena if this condition is not satisfied or if transport of active material within the pores of the sponge limits the rate of discharge.

DeHaan and Harshad have reported the use of an auxiliary electrode of stainless steel embedded at a precise depth in the lithium anode.[58] The stainless steel electrode is exposed to the lithium on the surface toward the outside of the anode assembly but otherwise insulated from the lithium. As the cell discharges and the lithium erodes away, the outside surface of the stainless steel electrode eventually becomes free of lithium. Because the stainless steel is then isolated from the lithium and in contact with the catholyte, its potential rises to that of the cathode.

The main problems with this approach are the need for an additional feedthrough in the battery for the reference electrode and the need to locate the reference at a precise depth within the lithium anode. Attainment of the required precision in a ductile medium such as lithium demands extraordinary control in manufacturing.

Coulometric circuits could also be built into the implantable devices. Because of uncertainties remaining in the determination of parasitic losses under application conditions and because other unanticipated modes of loss may occur, a positive indication of depth of discharge is a more desirable approach. Coulometric discharge of the cell within the device does not register losses because of accidental shorting of the cell or unanticipated parasitic losses.

Whether one of these approaches or some other approach to end-of-life indication emerges as clearly superior to the others remains to be seen. At this time any of those described seems to be viable.

5. VOLTAGE DELAY

5.1. Anode Passivation

Lithium/thionylchloride cells designed for application in implantable medical devices are expected to discharge for long periods at a low background rate with intermittent discharge at higher rates for short periods. The response of the battery voltage to the higher-rate demand is likely to be important throughout the entire discharge life of the battery. Thus, characterization of voltage response for most medical implantable applications is more complicated than that for single-rate discharge after a period of storage at low or zero background rate.

The effects of various design parameters such as electrolyte composition, anode coatings electrode area, and case polarity as well as demand parameters including quiescent period, device impedance, and pulse shape are important in determining the voltage response. In addition, the evolution of the voltage response with time, depth of discharge, background current, and other details of discharge history need to be known in order to predict the useful lifetime of a battery. Little work of this type of a general nature has been reported.

When a lithium/thionylchloride battery is placed under load after storage, the voltage may fall initially and then recover to its normal operating level. This phenomenon is termed *voltage delay* and is related to the growth of a protective, passivating film of lithium chloride on the anode during storage.

Two types of film have been identified.[12,59] The primary film is reported to be a thin, dense layer of lithium chloride immediately adjacent to the lithium, <500 Å thick. This type of film has been identified as the solid electrolyte interphase mentioned in Section 2.1.[12] The rate of further corrosion is limited by the kinetics of diffusion through this primary film layer. The diffusion has been suggested to involve a combination of ionic diffusion and electronic leakage through the increasing SEI layer[59]; molecular diffusion of chlorine or thionylchloride through the layer is also likely to contribute significantly to the growth of the layer. The impedance associated with the primary film is essentially ohmic.[59] Voltage recovery occurs through the rupture of the passivating layer and its reformation at a lower thickness determined by the discharge rate. For the primary film voltage recovery is rapid.

With time or elevated temperature, a secondary film, also consisting mainly of lithium chloride, has been found to grow over and reportedly from the primary film.[59] The secondary film is characterized as thick (up to a few micrometers), strong, and somewhat porous. The presence of the secondary film leads to polarization within its porous structure and slower voltage recovery when the cell is placed under load. It is the kinetics of these lithium chloride layers on the anode, their dynamic interaction with the anode and electrolyte, and ultimately their structures that determine the voltage response of a battery.

5.2. Alleviation of Voltage Delay

Most of the agents reported to reduce the anode corrosion rate in lithium/thionylchloride batteries are also reported to improve the voltage delay characteristics. This is not without exception in that acidic electrolytes, made with excess $AlCl_3$ over the amount of LiCl provided, are also reported to reduce voltage delay temporarily. Efforts to improve the voltage response have focused on the composition of the electrolyte and the application of polymeric films to the anode.

In the electrolyte the first consideration in the alleviation of voltage delay is purity of both salt and solvent. Iron salts are common contaminants of $AlCl_3$ that have been shown to contribute both to anode corrosion and to voltage delay.[60] Iron and other transition metal contaminants can be removed from solution by contact of the solution with lithium or by electrolysis. The hydrolysis products of thionylchloride, HCl and SO_2, can be removed from solution by refluxing. Hydrolysis products of the $AlCl_3$ are more difficult to remove. Careful materials preparation and handling techniques can minimize their presence in the electrolyte.

Effects of various electrolyte salts and additives in varying concentrations on voltage delay have been studied. Dey[11] carried out studies of the growth rate of the passivating film with varying $LiAlCl_4$ concentration. He found that the rate of film growth increased with increasing concentration over the range 0.25–2.0 *M*. Schlaikjer has reported that a considerable improvement in voltage delay can be achieved through the use of the polyhedral perchloroborate salts $LiB_{10}Cl_{10}$ and $LiB_{12}Cl_{12}$ instead of $LiAlCl_4$.[61] Others have indicated beneficial effects resulting from the use of solutions of Li_2O, Li_2S and alkaline earth oxides, and $AlCl_3$ in thionylchloride and in the use of lithium chloride salts of Group III chlorides other than $AlCl_3$.[62,63]

The presence of SO_2 in $LiAlCl_4$ electrolyte has been reported to reduce voltage delay up to concentrations of 5%.[64] As a reaction product, however, that level cannot be maintained as the cell discharges. Addition of BrCl to $LiAlCl_4$-$SOCl_2$ electrolyte is reported to improve the voltage delay properties.[27] Polymeric anode coatings,[65,66] mentioned earlier for the reduction of the rate of anode corrosion under both open-circuit and discharge conditions, are also reported to alleviate voltage delay.

6. SAFETY

Two properties that distinguish the lithium/thionylchloride system and make it attractive as a power source are high energy density and high power capability. These same two characteristics also create potential safety hazards in its use. Additional hazards arise from the caustic and corrosive nature of the active components, should they be released outside the battery case.

Several potentially hazardous conditions have been identified. These include short circuit, forced overdischarge, charging, casual storage, and improper disposal. These conditions can be avoided in many situations. When hazardous situations cannot be avoided, the hazard can in many cases be mitigated through design of the battery or the device in which the battery is used. The level of risk can then be quantified through evaluation and a decision can be made regarding the acceptability of the battery for the particular application.

Johnson[67] has described principles involved in the design of a cylindrical lithium/thionylchloride battery that was intended to be safe and that demonstrated a high degree of safety in extensive abuse testing. These include the following:

1. The cell contains a vent.
2. The cell has low-area electrodes and a central anode.
3. The cell is anode limited and is designed to give complete lithium utilization and uniform distribution of the reaction product.
4. The composition and purity of the cathode electrolyte solution are closely controlled during manufacturing.

6.1. Short Circuit

The primary hazardous effect of a short circuit in a thionylchloride battery is heating of the battery. This can lead to increases in pressure within the container and to venting or explosion if the chemical discharge reaction rate reaches a critical value. Even if venting or explosion does not occur as a result of a temporary short circuit, the attendent capacity loss could cause premature unanticipated failure of the battery through depletion of the limiting electrode.

Generally, it is thought that the explosion hazard in shorting a lithium/thionylchloride battery comes about from melting the lithium. Although explosion does not always follow the heating of a battery above the melting point of lithium, several studies have associated explosions with heating lithium above the melting point.[68]

A mechanical vent is often incorporated to minimize the danger if a short circuit does occur. The battery case can be designed to vent at moderate pressure, thus preventing thermal runaway. This could be an acceptable design alternative if the battery is contained within an outer package that is also hermetically sealed and capable of withstanding the contents of the battery long enough to permit identification and removal, as is usually the case with implantable medical devices.

The high inherent maximum rate capability of the lithium/thionylchloride battery system often exceeds the current demand in applications that require the high energy density. In such cases the battery can be designed with scaled-down rate capability to match the application requirement. The simplest principle of low-rate thionylchloride cell design is limitation of the electrode area.

Several approaches to limit further the current have been reported. A low-concentration–low-conductivity electrolyte could be used; however, as shown in Figure 2, the deliverable capacity declines significantly below 1 *M* $LiAlCl_4$. Howard *et al.*[40] describe a case-positive cell design in which contact between the carbon element and the positive terminal is made through ribs molded into the carbon element rather than through an internal metal screen current collector. Cells of this type discharged efficiently at constant rates up to about 400 $\mu A/cm^2$ but showed only limited swelling and no venting on short-circuit testing.[69] Short-circuit currents through a 50-mΩ resistor declined quickly to less than 1.0 mA/cm^2, in contrast to current densities greater than 300–500 mA/cm^2 for otherwise similar cells with metal screen current collectors internal to the carbon element. A third method of current limitation is the use of a fuse in series with the battery. This method only protects the battery from external short circuits.

Internal short circuits can arise through migration of carbon particles. The polymeric anode coatings that have been shown to be effective in reducing self-discharge rates also provide protection from this type of short circuit. Cells constructed using coated anodes and no other means separating the carbon and lithium discharged well and showed no evidence of shorting.[70]

External short circuits can be avoided through prudent design of external terminals and connectors on both battery and the device in which it is used. Attention must be given to the manufacturing process to ensure that cells cannot be shorted accidentally during assembly. Besides creating a manufacturing hazard, a battery that has seen an undetected short circuit may have undergone a significant capacity loss and fall short of the expected deliverable capacity.

6.2. Overdischarge

Forced overdischarge can come about when nonidentical cells are discharged in series. If one cell of the battery becomes depleted before the others, continued discharge of the battery results in the depleted cell being driven. The safety implications of overdischarge depend on the limiting component of the depleted cell.

In carbon-limited cells oxidation of lithium can remain the anode reaction, but a new cathode reaction is required. This can be reduction of lithium ion to lithium metal or reduction of some other component of the residual electrolyte solution. In lithium-limited cells reduction of thionylchloride can still occur at the cathode, but the anode reaction must also involve oxidation of some component of the same solution. The different cell reactions lead to different products and combinations of potential reactants within the cell.

Abraham *et al.* found no hazard in the forced overdischarge of small carbon-limited cells.[71] They report internal shorting of the cells resulting from the formation of lithium dendrites. The short poses no hazard because the main cell reaction is prevented by saturation of the carbon element. Although explosions were not observed on overdischarge in these experiments, others have found shock sensitivity to be a hazardous result of the formation of lithium dendrites in an overdischarged cell with a passivated carbon element.[72]

Shipman and McCartney[73] have proposed the prevention of lithium deposition on the saturated cathode of an overdischarged cell by the addition of the salt of an aliovalent metal, $NbCl_5$, to the electrolyte. Reduction of the salt to the lower oxidation state should be expected to occur before reduction of the lithium ion and thus prevent the deposition of lithium.

The forced overdischarge of anode-limited cells by several groups of workers has led to a variety of results and conclusions,[74–76] presumably reflecting differences in the details of cell design and of the experiments. While some workers have noted no hazard in discharging large (>2000-Ah) cells,[74,75] others have observed significant increases in anode potential accompanied by fluctuations, and in some cases explosions, in small (C size and smaller) anode-limited cells subject to overdischarge.[71] The fluctuations in anode potential were identified as due to intermittent contact of the anode current collector with isolated lithium.[71]

The explosions were attributed to the formation of unstable oxidation products in the electrolyte. Solutions subjected to forced overdischarge have been analyzed spectroscopically and electrochemically[71] with a number of products identified, including products of a cyclic regenerative process proposed to account for the steady-state passage of current, notably $AlCl_3$ and Cl_2. Some evidence has been presented for the presence of the unstable oxidant Cl_2O.[76] However, the cause of explosions observed on overdischarge of small anode-limited lithium/thionylchloride cells has not been identified definitively.

Marincic has described the end-of-life characteristics of cells that are thionylchloride limited.[54] Little has been reported about the safety characteristics of such cells. The loss of conductivity of the electrolyte as the thionylchloride becomes completely consumed could have beneficial safety consequences in a low-voltage battery with limited current capability; in a high-current, high-voltage battery, the opposite might be true.

Forced overdischarge of any lithium/thionylchloride battery must be regarded as potentially hazardous. The best design approach is to avoid a situation in which overdischarge can occur. If a series combination is required for the application the battery can be designed so that overdischarge is highly unlikely. One possibility is the careful matching of the capacity of the cells together with a positive end-of-life warning of sufficient duration to allow identification and removal of the device before any cell in the battery is depleted. Any proposed design of a multicell battery must be accompanied by rigorous and exhaustive testing.

6.3. Charging

Charging a lithium/thionylchloride battery is thought to lead to cyclical processes like those involved in the overdischarge of anode-limited cells.[71] Similar hazards must be presumed to be operative. Parallel cell combinations can be avoided in general by design of single cells with sufficient rate capability. If this is not possible the cells should be diode-protected. In this connection the calcium anode in a calcium/thionylchloride cell has been characterized as a chemical diode because of the anionic nature of ion transport within the solid electrolyte interphase.[77] This and the high melting point of calcium have suggested possible safety advantages in the use of that couple.

6.4. Casual Storage

Incidents of venting or explosion of lithium/thionylchloride cells on storage after discharge or partial discharge and without the general heating associated with a short circuit have been reported.[78] Because of the random nature of the reported explosions, the reports tend to be anecdotal. However, much work has been done to identify the likely cause or causes of such explosions.

One proposal has been the initiation of local melting of lithium at "hot spots" in cells in which the average temperature is well below the melting point of lithium. Dey[79] demonstrated that D cells could be made to explode through local heating using externally powered nichrome heaters built into the cells. While the "hot spot" hypothesis is a plausible explanation of the explosions, the existence of "hot spots" has not been demonstrated and the chemical origin remains a matter of conjecture. A few suggestions have been made. For example, lithium nitride, a likely impurity, has been shown to react violently with thionylchloride electrolyte under some conditions.[73]

Others have suggested the formation of unstable reaction intermediates or the deposition of sulfur and lithium in proximity to one another as causes.[80]

One safety advantage proposed for the addition of bromine chloride to thionylchloride electrolyte is the prevention of intimate mixing of lithium and solid sulfur by increasing the solubility of the sulfur in the electrolyte.[27] Cells of this type have shown excellent safety in abusive testing.[27] The addition of copper to the carbon element has been claimed to improve the safety of a lithium/thionylchloride cell by reacting with sulfur as it is formed and thus preventing the buildup of sulfur in the presence of lithium.[81]

Reports of random explosions of lithium/thionylchloride cells may raise concerns about the use of this battery system in any application where another power source could be used instead. Yet in our construction and normal-use testing over a period of seven years of several thousand small lithium/thionylchloride batteries designed to power implantable medical devices, not a single instance of explosion or venting has occurred. Further, the thionylchloride cells employed by ARCO to power several thousand pacemakers have not been involved in any reported incidents, nor have any of the medical BCX batteries manufactured by Wilson Greatbatch Ltd.

We conclude that small, low-rate designs may be used safely if appropriate precautions are taken in both battery and device design to preclude abusive conditions.

6.5. Disposal

Safe disposal of the batteries used in implantable devices should be provided. Explanted devices should be transferred to a qualified battery disposal service or to the manufacturer for disposal of the battery. Techniques for the safe disposal of small lithium/thionylchloride batteries have been described.[82] Incineration of any sealed container can result in explosion leading to injury and/or property damage. Casual incineration of any implantable medical electronic device should be avoided.

6.6. Future

The lithium/thionylchloride cell system has been developed into a promising power source for medical applications. Looking to the future, one can see extension of the domain of efficient discharge to lower discharge rates, the alleviation of voltage delay both through the use of anode coatings or new electrolyte salts, for example, and the clarification of the chemical basis of cell discharge as milestones that when passed, will encourage wider use of lithium/thionylchloride and other liquid oxidant systems as power sources for implantable medical devices.

REFERENCES

1. K. R. Brennen and B. B. Owens, in: *Lithium Battery Technology* (H. V. Venkatasetty, ed.), pp. 139–158, Wiley-Interscience, New York (1984).
2. A. Schneider, J. Moser, T. Webb, and J. Desmond, in *Proc. 24th Power Sources Symposium*, pp. 27–29, PSC Publications Committee, Red Bank, New Jersey (1970).
3. K. Fester, W. D. Helgeson, B. B. Owens, and P. M. Skarstad, *Solid State Ionics 9&10*, 107–110 (1983).
4. G. E. Blomgren and M. L. Kronenberg, German Patent 2,262,256 (1972).
5. M. J. Harney and S. Brown, in: *Proceedings of Symposium on Power Sources for Biomedical Implantable Applications, and Ambient Temperature Lithium Batteries* (B. Owens and N. Margalit, eds.), pp. 102–109, The Electrochemical Society, Princeton, New Jersey (1980).
6. J. P. Slack, P. Hurzeler, and D. Morse, *Pace 5*, 567–570 (1982).
7. C. R. Schlaikjer, in: *Lithium Batteries* (J. P. Gabano, ed.), pp. 301–370, Academic Press, New York (1983).
8. J. J. Auborn and H. V. Venkatasetty, in: *Lithium Battery Technology* (H. V. Venkatasetty, ed.), pp. 127–137, Wiley-Interscience, New York (1984).
9. E. Peled, in: *Lithium Batteries* (J. P. Gabano, ed.), pp. 43–72, Academic Press, New York (1983).
10. T. G. Hayes, B. B. Owens, J. B. Phipps, P. M. Skarstad, and D. F. Untereker, *Extended Abstracts*, Vol. 82-1, Abstract 764, The Electrochemical Society, Pennington, New Jersey, (1983).
11. A. N. Dey, *Thin Solid Films 43* 131 (1977).
12. E. Peled, *J. Electrochem. Soc. 126*, 2047–2051 (1979).
13. J. P. Gabano, French Patent 2,079,744 (1971).
14. J. J. Auborn, K. W. French, S. I. Lieberman, V. K. Shah, and A. Heller, *J. Electrochem. Soc. 120* 1613–1618 (1973).
15. W. K. Behl, J. A. Christopulos, M. Ramirez, and S. Gilman, *J. Electrochem. Soc. 120* 1619–1623 (1973).
16. G. E. Blomgren, V. Z. Leger, M. L. Kronenberg, T. Kalnoki-Kis, and R. J. Brodd, in: *Power Sources 7, Proc. 11th International Power Sources Symposium* (J. Thompson, ed.), pp. 583–593, Academic Press, New York (1978).
17. A. N. Dey, *J. Electrochem. Soc. 123*, 1262–1264 (1976).
18. C. R. Schlaikjer, F. Goebel, and N. Marincic, *J. Electrochem. Soc. 126*, 513–522 (1979).
19. J. R. Driscoll, G. L. Holleck, and D. E. Toland, *Proc. 27th Power Sources Symposium*, pp. 28–30, PSC Publications Committee, Red Bank, New Jersey (1976).
20. J. C. Baily and J. P. Kohut, in: *Power Sources 8, Proc. 12th International Power Sources Symposium* (J. Thompson, ed.), pp. 17–26, Academic Press, New York (1981).
21. W. L. Bowden and A. N. Dey, *J. Electrochem. Soc. 127*, 1419–1426 (1980).
22. H. V. Venkatasetty, *J. Electrochem. Soc. 127*, 2531–2533 (1980).
23. B. J. Carter, R. M. Williams, F. D. Tsay, A. Rodriquez, S. Kim, M. M. Evans, and H. Frank, *J. Electrochem. Soc. 132*, 525–528, (1985).
24. H. V. Venkatasetty and D. J. Saathoff, *J. Electrochem. Soc. 128*, 773–777 (1981).
25. P. M. Skarstad and T. G. Hayes, U.S. Patent 4,246,327 (1980).
26. V. Feiman and E. Luksha, U.S. Patents 4,247,609, 4,258,420 and 4,263,378 (1981).
27. C. C. Liang, P. W. Krehl, and D. A. Danner, *J. Appl. Electrochem. 11*, 563–571, (1981).
28. T. G. Hayes, D. R. Merritt, and P. M. Skarstad, unpublished results.
29. A. N. Dey, *J. Electrochem. Soc. 126*, 2052–2056 (1979).
30. K. A. Klinedinst, *Extended Abstracts*, Vol. 84-2, Abstract 136, The Electrochemistry Society, Pennington, New Jersey (1984).

31. N. Doddapaneni, *Extended Abstracts,* Vol. 81-2, Abstract 83, The Electrochemistry Society, Pennington, New Jersey (1981).
32. K. A. Klinedinst, *J. Electrochem. Soc. 128,* 2507–2512 (1981).
33. R. C. Weast, ed., *Handbook of Chemistry and Physics,* 59th Ed., CRC Press, W. Palm Beach, Florida (1978).
34. R. Selim and P. Bro, *J. Electrochem. Soc. 118,* 829–831 (1971).
35. K. A. Klinedinst, *Proc. 29th Power Sources Conference,* pp. 118–121, The Electrochemical Society, Pennington, New Jersey (1981).
36. K. C. Tsaur and R. Pollard, *J. Electrochem. Soc. 131,* 975–984, 984–990 (1984).
37. N. Marincic, *J. Appl. Electrochem. 5,* 313–318 (1975).
38. K. A. Klinedinst, *J. Electrochem. Soc. 131,* 492–499, (1983).
39. B. K. Patel, P. M. Skarstad, and D. F. Untereker, in: *Proc. Symp. on Lithium Batteries,* Vol. II (A. N. Dey, ed.), pp. 221–230, The Electrochemical Society, Pennington, New Jersey (1984).
40. W. G. Howard, R. C. Buchman, B. B. Owens, and P. M. Skarstad, to be published in *Proc. 14th Int. Power Sources Symposium,* Brighton, England, 1985.
41. K. R. Brennen, K. E. Fester, B. B. Owens, and D. F. Untereker, *J. Power Sources 5,* 25–34 (1980).
42. P. Bro, in: *Power Sources 7, Proc. 11th International Power Sources Symp* (J. Thompson, ed.), pp. 571–582, Academic Press, New York (1979).
43. H. F. Gibbard, in: *Proc. Symp. Power Sources for Biomedical Implantable Applications and Ambient Temperature Lithium Batteries* (B. B. Owens and N. Margalit, eds.), pp. 54–63, The Electrochemical Society, Princeton, New Jersey (1981).
44. H. F. Gibbard, *J. Electrochem. Soc. 125,* 353–358 (1978).
45. D. F. Untereker, *J. Electrochem. Soc. 125,* 1907–1912 (1978).
46. G. N. Lewis and M. Randall, *Thermodynamics,* revised by K. S. Pitzer and L. Brewer, 2nd Ed., p. 164, McGraw-Hill, New York (1961).
47. N. Godshall and J. Driscoll, *J. Electrochem. Soc. 131,* 2221–2226 (1984).
48. C. R. Schlaikjer, F. Goebel, and N. Marincic, *J. Electrochem. Soc. 126,* 513–522 (1979).
49. M. H. Miles, *J. Electrochem. Soc. 126,* 2168 (1979).
50. B. K. Patel and P. M. Skarstad, unpublished results.
51. R. C. Buchman, K. Fester, B. K. Patel, P. M. Skarstad, and D. F. Untereker, in: *Proc. Symp. Lithium Batteries,* Vol. II (A. N. Dey, ed.), pp. 212–220, The Electrochemical Society, Pennington, New Jersey (1984).
52. R. C. Buchman and K. Fester, Personal communication.
53. L. D. Hansen and H. Frank, in the 1983 Goddard Space Flight Center Battery Workshop (D. Baer and B. Morrow, eds.), pp. 109–116, NASA Conference Publication 2031 (1984).
54. N. Marincic, U.S. Patent 4,293,622 (1981).
55. F. Goebel and R. C. McDonald, U.S. Patent 4,416,957 (1983).
56. P. Kiester, J. M. Greenwood, C. F. Holmes, and R. T. Mead, Abstracts of 2nd International Meeting on Lithium Batteries, Paris, April, 1984, Abstract 32, pp. 70–71.
57. R. P. Elliot, *The Constitution of Binary Alloys,* 1st Supplement, p. 242, McGraw-Hill, New York (1965).
58. A DeHaan and H. Tataria, U.S. Patent 4,388,380 (1983).
59. R. V. Moshtev, Y. Geronov, and B. Pureshevea, *J. Electrochem. Soc. 128,* 1851–1857 (1981).
60. E. Peled and H. Yamin, *Israel J. Chem. 18,* 131 (1979).
61. C. R. Schlaikjer, U.S. Patent 4,020,240 (1979).
62. J. P. Gabano and P. Lenfant *Proc. Symp. Battery Design and Optimization* (S. Gross, ed.), Vol. 79-1, pp. 348–355. The Electrochemical Society, Princeton, New Jersey (1979).
63. J. P. Gabano and J. Y. Grassien, U.S. Patent 4,228,229 (1980).
64. D. Chua, W. C. Merz, and W. S. Bishop, *Proc. 27th Power Sources Symposium,* pp. 33–37, The Electrochemical Society, Princeton, New Jersey (1977).

65. T. Kalnoki-Kis, U.S. Patent 4,278,741 (1981).
66. V. O. Catanzarite, U.S. Patent 4,170,693 (1979).
67. D. H. Johnson, in: *Proc. 1982 Goddard Space Flight Center Battery Workshop,* NASA Conference Publication 2263 (G. Halpert, ed.), pp. 75–84 (1983).
68. M. Babai and U. Zak, *Proc. 29th Power Sources Symposium,* p. 150, The Electrochemical Society, Pennington, New Jersey (1981).
69. R. C. Buchman, personal communication.
70. W. G. Howard, personal communication.
71. K. M. Abraham and R. M. Mank, *J. Electrochem. Soc. 127,* 2091–2096 (1980).
72. S. Dallek, S. D. James, and W. P. Kilroy, *J. Electrochem. Soc. 128,* 508–516 (1981).
73. W. Shipman and J. F. McCartney, U.S. Patent 4,307,160 (1981).
74. V. F. Garoutte and D. L. Chua, in: *Proc. 29th Power Sources Conference,* pp. 153–157, The Electrochemical Society, Pennington, New Jersey (1981).
75. N. Marincic and F. Goebel, *J. Power Sources 5,* 73–82 (1980).
76. D. J. Salmon, M. E. Adcamczyk, M. E. Henricks, L. L. Abels, and J. E. Halls, in *Proc. Symp. Lithium Batteries,* (H. V. Venkatasetty, ed.), pp. 64–77, The Electrochemical Society, Pennington, New Jersey (1981).
77. A. Meitav and E. Peled, *J. Electrochem. Soc. 129,* 451–457, (1982).
78. A. N. Dey, *J. Power Sources 5,* 57–72 (1980).
79. A. N. Dey, Report DELET-TR-74-0109-F, Fort Monmouth, New Jersey (1978).
80. A. N. Dey, *Proc. 28th Power Sources Symposium,* pp. 251–255, The Electrochemical Society, Pennington, New Jersey (1978).
81. L. R. Giattino and V. O. Catanzarite, U.S. Patent 4,167.608 (1979).
82. W. V. Zajac, Jr., *Extended Abstracts 81-2,* Abstract 13, The Electrochemical Society, Pennington, New Jersey (1981).

9

Mercury Batteries for Pacemakers and Other Implantable Devices

ALVIN J. SALKIND and SAMUEL RUBEN

1. BACKGROUND

The first pacemakers powered by primary cells, circa 1958, utilized the zinc-alkaline electrolyte-mercuric oxide cell previously developed[1] for military and other special high-reliability requirements. These were commonly called RM cells for Ruben–Mallory, after the inventor Samuel Ruben and the principal manufacturer P. R. Mallory and Co., Inc. (now Duracell). Because of military needs during the early 1940s, these cells were also manufactured by Ray-O-Vac and Sprague and were labeled RMR and RMS. Although these cells were sealed, the seal was a plastic grommet that permitted hydrogen to vent from the cell. These cells had been designed to achieve high energy density per unit volume, a long storage (stand) life at elevated temperatures, and a constant flat discharge voltage. Figure 1 is a photograph of the three mercury zinc pacemaker cells that were manufactured by Mallory, LeClanché, and General Electric. The stability of this electrochemical battery system is illustrated by the data (Fig. 2), which plots open-circuit voltage versus time for batteries that were stored at 21°C, over a five-year time interval.[2] These design factors matched the electrical requirements of the early pacemakers, except for voltage, and five or six cells were used in series to achieve circuit voltages between 6 and 8 V.

ALVIN J. SALKIND • Department of Surgery, Bioengineering Section, UMDNJ–Rutgers Medical School, Piscataway, New Jersey 08854   SAMUEL RUBEN • Reed College, Portland, Oregon 97202.

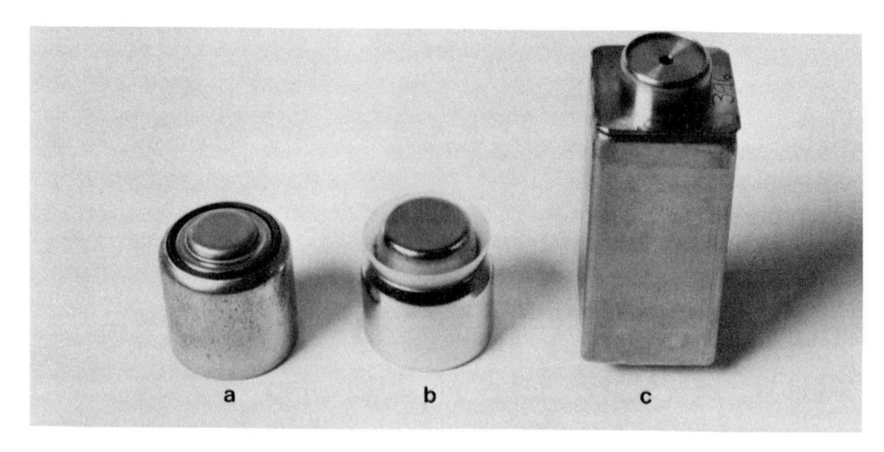

FIGURE 1. Mercury Pacemaker Cells: (a) Mallory RM-1, (b) LeClanché, (c) General Electric.

2. CHEMISTRY

A simplified overall reaction of the system is shown in Eq. (1).

$$Zn_{(s)} + HgO_{(s)} \underset{H_2O}{\overset{KOH}{\rightarrow}} ZnO_{(s)} + Hg_{(l)} \tag{1}$$

Some forms of zinc hydroxide may also be present.

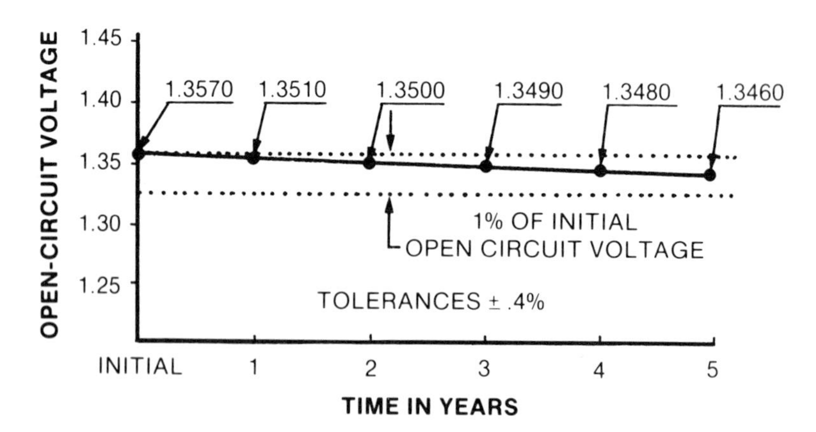

FIGURE 2. The Time Dependence of the Cell Voltage of RM-12R Mercury Cells After Indicated Storage Time, at 21°C.

The chemistry and atomic structures of the materials in the mercuric oxide positive electrode are such that the electrode is stable, as its voltage is lower than the reversible oxygen electrode in alkaline electrolyte, illustrated in the Pourbaix diagram (Fig. 3), and the thermodynamic voltage on discharge does not change from the open-circuit potential, since the reactant, mercuric oxide, and the product, mercury, are in different crystal habitats and the electrochemical activity is unity. The only observed difference in the voltage of the mercuric oxide electrode on discharge is from ohmic changes in the electrode, and these are very small.

The zinc electrode potential, on the other hand, is more negative than that of the reversible hydrogen electrode and depends on the local hydroxyl ion and zinc ion activity. The electrode and cell potentials are given in Figure 4.

The zinc electrode is thermodynamically unstable in alkaline electrolyte and tends to interact with the electrolyte, self-discharging and releasing hydrogen. This instability is greatly reduced by alloying the zinc active material with mercury, and approximately 10 wt % of mercury[3] was used in the pacemaker cells.[10] This raised the hydrogen overvoltage by approximately 200 mV (Fig. 5).

Although the voltage of the zinc electrode is dependent on the electrolyte concentration, the latter changes little during discharge, the overall cell voltage

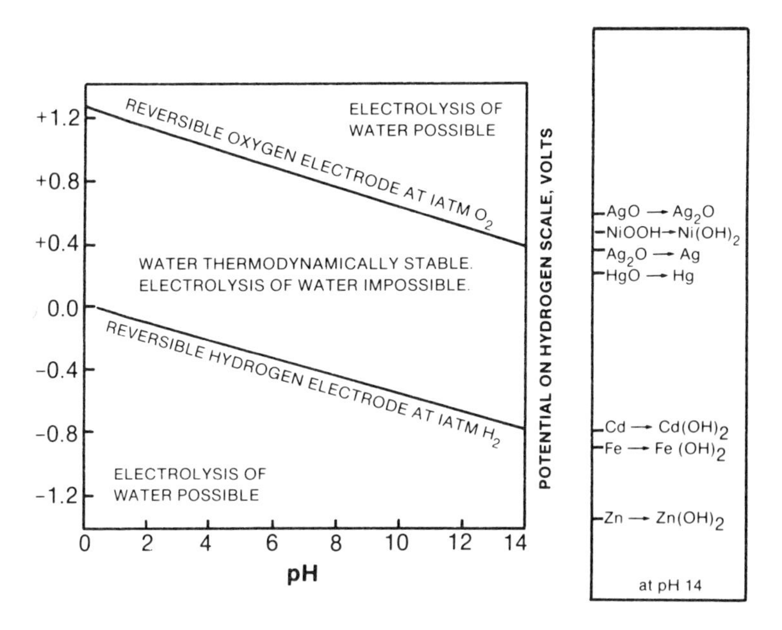

FIGURE 3. Pourbaix Diagram for Electrolysis of Water.

ELECTRODE POTENTIALS—The half cell potential in reference to the hydrogen electrode, expressed by the Nernst equation, is $E = E° - \frac{RT}{nF} \ln Q$, where Q is the product of the activities (a) of the resulting substances divided by the activities of reacting substances, each raised to that power whose exponent is the coefficient of the substances in the chemical equation. Applying this:

Anode: $E_{Zn;Zn^{++}} = E° - \frac{RT}{nF} \ln a_{zn^{++}}$. Since activity product $(a_{Zn^{++}})(a_{OH^-})^2$ is 4.5×10^{-17}, OH^- molality 7.7 with activity coefficient of 3.75 and n = 2, at 25° C and $\log_{10}$;

$$E_{Zn;Zn^{++}} = 0.76 - 0.0296 \log \frac{4.5 \times 10^{-17}}{(28.9)^2} = 1.332 \text{ volts.}$$

Cathode: $E_{HgO;OH^-} = E^o_B + \frac{RT}{nF} \log a^2_{OH^-}$

$E_B = Hg + 2OH^- \rightleftharpoons HgO + H_2O + 2_e- = -0.098$, HgO/OH^- activity = 1, n = 2. $E_{HgO;OH} = E^o_B + \frac{RT}{F} \log a_{OH^-} = -0.098 + 0.0592 \log 28.9 = -0.012$ volt. $E_{CELL} = E_{Zn;Zn^{++}} - E_{HgO;OH} = 1.332 - (-0.012) = 1.344$ volts, which is in close agreement with observed commercial cell potentials.

FIGURE 4. Electrode Potentials of the Zinc, Mercuric Oxide System.

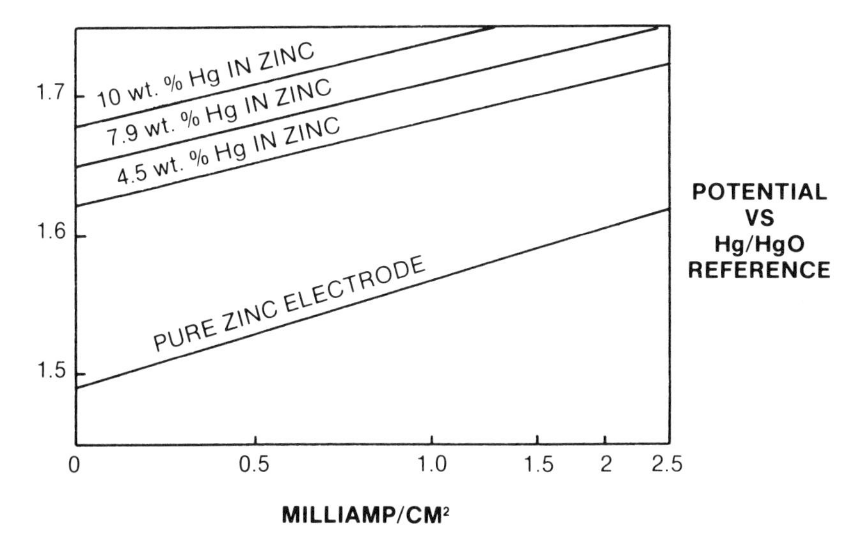

FIGURE 5. Hydrogen Overvoltage on Zinc–Amalgam Electrodes; All Pacemaker Cells Used 10 wt % Hg in Zinc.

was virtually constant during discharge, and batches of cells were extremely reproducible.

This is different from the behavior of other common cells, such as those with manganese dioxide (MnO_2), or nickel hydroxide ($NiOOH$) positive active materials, used, respectively, in Leclanché dry cells or nickel–cadmium cells. In these materials the discharge reaction involves mainly a proton transfer, and there is a decrease of the thermodynamic voltage during discharge, because the activity of the components in the Nernst equation is changing during reaction.

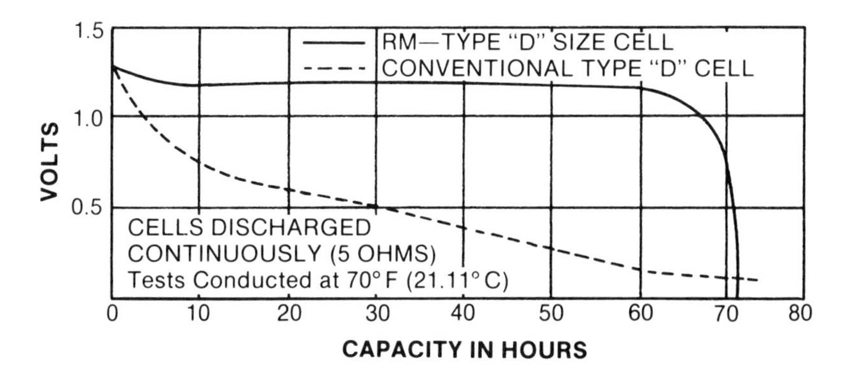

FIGURE 6. Discharge Curve of RM-1 Cell, Compared to Conventional-D Cell.

A discharge curve characteristic of a concentration cell is obtained and the open-circuit voltage decreases linearly with amount of discharge, as shown by comparison of the two types of discharge curves in Figure 6.

3. CELL DESIGN AND PERFORMANCE CHARACTERISTICS

To produce a sealed cell with alkaline electrolyte several important design factors were considered:

1. The electrolyte was saturated with zinc oxide to minimize local chemical action at the zinc electrode.
2. Contact materials and insulators were selected to provide good electrical properties without any consummable metals.

One problem inherent in a sealed alkaline cell structure is that if the depolarizer is exhausted before the anode is completely consumed but the cell is still connected to the load, there will be generation of hydrogen gas at the cathode (since the zinc anode potential of 1.32 V is more negative than the hydrogen deposition voltage of 0.989 V), with attendant high-pressure buildup in the cell. To overcome this problem, Ruben[7] developed the balanced cell, where the coulombic capacities of the mercuric oxide cathode and the zinc anode are equal, so that on completion of the cell's discharge there was no residual unoxided zinc. This was a very important consideration for sealed cells that would be encapsulated in a resin and implanted in the bodies of cardiac pacemaker recipients.

Other special factors had to be considered for pacemaker applications, specifically (1) means for preventing internal current-bridging paths caused by minute mobile droplets of mercury produced by low current density at the mercuric oxide cathode, and (2) semiconductive interelectrode paths of zinc oxide produced at the anode.

The immobilization of mercury was accomplished by adding manganese oxide and graphite to the mercuric oxide, the intergranular spaces of which served as a retainer for mercury, or, in later cells, by the complete immobilization of the mercury by adding silver to the mercuric oxide, forming a solid silver amalgam during operation. The prevention of zinc oxide semiconductive paths required the sealing off of all electrolytically conductive paths except that between the cathode and anode surfaces, which were separated by an ionically conductive spacer.

The combination of these factors provided the pacemaker cell of Figure 7 having internal chemical stability within a sealed container with freedom from atmospheric effects, pressures, or humidity, including the effect of carbon dioxide on the alkaline electrolyte. Two important mechanical features of this "Certified Cell" were the double cap terminal for improved mechanical scaling and

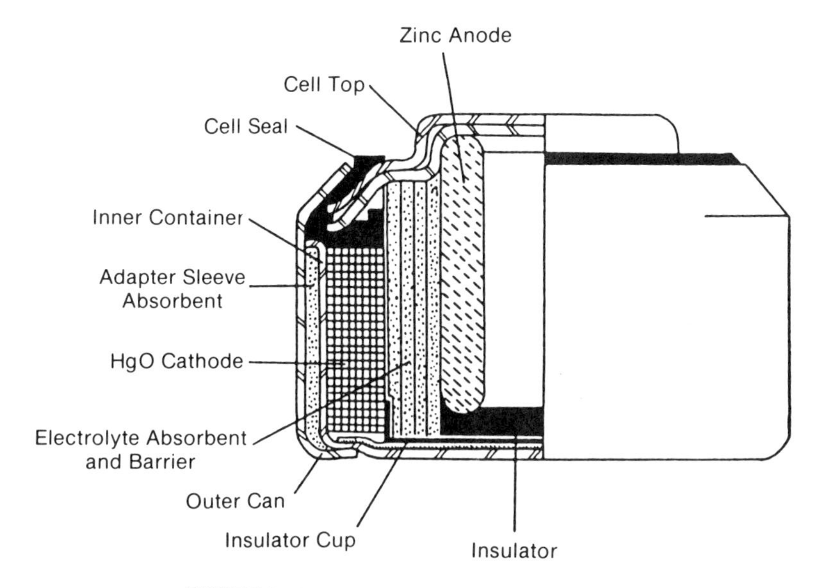

FIGURE 7. Cross Section of Certified Mallory Cell.

the cardboard sleeve used as an absorber between the outer can and the inner container. In a nonencapsulated application the double cap with grommet also acted as a pressure relief valve at a few atmospheres.

In the first 16 years of the wide use of implantable heart pacers (1960–1976), mercury cells were used as the power source in 99% of all pacers. Over 3 million cells were fabricated in a special high-reliability "Certified Cell" line by the Mallory Company (now called Duracell), principally with one size, the RM-1, cell (Fig. 1a and 7). Each cell was cylindrical, with the height and diameter both being close to 1.6 cm. These dimensions determined the thickness of the pulse generators, resulting in a total thickness of 2.0–2.5 cm. The weight of each cell was a little more than 12 g, and a battery of four to six cells typically represented about half the total weight of the pacemaker. Each cell had a voltage of 1.35 and nominal capacity of 1 Ah (for a load of 1250 Ω with a cutoff voltage of 0.9 at a temperature of 37°C).

Various improvements were incorporated in the cells over the years to improve their performance in pacemaker applications. They included:

1. The inclusion of a special insulator on the bottom to provide a long leak path.
2. The addition of silver rather than MnO_2 to the mercuric oxide, to immobilize completely the mercury formed on discharge (1969).
3. A welded connection between the inner and outer can (late 1969).
4. An improved double wrap separator (1973).

A problem during the initial growth period of pacemakers, when discrete electronic components were used, was the effect of epoxy impregnation on the battery. Vacuum impregnation, which was commonly used in pacer fabrication, led to epoxy intrusion between the inner and outer cans, hampering proper contact. The welded double-can cell, introduced in 1969, corrected this problem. New separator materials were also studied and the double-sealed Acropor(TM) and Permion,(TM) introduced in 1973, were the last significant changes. This cell is still used in some nonpacer applications where reliability and shelf-life are required factors. Its discharge characteristics at body temperature and 11,000 Ω are shown in Figure 8, compared with earlier cells with the $HgO/MnO_2/C$ cathode. The curves are virtually identical except that the HgO/Ag cathode does not have the early high-voltage area caused by MnO_2 and has a more uniform voltage and discharge life.

Pacemaker life increased with these changes and improvements in electronic circuitry and packaging, and by the mid-1970s, four to five yers of life was achieved with some designs.

According to Parker's 1978 review of early pacemakers,[6]

> designers believed that the battery would function in pacemaker applications for more than five years. This belief followed from a supposition that a 20 microampere drain on a cell with a capacity of 1 Ah represented a life of 50,000 hours, which corresponds to 5.7 years. Yet, it quickly became apparent in the early pacemakers, that pulse generator lifetimes were much shorter than the projected five years. Throughout the mercury era there was a progressive improvement in the mean time to failure from about 18 to 36 months. Manufacturers' reports on returned defective pulse generators

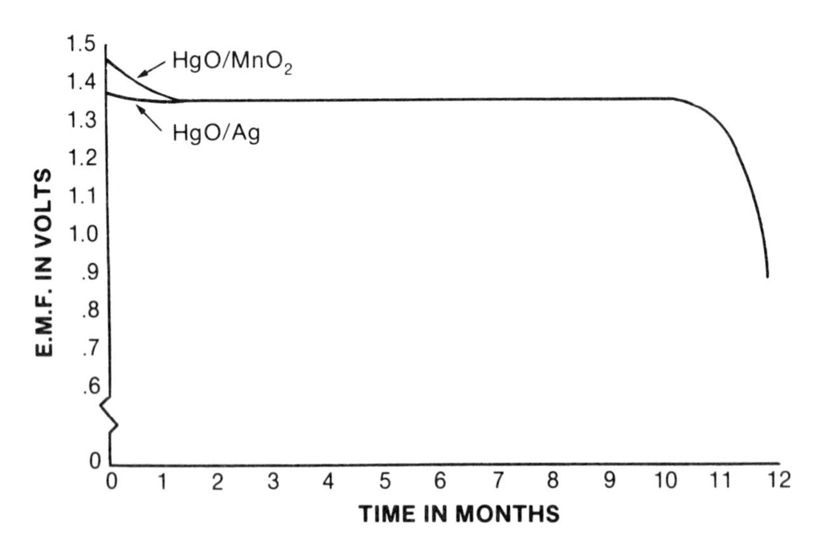

FIGURE 8. Discharge Characteristics Through 11 k Ω at 37°C, of HgO/Ag Cathode Cell Compared with HgO/MnO_2 Cell.

almost invariably attributed malfunction to premature battery failure. Failures due to other causes were quoted at less than three percent. Yet, there were puzzling aspects to this convenient "diagnosis." One discrepancy was the fact that representative cells kept at 37°C with a 67,000 ohm (20 microampere) load all seemed to last about five years. Also, while it is difficult for the user to determine whether pulse generator malfunction is accompanied by battery failure, one user study concluded that, of 175 defective pulse generators removed in 1973 and 1974, over 37 percent had good batteries. It is even more difficult for the user to determine whether a battery failure is secondary to some other failure. In only a few rare circumstances can non-destructive tests demonstrate the primary failure was not the battery. Yet, sufficient numbers of these rare circumstances have been noted, leading to the suspicion that battery failure was commonly secondary to another defect. Dr. F. Tyers, in particular, felt that plastic packaging was unsatisfactory and that it had been a major cause of premature pulse generator failure. Another anomaly is the case of at least one model of mercury-powered pulse generator which, so far, has shown a longevity similar to lithium pacemakers. Thus, there is considerable justification for the hypothesis that pulse generator malfunction is probably not due to battery failure, if it occurs earlier than one year before the expected end of battery life, based on coulombic calculations.

The earlier epoxy resins used to encapsulate pacer components and batteries (Fig. 9) became saturated with biological fluids. This provided a parasitic drain

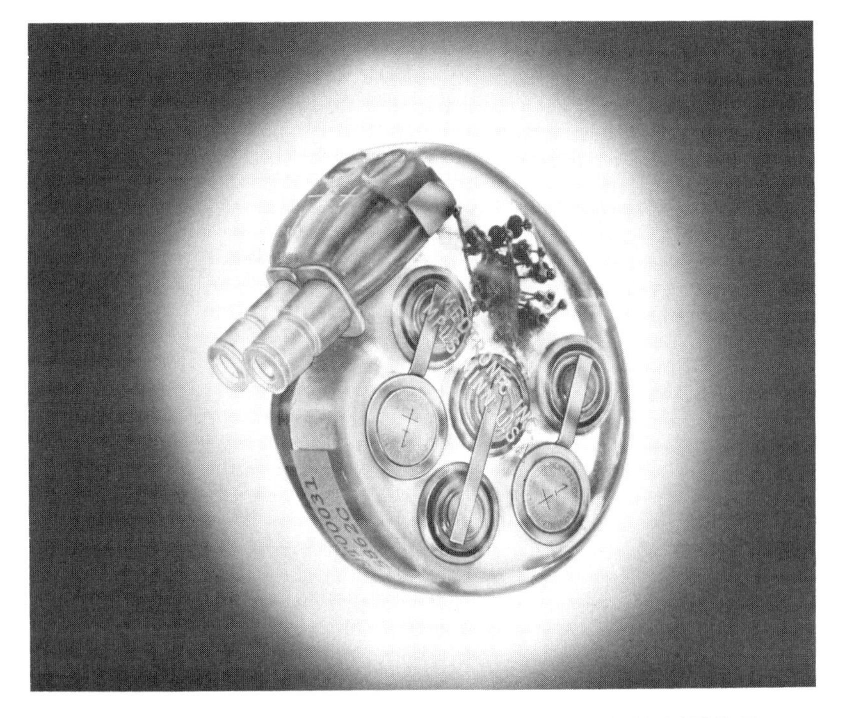

FIGURE 9. Epoxy Encapsulated Pacer with Cells (Medtronic Model 5862 C).

on the pacers equal to their working drain. In some designs the components were coated with a nonwater-permeable resin to minimize this phenomenon.

In the early 1970s, two other designs of mercuric oxide-zinc cells were fabricated for implantable pacemakers, one by H. Cataldi and colleagues at General Electric Company,[5] and the other by P. Ruetschi and associates at LeClanché S. A. (Switzerland).[4]

In the G.E. design (Fig. 10 and Fig. 1c), additional steps are taken to prevent internal short-circuiting. First, a chemical-resistant barrier material was used to completely wrap the anode (including top and bottom) to eliminate direct-shorting paths, such as around the open ends of a spiral wrapping (as did the

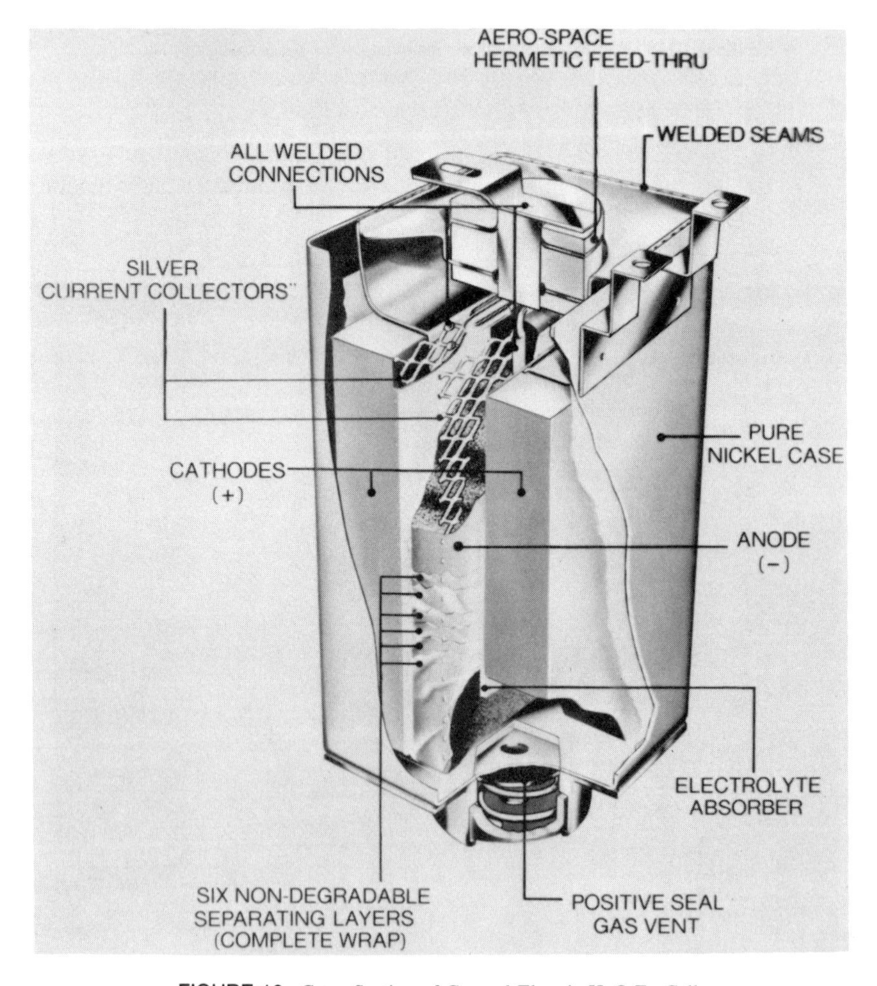

FIGURE 10. Cross Section of General Electric HgO/Zn Cell.

special pacemaker versions of the HgO/Zn cell made by Mallory). Six layers of this material between the electrodes minimized the possibility of penetration by liquid mercury or conducting zinc oxide crystals.[8] Finally, a corrosion-resistant Ni/Ti braze eliminated the possibility of shorting by corrosion products across the ceramic insulated feedthrough. Other design features were a spring-loaded valve that eliminated venting, at normal pressures, throughout the life of the cell and a nickel case and cover to preclude the accumulation of iron contaminant in the electrolyte.

The performance characteristics of prototype G.E. cells are given in Figure 11. With the modified design, the average rate of self-discharge at 37°C was reduced to 1–2% per year, and the projected life of implanted G.E. mercury cells was over eight years. Several hundred pacemakers manufactured by G.E. used this design with excellent life characteristics.[13]

Although the stable voltage of the mercury/zinc cell is a very desirable feature, without special precautions battery exhaustion can occur with little advance notice. To provide a warning of battery exhaustion, G.E. added lead to the zinc anode. (Lead also increased shelf life, as mentioned by Ruben.[1]) When zinc is the limiting electrode, after the zinc is consumed the cell voltage drops to about 0.7 V, the HgO/Pb plateau. The lower circuit voltage results in a corresponding change in the pacemaker rate of five to six beats per minute.

Ruetschi[4] identified internal side reactions and some external cell processes that were responsible for self-discharge. The four main causes of capacity loss are (1) solubility of mercuric oxide in alkaline electrolytes and diffusion of dissolved mercuric oxide species toward the zinc electrode; (2) hydrogen evolution at the zinc electrode; (3) external electrolysis in moisture films on cell

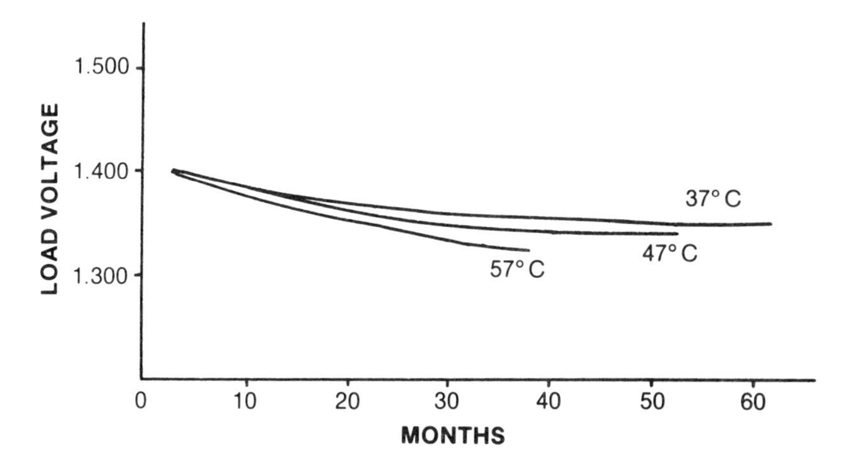

FIGURE 11. Performance Characteristics of General Electric Implantable HgO/Zn Cell at an Average Current of 37 μA.

surfaces or leads, including anodic oxygen evolution or anodic corrosion of metal parts, and cathodic hydrogen evolution; (4) external oxygen cycle, resulting from electrochemical reduction of oxygen on negative terminal parts and oxygen evolution on positive terminal parts.

In the LeClanché MR81 cells (Fig. 1b and Fig. 12), the electrodes were pellets, and the separators were locked at the edge between a grommet seal and a supporting steel ring. This design is comparable in size and capacity to the Mallory RM-1 cells, but round button cells and sealed cells with a glass feed-through were also fabricated (Fig. 13). With these special designs, self-discharge of 1–2% yearly at 37°C was claimed.[9]

In the early 1980s a new pacer fabrication facility was constructed in East Germany, based on the use of a mercury/zinc battery. At the time of this writing (1984), this remains the only pacer in production utilizing mercury batteries.

The zinc/mercuric dioxysulfate cell was another system developed by Ruben,[11,12] for lower current requirements, that has some of the desirable flat voltage discharge characteristics of the zinc/mercuric oxide alkaline cell while allowing the use of the less expensive conventional dry cell construction. In the past, attempts have been made to use mercury compounds in the cathode of a

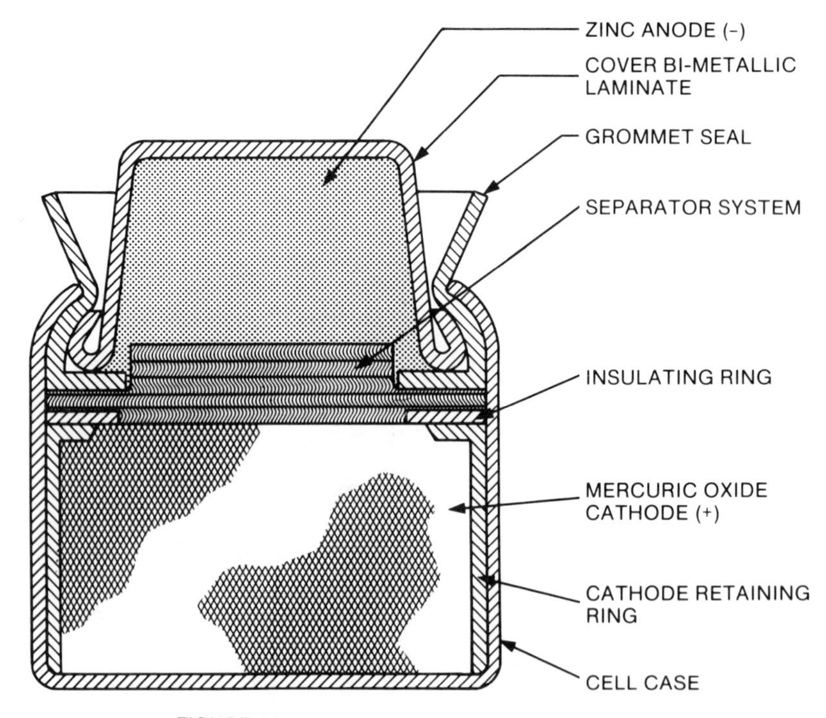

FIGURE 12. Cross Section of LeClanché Cell (MR 81).

FIGURE 13. LeClanché Mercury Cells Hermetically Sealed.

nonalkaline dry cell, but these have failed because of inherent limitations of the materials used. In this cell the cathode reactant is the basic mercuric sulfate, or mercuric dioxysulfate, and the electrolyte is a zinc sulfate solution in which the cathode is stable. The anode is amalgamated zinc. The electrochemical system in the presence of an aqueous solution of zinc sulfate can be expressed as

$$Zn/ZnSO_4/HgSO_4,\ 2HgO + C \quad (2)$$

with an overall reaction on discharge as

$$3Zn + HgSO_4 \cdot 2HgO \rightarrow ZnSO_4 + 2ZnO + 3Hg \quad (3)$$

The potential of the cell is 1.36 V. The theoretical capacity of $HgSO_4 \cdot 2HgO$ is 0.2204 Ah/g, and the practical capacity of the carbon mix is 0.15 Ah/g.

While this system can be used in a standard dry cell structure, one of the important characteristics of the zinc/mercuric dioxysulfate electrochemical system is that it allows the use of stainless steel containers for contact to the cathode. This permitted production of thin wafer-type cells of the structure shown in Refs. 11 and 12. The case was a shallow stainless steel cup. The cathode depolarizer was a pressed disc of mercuric dioxysulfate, Shawinigan carbon, and zinc sulfate solution containing 1% potassium dichromate as an inhibitor, in contact with the bottom of stainless steel. The spacer was laminate of paper and cellophane that separates the cathode from the amalgamated zinc disc anode. The anode had a polyethylene grommet around its edge, insulating it from the cathode container that is crimped against it for sealing the cell. The physical specifications of the Mallory WD5 wafer type cells were 2.54 cm diameter, 0.272 cm height, 1.37 cm^3 volume, and 4.9 g weight. The electrical characteristics were a capacity of 230 mAh and initial flash currents of approximately 0.5 A. Cell impedance was in the 10-Ω range at 1000 Hz. The system has excellent shelf life; experimental samples 10 years old show good voltage and flash current characteristics.

ACKNOWLEDGMENT. The authors want to express their appreciation for the generous cooperation in providing historical construction data and photographs to several individuals: Mr. Mike Firshein, Duracell Inc., Tarrytown, New York; Dr. Horace Cataldi, General Electric Medical Systems Business Group, Milwaukee, Wisconsin; Dr. Paul Ruetschi, LeClanché S.A., Yverdon, Switzerland; Dr. Jacques Mugica, Chirurgical Val d'Or, Saint-Cloud, France.

REFERENCES

1. S. R. Ruben, U.S. Patent 2,422,045 (June 10, 1947).
2. S. R. Ruben, Sealed zinc–mercuric oxide cells for implantable cardiac pacemakers, in: Advances in Cardiac Pacemakers, *Ann. N.Y. Academy of Science 167,* pp. 627 (1969).
3. A. J. Salkind and W. J. Raddi, Primary and secondary cells, *Ann. N.Y. Academy of Science 167,* pp. 636.
4. P. Ruetschi, Alkaline aqueous electrolyte cells for biomedical implantable applications, *J. Electrochem. Soc. 127,* 1667, (1980).
5. H. Cataldi, A mercuric-oxide-zinc cell for implantable cardiac pulse generator, *Proc. 27th Annual Power Sources Symposium* PSC Publications Committee, Red Bank, New Jersey (1976).
6. B. Parker, Obituary—a vindication of the zinc–mercury pacemaker battery, *Pace 1,* 148 (1978).
7. S. Ruben, Balanced alkaline dry cells, *Trans. Electrochem. Soc. 92,* 183 (1947).
8. H. Cataldi, private communication to A. J. Salkind (1983).
9. P. Ruetschi, private communication to A. J. Salkind, March 23, 1983.
10. M. Firshein, private communication to A. J. Salkind, March 7, 1983.
11. S. Ruben, Sealed Mercurial Cathode Cells, Proceedings CITCE 17 meeting, Tokyo, Japan (1967).
12. S. Ruben, Zinc–mercuric dioxysulfate dry cell, *J. Electrochem. Soc. 106,* 77 (1959).
13. J. Mugica, Chirurgical Val d'Or, private communication to B. Owens (1983).

10

Rechargeable Electrochemical Cells as Implantable Power Sources

GERHARD L. HOLLECK

1. INTRODUCTION

The concept of using a rechargeable cell for applications requiring delivery of a high energy per weight or volume over an extended time is both attractive and technically sound. Even a fairly low-energy-density Ni/Cd cell needs to be recharged only about 10 times to deliver the same energy as a modern high-energy-density lithium cell. One may ask, then, why there are no secondary cells used in any of today's pacemakers. The reason can be found in the miniaturization of electronic circuits and the reduction in required power levels coupled with the development of high-energy-density, long-lived reliable electrochemical primary cells that are able to satisfy the majority of present needs.

If power and energy requirements increase again as implantable electronic devices become more diversified and more ambitious in their functions, the use of rechargeable power sources may be reconsidered, in particular, if a reliable long-lived high-energy-density secondary cell (e.g., based on a rechargeable lithium system) can be developed. Unfortunately, the chemical and physical requirements for the components of secondary cells are much more restrictive than those needed for primaries. Therefore, the list of viable candidates is very narrow. In fact, to date only two rechargeable systems, nickel oxide/cadmium cells and mercuric oxide/zinc cells have been successfully implanted in human patients.

GERHARD L. HOLLECK • Battery Division, EIC Laboratories, Inc., Norwood, Massachusetts 02062.

In the following these two systems and their applications will be discussed in more detail.

2. NICKEL OXIDE/CADMIUM CELLS

2.1. Brief History

The first totally implantable cardiac pacemaker, developed and clinically used by Senning and Elmquist in 1958, was powered by nickel oxide/cadmium cells.[1] The 60-mAh two-cell battery was inductively recharged through the intact skin. However, despite rechargeability it had only a short functional life. A moving account of this early phase of pacing has been given by Lagergren.[2] In the following years several additional attempts were made to develop implantable pacemakers with rechargeable nickel oxide/cadmium cells as power sources.[3–5] These devices employed commercially available nonhermetically sealed cells that proved unsuitable for operation at body temperature at which they suffered rapid capacity degradation. During the late 1960s, taking advantage of the development of sealed nickel oxide/cadmium cells for use in satellites, members of the Johns Hopkins University Medical Institutions and the Applied Physics Laboratory adopted this technology for use at body temperature.[6–8] This led to the formation of Pacesetter Systems, Inc., which developed rechargeable nickel oxide/cadmium-based pacemakers and produced devices for clinical use. Approximately 5000 rechargeable pacemakers have been successfully implanted of which about 1500 remain still in patients.[9] Fabrication of rechargeable pacemakers has been discontinued. However, this decision appears to be based maninly on marketing considerations. A clinical evaluation of the rechargeable pacemaker system covering 66 patients and 194 person years found an overall failure rate of 3% per year and a difficulty in some patients to accept the recharging concept.[10] This failure was judged normal for the time and patient screening was used to counter the acceptability problem. There appeared to have been no technical problems with the power source itself.

2.2. General Nickel Oxide/Cadmium Cell Characteristics

Nickel oxide/cadmium cells and batteries are fabricated in a great variety of design types and sizes. This includes vented and sealed configurations for such diverse applications as airplane starting to powering of small consumer electronics devices. A comprehensive description of the system can be found in the monograph by Falk and Salkind[11] and in review articles.[12] The following discussion will be restricted to sealed cell configurations and to the design and operating tradeoffs relevant for its use as a power source in biomedical implantations.

2.2.1. Cell Reactions

The cell reactions, resulting in a nominal discharge voltage of 1.25 V, can be described by the following simplified equation:

$$2NiOOH + Cd + 2H_2O \underset{\text{charge}}{\overset{\text{discharge}}{\rightleftharpoons}} 2Ni(OH)_2 + Cd(OH)_2$$

In reality, the processes are considerably more complicated as shown in the review by Milner and Thomas.[13] The NiOOH and $Ni(OH)_2$ contain differing amounts of hydrated or absorbed water and electrolyte. Also, the oxidation states do not precisely vary between 3 and 2. Extensive crystallographic studies revealed at least two reactions with different starting materials and different products that were not in equilibrium with each other.[14] Thermodynamically, this means that neither reaction series can be expressed by a simple, reversible potential. This is reflected in a variation of the nickel oxide potential with the state of charge and in a considerable hysteresis in the voltage between charge and discharge.

The chemistry, electrochemistry, and crystal structure of the cadmium electrode is much simpler. The cadmium electrode maintains a constant E_o potential throughout the discharge, as expected from electrodes where the charged and discharged states form independent distinct crystalline forms. Contrary to the nickel oxide electrode, there is, however, a soluble intermediate involved in the formation of $Cd(OH)_2$. This leads to an appreciable cadmium solubility in the order of $2 \cdot 10^{-4}$ mol/l of 8 M KOH at room temperature in the form of complex hydroxides resulting in relatively large $Cd(OH)_2$ crystals and in cadmium migration into the separator matrix.

2.2.2. Cell Construction

To construct a sealed system, it is necessary to consider not only the discharge–charge reaction but also the reactions that occur on overcharge and overdischarge. Sealed nickel oxide/cadmium cells operate on an oxygen cycle by appropriate adjustment of the anode and cathode capacities as well as their state of charge prior to cell sealing. Typical anode–cathode capacity ratios are 1.5–1.8. Thus, oxygen is generated at the cathode at the end of charge that can be reduced at the cadmium electrode. A precharge or $Cd(OH)_2$ addition to the nickel-oxide electrode can be used to assure that overdischarge leads also to an oxygen cycle. Significant hydrogen generation has to be avoided, since it does not react fast enough on either electrode and would lead to pressure buildup. Rapid oxygen transport from the cathode to the anode requires open (nonelectrolyte filled) pores in the separator matrix. Thus, careful separator selection and adjustment of the electrolyte level is necessary. Some oxygen evolution occurs already during charge of the nickel oxide electrode and at low to moderate charge rates, no clear transition in potential is observed between charge and overcharge. Recharge ratios between 1.1 and 2 are common. A high-quality sealed cell may consist of the following: (1) highly porous nickel sinter matrix electrodes into which the anode and cathode active materials are deposited, (2) nonwoven separator mats of nylon or polypropylene, (3) potassium-hydroxide-based electrolyte to fill the pore volume of the electrodes and approximately 70–80% of the pores in the separator, (4) a welded stainless steel case with ceramic seal terminal.

2.2.3. *Cell Operation at Body Temperature*

An ideal operating temperature for NiCd cells is 10–25°C. The elevated temperature of 37°C results in reduced charge acceptance, accelerated self-discharge, larger $Cd(OH)_2$ crystals, and gradual hydrolytic degradation of nylon separators. Such shorcomings can be counteracted by specialized cell design.

The presence of cobalt hydroxide in the cathode and in particular additions of Li ions to the electrolyte improve charge acceptance at elevated temperature.[15,16] Self-discharge can be minimized by careful exclusion of impurities during manufacture that could lead to chemical shuttles. For example, nitrate ions can be electrochemically reduced at the Cd electrode to nitrite, which then diffuses to the nickel oxide electrode, where it is reoxidized. Thus, a discharge load is imposed on the cell, its magnitude depending on the diffusional transport of the redox species. Even in well-built cells, observed self-discharge rates are still appreciable (e.g., 0.3–0.7% per day at room temperature and three to four times as high at 37°C). This is not detrimental for a rechargeable cell, but it places a limit on the maximum operating time between recharges.

2.3. The Nickel Oxide/Cadmium Pacemaker Cell

The rechargeable nickel oxide/cadmium cell successfully used by Pacesetter Systems, Inc. to power cardiac pacemakers has been specifically designed and constructed for long-term use at temperatures of 38°C.[6–8,17] Cell characteristics are summarized in Table 1.

The cell is operated at a very mild depth of discharge, which assures a considerable margin on cycle life and reserve capacity. A weekly recharge for 10 years amounts only to 520 cycles. Thus, it is probable that time-dependent rather than cycle-number-dependent changes eventually will lead to performance degradation. In accelerated and real-time testing over a four-year period no significant degradation in either end of discharge voltage or capacity has been detected.[17] At body temperature the cell can be safely overcharged at the relatively high five-hour rate. Prior to implantation the cell is maintained in the fully charged state by a lower trickle charge current. Recharge *in vivo* occurs via an alternating magnetic field of 25 kHz through the skin. Without recharge the cell can operate the pacemaker for eight weeks before its energy is depleted.

TABLE 1 Specifications for the PSI Power Cell

Nominal discharge voltage	1.265 V
Nominal capacity	190 mAh at 38°C
Charge rate	40 mA
Recommended recharge cycle	60 min/week
Nominal depth of discharge	15%
Cell internal impedance	0.050 Ω
Cell size	2.4 cm diameter
	0.95 cm thick
Cell volume	4.3 cm^3
Cell weight	12 g
Seal type	Hermetic
Energy density (volumetric)	
15% depth	0.0084 Wh/cm^3
100% depth	0.056 Wh/cm^3

This time limit is, in part, due to internal cell self-discharge. The fraction of used energy over energy lost via self-discharge becomes less favorable as the time between recharges is increased. A significant extension of the recharge interval is therefore limited to the ability to control the rate of self-discharge. Pacesetter Systems, Inc. achieved values of ~0.3%/day at body temperature, which allowed extension of the recharging period for a 1-Ah cell to once every six months.[18] Under the controlled conditions of pacemaker operation, cell voltage can be used as a state of charge monitor. In general, for randomly used nickel oxide/cadmium cells, this is not a reliable parameter. To avoid potential complications resulting from mismatched series connected cells, Pacesetter Systems, Inc., used only a single cell. Voltage amplification of the pulse was achieved by an output transformer in the pacing circuit.

3. RECHARGEABLE MERCURIC OXIDE/ZINC CELLS

3.1. Brief History

Generally, mercuric oxide/zinc is thought of as a primary system, but by modifications in the electrode structure it can be recharged.[18,19] Since the recharge is not without difficulty and has to be closely controlled, it has found only limited commercial application. Serious consideration for use in biomedical applications became feasible only after development of an improved high-reliability cell by Fagan.[20] Experimentation to explore recharging of mercuric oxide/zinc cells to extend their use as power sources for heart pacemakers began in 1969 by a medical team at the Pennsylvania State University.[21] The initial preclinical test involved unmodified cells that were soon replaced by improved cells optimized for low-drain application. Laboratory tests were conducted on about 130 cells in various accelerated regimes.[22,23] This was followed by development of hermetically sealed units with control circuitry for pacing recharging and monitoring telemetry[24,25] and by clinical evaluation. In a recent interview, Tyers *et al.* reported that cell implanted in dogs and tested in biological simulators have functioned for 10 years and delivered in excess of 20 Ah at potentials between 1.2 and 1.6 V.[26]

Ten clinical prototype units constructed at the Pennsylvania State University were implanted, beginning in 1974, in patients ranging in age from their teens to their 80s. After five years, all 10 units continued to function normally and nine of 10 patients were alive, seven paced by their original rechargeable unit. One of the mercuric oxide/zinc rechargeable units was explanted following an unrelated patient death, a second at the time of a lead break and the third because of threshold problems. The three explanted units were implanted in dogs with surgically created complete heart block and continued to function normally.

In spite of this demonstration of performance and reliability, rechargeable mercury oxide/zinc-cell-powered pacers were never introduced commercially, probably because of the outstanding competitive simplicity and reliability of the primary lithium anode cells.

3.2. Cell Chemistry and Construction

The cell reaction in the rechargeable cells is the same as that in the primary and can be described in simplified form by

$$HgO + H_2O + Zn \underset{\text{charge}}{\overset{\text{discharge}}{\rightleftharpoons}} Hg + Zn(OH)_2 \qquad E_o = 1.343 \text{ V}$$

To allow the cell to be recharged the agglomeration of elemental mercury has to be prevented and a conductive matrix for the replating of zinc has to be provided. In the cell configuration patented by Fagan,[20] finely divided silver power (~30% by weight) is added to the mercuric oxide. This mixture provides excellent conductivity, and as the mercuric oxide is reduced to mercury it is held in place by amalgamation with the silver. On charge the mercury is easily reconverted to mercuric oxide without change in electrode dimension or structure. The zinc electrode contains also approximately 10 wt % silver powder in a conductive epoxy matrix, providing a high area surface for replating zinc during charge. The zinc is amalgamated to suppress the thermodynamically favored H_2 gas evolution. To prevent hydrogen evolution and to provide good electrical contact the inner cell can surfaces are also silver plated. (Silver has a high hydrogen overvoltage.) The two electrodes are separated from each other by a multilayer arrangement of ionically permeable, electronically nonconductive microporous separators. A typical sequence may be as follows: microporous polyvinyl chloride, irradiated polyethylene (e.g., Permion), regenerated cellulose, irradiated polyethylene, regenerated cellulose. Such multilayer separators are necessary because the required properties of electrolyte absorption—impermeability to particulate material and long-term stability toward oxidative chemical attack—are not obtained with any one material alone. The separators and electrodes are saturated with potassium hydroxide electrolyte.

In practical cells the material balance is adjusted such that the cell is anode (Zn electrode) limited on discharge and cathode (HgO electrode) limited on charge. This can be achieved with a Zn/ZnO anode mixture in which the Zn equivalent is less and the combined Zn and ZnO equivalent is greater than the HgO equivalent. Continuous overcharge has to be avoided, since this would lead to gas evolution, which cannot be accommodated in the cell without overpressurization. In controlled charging applications this does not present a problem, since full charge is indicated by a clear voltage change. During charge the cell voltage raises from 1.4 to 1.6 V. After all Hg is converted to HgO, continued charge would convert Ag to Ag_2O, which occurs at an increased voltage of about 1.7 V in this system, and eventually oxygen would be generated. To avoid this, charging has to be voltage limited.

3.3. Cell Performance

The cells used in the investigations of Tyers *et al.* were 36 mm in diameter and 6 mm thick of the type described by Fagan.[20] The long time performance of these rechargeable mercuric oxide/zinc cells in low-drain operation at elevated temperature (37°C) is impressive. The energy density of the cells is 0.5 Wh/cm^3, and over a seven-year period 9 Wh/cm^3 have been delivered.[(27)]

The self-discharge rate of the rechargeable mercury oxide/zinc cells is equal to that of the primary and may be as low as 1 to 2% per year. Ruetschi[(28)] gave an excellent account of various internal and external processes that contribute to the overall observed self-discharge rate. One of the mechanisms is hydrogen evolution at the zinc electrode, since its potential is considerably more negative than that of the hydrogen electrode. In heavily amalgamated pure zinc electrodes this can be as low as 1 cm^3 H_2 g^{-1} $year^{-1}$. Tyers measured on his cells the generation of 6–7 cm^3 H_2/year,[(25)] which slowly diffused through the crimped seal. To provide a hermetically sealed unit the electrochemical cell was packaged together with palladium sponge as hydrogen absorber into a soldered metal can. Approximately 185 cm^3 of H_2 can be bound by 1 g Pd sponge ($\sim$0.2 cm^3).

Extended cycle testing of 150 cells in laboratory simulators under accelerated conditions was reported by Tyers *et al.*[(27)] Accelerations of 8 and 250 times translate into actual currents of 200 μA and 6.4 mA, respectively. Cells discharged at 200 μA were recharged every 22 h or after complete discharge (five to six months) and continued to function for over seven years, delivering in excess of 12 Ah. The possibility of recharging the cells at relatively high currents (20–40 mA) permits the use of short recharge times. Two group of cells discharged at 6.4 mA were recharged after 40 min and 43.2 h, corresponding to 0.43% and 28% depth of discharge, respectively. The first failure in the latter group occurred after 2.7 months, with a mean life of 6.7 months. The cells cycled to shallow depth remained on test 1.3 to 2 years, after which time the capacity had decreased to about 25% of its original value. Over this time a cumulative capacity of 100 Ah had been delivered. The failure mode appeared to be a gradual loss in usable capacity.

The results suggest that the mercuric oxide/zinc cells can be recharged repeatedly after full discharge and a large number of cycles (10,000 to 20,000) can be realized at low depth of discharge. With a pacemaker current requirement of 25 μA, Tyers *et al.* extrapolate 50 to 500 years of pacing equivalent. Together with the seven-year real-time test experience a life expectancy of 10 to 15 years appears realistic.

4. PROSPECTS FOR FUTURE USE OF RECHARGEABLE CELLS

At present only primary electrochemical cells are used in biomedical implantable devices. It appears, however, that except for the inconvenience of periodic recharging the Pacesetter-type nickel oxide/cadmium cell and the Fagan–Mallory mercuric oxide/zinc cell can achieve equal lifetimes and reliabilities. Among the advantages of the secondaries are a smaller physical size in a given application, since the capacity can be used repeatedly, a low impedance, and a relatively high current capability. Cumulative delivered energies per unit volume are significantly higher than those for any primary cell. This may become important if higher power levels are needed for new applications or more ambitious diagnostic monitoring via telemetry from implanted sensors.

Most other conventional rechargeable systems appear unsuitable for long-lived implantable biomedical devices for various reasons, with the possible exception of the mercuric oxide/cadmium couple. A relatively low cell voltage (0.92 V) and the need for cell development and performance demonstration will probably inhibit its practical use.

In the future, rechargeable lithium anode systems (e.g., TiS_2/Li cells), which are presently under active development mainly with orientation toward military application, will become available and may prove suitable for implantable biomedical applications. They promise a high energy density and low self-discharge rate. A current and comprehensive review of secondary lithium cells by Abraham *et al.* has been published recently.[29] The discussion of the more conventional rechargeable systems demonstrated the need for extensive specific cell development to achieve the performance and reliability parameters required for long-term biomedical applications. The need for a similarly extensive effort can be expected for the development of a suitable rechargeable lithium cell. Given this the prospects appear quite promising.

REFERENCES

1. R. Elmquist and A. Senning, An implantable cardiac pacemaker, *Medical Electronics* (N. Smythe, ed.), pp. 253, Iliff & Son, London (1960).
2. H. Lagergren, *Pace 1,* 140–143 (1978).
3. H. Siddons and O'N. Humphries, *Proc. Roy. Soc. Med. 54,* 237 (1961).
4. S. Furman, W. Raddi, P. Escher, J. Schnebel, and S. Horwitt, *Arch. Surg. 91,* 796 (1965).
5. A Silver, G. Root, F. X. Byron, and H. Sandberg, *Ann. Thorac. Surg. 1,* 380 (1965).
6. P. W. Barnhart, R. E. Fischell, K. B. Lewis, and W. E. Radford, A fixed-rate rechargeable cardiac pacemaker, Johns Hopkins University, *Appl. Phys. Lab. Techn. Dig. 9,* 2 (1970).
7. K. B. Lewis, R. E. Fischell, and J. W. Love, *Circulation 40* (Suppl. III), 132 (1969).
8. J. W. Love, K. B. Lewis, R. E. Fischell, and J. Schulman, *Ann. Thorac. Surg. 17,* 152 (1974).
9. J. H. Schulman, Pacesetter Systems, Inc., personal communication.
10. S. H. Stertzer, N. P. DePasquale, L. J. Cohn, and M. S. Bruno, *Pace 1,* 186–188 (1978).
11. S. U. Falk and A. J. Salkind, *Alkaline Storage Batteries,* Wiley, New York (1969).
12. A. J. Salkind, Batteries, Secondary (Alkaline Cells), *Encyclopaedia of Chemical Technology* Vol. 3 (Kirk-Othmer, ed.), Wiley, New York (1978).
13. P. C. Milner and U. B. Thomas, The nickel–cadmium cell, in: *Advances in Electrochemistry and Electrochemical Engineering,* Vol. 5 (P. Delahay and C. W. Tobias, eds.) Interscience, New York (1967).
14. H. Bode, K. Dehmelt, and J. Witte, Nickel Hydroxide Hydrate, paper presented at CITCE Meeting, Strasburg, France (1965).
15. E. J. Rubin and R. Baboian, *J. Electrochem. Soc. 118,* 583 (1970).
16. M. Oshitani, M. Yamane, and S. Hattori, *Power Sources,* Vol. 8 (J. Thompson, ed.), Academic Press (1980).
17. R. E. Fischell and J. H. Schulman, *Proc. 11th IECEC, American Inst. of Chemical Engineers,* pp. 163, New York (1976).
18. S. Ruben, U. S. Patent 2,554,504 (May 29, 1951).
19. S. Ruben, The mercuric oxide–zinc cell, in: *The Primary Battery* (G. W. Heise and N. C. Cahoan, eds.), pp. 207–222, Wiley, New York (1971).
20. F. G. Fagan, U.S. Patent, 3,824,129 (March 14, 1973).
21. G. F. O. Tyers, R. H. Foresman, C. K. Park, E. H. Lerner, H. A. Torman, and J. A. Waldhausen, *J. Thorac. Cardiovasc. Surg. 62,* 763–768 (1971).
22. G. F. O. Tyers, R. A. Foresman, Jr., R. R. Brownlee, C. Volz, N. J. Manley, and J. B. Dixon, *J. Surg. Res. 16,* 262–267 (1974).
23. F. G. O. Tyers, R. R. Brownlee, H. C. Hughes, Jr., C. Volz, N. J. Manley, and J. A. Waldhausen, *28th Proc. Am. Conf. Eng. Med. Biol.* (C. Chase, ed.), pp. 92, New Orleans, Louisiana (1975).
24. R. R. Brownlee, G. F. O. Tyers, C. Volz, Sr., U.S. Patent 4,014,346 (March 29, 1977).
25. G. F. O. Tyers, R. R. Brownlee, H. C. Hughes, Jr., J. H. Danadii, and C. Volz, *J. Surg. Res. 20,* 405–411 (1976).

26. G. F. O. Tyers and R. R. Brownlee, *Prop. Cardiovasc. Diseases* 2, 421–434 (1981).
27. G. F. O. Tyers, H. C. Hughes, Jr., R. R. Brownlee, N. J. Manley, and I. N. Goreman, *Am. J. Cardiol. 38,* 607–610 (1976).
28. P. Ruetschi, *Power Sources,* Vol. 7, pp. 533 (J. Thompson, ed.), Academic Press, London (1979).
29. K. M. Abraham and S. B. Brummer, Secondary lithium cells, in: *Lithium Batteries* (J. P. Gabano, ed.), Academic Press, New York (1983).

11

Nuclear Batteries for Implantable Applications

DAVID L. PURDY

1. GENERAL DESCRIPTION OF NUCLEAR BATTERIES

1.1. Description of Isotopic Decay

The nuclear battery is so named because its source of energy is derived from energy stored in the "nucleus" of the atoms of the fuel, rather than in the electrons that surround the nucleus and that are the fundamental source of energy for the chemical batteries described elsewhere in this book. Since the energy stored in the atom's nucleus is immense compared with the chemical energy stored in the electron shells around the nucleus, the nuclear battery promises to be very powerful, small, and light compared with its chemical counterpart.

There are three phenomena that provide access to the energy stored in the nucleus: fission of the nucleus, fusion of the nucleus, and radioactive decay of the nucleus. Fission of the nucleus provides the source of heat for commercial nuclear reactors and for atomic weapons; fusion is being studied as a commercial source of heat and has been used for atomic weapons; and radioactive decay has been used as a source of heat for a variety of nuclear batteries for aerospace, oceanographic, and medical applications. Both fission and fusion reactors require large amounts of nuclear material to sustain the reactions, and thus are unsuitable for small, nuclear batteries. Radioactive decay is the only energy-producing process applicable to implantable devices.

DAVID L. PURDY • Coratomic, Inc., Indiana, Pennsylvania 15701-0434.

Radioactive decay is the process by which unstable atomic nuclei disintegrate into other nuclei, with the release of subatomic particles. To understand the process a brief review of atomic structure and types of radioisotopic decay is in order.

The nucleus of the atom is composed of two subatomic particles, protons and neutrons. The proton is positively charged, the neutron is neutral, and both have approximately the same mass. The number of positively charged protons in the nucleus is the same as the number of negatively charged electrons that surround the nucleus in successive shells. This number is the element's atomic number. The number of electrons in the outer shell determines the chemical properties of the atom, and definitive rules govern the number of shells and the number of electrons in each shell.

The number of neutrons in the nucleus of the atom of each element can vary, and the term *isotope* is used to define the different species belonging to the same element. For example, helium, with atomic number 2, has two protons in its nucleus but can have from one to four neutrons. The most common form of helium comprising almost 100% of the element found in nature has two neutrons. Oxygen has eight protons in its nucleus but can have from six to 11 neutrons. Of the oxygen found in nature, 99.76% has eight neutrons, with the eight protons, this oxygen has an atomic weight of 16. This isotope of oxygen is oxygen 16 and is designated $_{8}O^{16}$.

Isotopes are either stable, which means they do not change their nuclear composition with time, or radiactive. The degree of stability is related to the number of neutrons in the nucleus; if too many or too few neutrons exist, the nucleus is unstable and the nucleus disintegrates, giving off subatomic particles and changing into another nucleus and, hence, another element. This is the process of radioactive decay.

Radioactive isotopes can occur in nature or can be produced artificially in nuclear reacators or particle accelerators. Such isotopes as radium or uranium occur naturally, those such as promethium or plutonium are the results of reactions in a nuclear reactor. When radioisotopes decay, they do so by the release of subatomic particles, most commonly alpha particles, beta particles, or positrons. Alpha particles are helium nuclei, or two protons and two neutrons; beta particles are electrons; and positrons are positively charged particles with the mass of an electron. Often gamma photons, a form of electromagnetic radiation, are also emitted. When the decay is by means of an alpha particle, the nucleus loses two neutrons and two protons. Thus, its atomic number decreases by two because of the loss of the protons, and its mass number decreases by four because of the loss of both the neutrons and protons. Thus, the isotope of plutonium $_{94}Pu^{238}$, which decays by alpha emission, becomes $_{92}U^{234}$, which is a naturally occurring radioisotope of uranium. If the decay is by means of a beta particle, the atomic number increases by one, but its mass number stays the same. The isotope of prometheum, $_{61}Pm^{147}$, which decays by beta emission, becomes $_{62}Sm^{147}$, which is a naturally occurring radioisotope.

1.2. Types of Nuclear Batteries

When the radioisotope's nucleus decays and emits particles, energy is released in the form of the kinetic energy of the particles. This energy can then be used to create electric power in two ways. The first is by the use of the betavoltaic effect, in which the electrons coming from the nucleus of a beta emitter are caused to impinge on a semiconductor, giving off electrons, which are then converted to electron current. This current provides electric power to drive the circuitry of an implantable device. One such battery, manufactured by the Donald W. Douglas Laboratories and termed the Betacel, has been used clinically to power implantable pacemakers.

The second method by which the nuclear disintegration energy can be utilized is to allow the kinetic energy of the emitted particles to be converted into thermal energy and then to convert this thermal energy into electrical power by means of the Seebeck effect, or thermoelectric conversion.

The Seebeck effect is in evidence when two dissimilar metals or semiconductors are joined together at two junctions and the two junctions are held at different temperatures. When this occurs a voltage is created that is proportional to the difference in temperature of the two junctions, the proportionality constant being different for different materials and being much higher for semiconductors than for metals. This Seebeck voltage or potential voltage gradient will cause current to flow in the dissimilar materials, the amount of current flowing being a function of the geometry of the materials and the electrical resistivity of the materials. The Seebeck effect is most well known in its application in thermocouples for the measurement of temperature, some of the common thermocouples being chromel/alumel, copper/constantan, platinum/rhodium, and tungsten/rhenium.

The semiconductors that have been used for energy conversion in thermoelectric generators are bismuth telluride, lead telluride, and silicon/germanium.

Several different nuclear batteries for implantable generators have been developed using the heat of decay of the plutonium-238 isotope and thermoelectric conversion of this heat to electricity. Of the batteries implanted in humans, the battery developed by the Nuclear Materials and Equipment Corporation for the Atomic Energy Commission utilized as its thermocouple material tophel special cupron special, a modified chromel/alumel material. The batteries developed by Alcatel, Hittman, and Coratomic utilize bismuth telluride as the thermoelectric material.

2. ISOTOPE SELECTION

2.1. General Parameters

In selecting a radioisotope for implantable nuclear batteries, a number of characteristics are desirable. The isotope should be high in radioactive decay energy density, or watts per cubic centimeter, low in penetrating radiation to minimize the radiation level the patient will receive, fairly constant in power delivered as a function of time, and capable of being contained, should an accident, such as fire or mechanical crushing occur. This latter requirement delayed the introduction of the nuclear pacer for many years, until adequate standards of safety and credible accident definitions could be prepared and until devices could be designed to withstand hypothetical credible accidents. A release of the material to the biosphere, because of its toxicity, is not desirable. When all these factors are considered, the isotopes available for implantation use are very restricted.

2.2. Isotope Longevity

If a radioactive element's activity or intensity of radiactivity is examined in time, it is found that the experimental decay or activity can be expressed as a function of time by the equation

$$A(t) = A_0 e^{-\lambda t} \tag{1}$$

where A_0 is the initial activity of the element, $A(t)$ is the activity after a time t, and λ is the disintegration constant. Since the activity is proportional to the number of atoms that disintegrate per unit time, the activity may be replaced by the number of atoms N and Eq. (1) may be written as

$$N(t) = N_0 e^{-\lambda t} \tag{2}$$

where N_0 are the initial number of radiactive atoms and $N(t)$ are the number of radioactive atoms after a time t.

If both sides of this equation are differentiated, the resulting expression is

$$-\frac{dN}{dt} = \lambda N \tag{3}$$

This equation states that the decrease per unit time in the number of atoms due to radiactive disintegration is proportional to the number of atoms of the element remaining that have not yet disintegrated. The proportionality factor is the disintegration constant λ, and this constant is different for each radioactive isotope. The higher the disintegration constant, the higher the rate at which the isotope decays.

Equation (3) is the fundamental equation of radioactive decay, and all species follow this law.

A common quantity, used to characterize a radioisotope, is its half-life T. This is the time needed for one-half of the original atoms N_0 to disintegrate, and the quantity has more conceptual meaning than the disintegration constant λ. To relate T and λ, after one half-life, $N/N_0 = 0.5$, and from Eq. (2),

$$0.5 = e^{-\lambda T}$$

Taking the natural log of both sides of Eq. (4), we have

$$\ln 0.5 = -\lambda T \ln e$$

$$0.693 = \lambda T$$

or the half-life,

$$T = \frac{0.693}{\lambda} \tag{5}$$

Thus, the half-life is inversely proportional to the disintegration constant. The half-life is very important in the selection of the isotope, since it determines how rapidly the isotope decays and, ultimately, the decay characteristic of the electric power resulting from the conversion of the isotopic power to electrical power.

The decay of the isotope is actually a result of the statistical probability that an atom will disintegrate at any time. With a very large number of atoms, the disintegrations per second become greater, and hence a uniform, or smooth decay, of the isotope is observed. For example, there are 6.02×10^{23} atoms per gram mole (Avogadro's number). In the case of plutonium-238 there would thus be 6.02×10^{23} atoms in 238 g, or 2.5×10^{21} atoms per gram. The half-life of Pu^{238} is 87.8 years, or 2.769×10^9 s. From Eq. (5)

$$T = 2.769 \times 10^9 = \frac{0.693}{\lambda}$$

and therefore

$$\lambda = 0.693 \div 2.769 \times 10^9 = 250.3 \times 10^{-12}\ \text{s}^{-1}$$

and from Eq. (3), for 1 g of Pu^{238}

$$-\frac{dN}{dt} = \lambda N = (250.3 \times 10^{-12})\ (2.5 \times 10^{21})$$

$$= 625.7 \times 10^{9} \text{ atoms/s}$$

Thus, when 1 g of Pu^{238} isotope is disintegrating, it does so at the rate of 626 billion atoms each second. Since the number is so large, even though a single atom may disintegrate by chance at any time from zero to infinity, the very large number of atoms involved results in what appears, within the limit of measurement, to be a continuous decay process.

The half-life of the isotope is important not only from the effect that it has on the power produced by the battery, but also from the effect it has on the safety of the biosphere, since as the isotope decays, its hazard to the biosphere, if it were accidentally released, is reduced. Since after each half-life the amount of radioactive isotope is reduced by half, after two half-lives it is reduced by $\frac{1}{2} \times \frac{1}{2}$; after three half-lives, by $\frac{1}{2} \times \frac{1}{2} \times \frac{1}{2}$; or after n half-lives, by $(\frac{1}{2})^n$. Thus, after 10 half-lives, the amount of isotope is reduced by $(\frac{1}{2})^{10}$, or 1/1024. Thus, for Pu^{238}, after 10 × 87.8, or 878, years, a gram of Pu^{238} would have been reduced to 1 × 1/1024, or 0.000976 g, which presents a substantially reduced risk.

2.3. Isotope Comparisons

Shown in Table 1 is a list of all 14 isotopes that can be considered as "fuel" for nuclear batteries. They are arranged in order of their half-life, with amerecium having the longest half-life and thulium having the shortest.

In the case of an isotope to be used for human implantation, the first criteria for selection is to obtain an isotope that emits a low level of radiation and hence requires a small amount of shielding (or none). In the column referring to radiation level, only four isotopes have low levels of radiation, the isotopes americium-241, plutonium-238, promethium-147, and polonium-210.

The half-life of polonium-210 is seen to be 0.38 years, and its ratio of power after 10 years to initial power, calculated from equation (1), is

$$\frac{A}{A_0} = e^{-\lambda t} = e^{-\frac{0.693}{0.38}(10)}$$

$$= 0.000000012$$

From this calculation it is obvious that for a system to last 10 years, no power would remain in the isotope to provide electrical power output, and $polonium^{210}$ must be ruled out because of its short half-life.

Of the three remaining isotopes—americium-241, plutonium-238, and pro-

TABLE 1 Isotope Properties[a]

Isotope (in Order of Decreasing Half-life)	Compound	Radiation[b]	Watts per Gram of Compound	Production Density of Compound (g/cm^3)	Power Density (Thermal Watts per cm^3)[c]	Radiation Level	Half-Life (years)	Curies per Thermal Watt	Power After 10 Years/Initial Power
Americium, ^{241}Am	Metal	a	.1	11.7	1.17	Low	458	30	.984
Plutonium, ^{238}Pu	PuO_2	a	.39	10.0	3.9[c]	Low	87.8	30	.924
Uranium, ^{232}U	UO_2	aγ	3.3	10.0	33.0	High	74	26	.910
Cesium, ^{137}Cs	Borasilicate glass	βγx	.067	3.2	0.215	High	30 ± .3	207	.796
Strontium, ^{90}Sr	SrO	βγx	000.334	4.5	1.5	Medium	27.7	150	.781
Strontium, ^{90}Sr	$SrTiO_3$	βγx	.23	4.6	1.05	Medium	27.7	150	.781
Curium, ^{224}Cm	Cm_2O_3	aγn	2.3	11.75	27	Neutron shield–medium	18.4	30	.635
Cobalt, ^{60}Co	Metal	βγ	1.7	8.9	15.5	High	5.24	65	.266
Thallium, ^{204}Tl	Tl_2O_3	β	.12	9.0	1.08	Medium–low	3.9	640	.176
Prometheum, ^{147}Pm	Pm_2O_3	β	0.27	6.6	1.8	Low	2.6	2778	.069
Thorium, ^{228}Th	ThO_2	aγ	141	9	1270	High	1.9	24	.026
Cerium, ^{144}Ce	CeO_2	βγx	3.8	6.4	24.5	High	.78	124	.0001
Curium, ^{242}Cm	Cm_2O_3	aγn	98	11.75	1150	Neutron shield–medium	.445	28	.000000012
Polonium, ^{210}Po	Metal	a	134	9.3	1210	Low	.38	32	—
Thulium, ^{170}Tm	Tm_2O_3		1.03	7.7	7.9	Medium	.25	385	—

[a] References: (1) Oak Ridge National Laboratory and Stable Isotopes Catalog, April 1963;
(2) Extension of Table VI, page 52 of *Radioisotope Heat Sources*, by C. A. Rohrmann, Battelle-Northwest, December 1, 1965;
(3) Isotope Power Date Sheets compiled by S. J. Rimshw, Oak Ridge, Tennessee.

[b] a, alpha; β, beta; γ, gamma; n, neutron; x, bremsstrahlung.

[c] Does not include void volume.

metheium-147—plutonium-238 is superior to amerecium because of its higher power density, that of plutonium-238 being 3.9 W of thermal power per cm^3 versus 1.17 W/cm^3. This higher energy density is important, since it results in a smaller fuel capsule that, as will be seen later, results in fewer thermal losses, with a resultant higher efficiency and smaller nuclear battery.

Prometheium-147, since it is a beta emitter, can be used to power a nuclear battery by utilizing the beta voltaic effect. Its major disadvantage is that with a half-life of 2.6 years, a large amount of the isotope must be used, and its upper limit of usefulness is approximately 10 years, at which point its power has decayed to 7% of its initial value.

3. DETAILED CHARACTERISTICS OF THE PLUTONIUM-238 ISOTOPE

The plutonium-238 isotope meets all the requirements for a heat source, being high in energy density, being low in radiation level, and having a long half-life. As with all radioactive isotopes, it must be contained within a sealed capsule that prevents its release into the biosphere.

3.1. Fuel Form

The plutonium, which is a metal, can be used in its metal form, or as an oxide or nitride. In the early development of the nuclear pacer, the metal form was used because of the fact that its thermal energy density is higher than the oxide or nitride form. As research continued on the safe containment of the isotope during potential accidents, the oxide form was found to be much safer, particularly at the high temperatures that could be encountered in an industrial or accidental fire or that would be encountered if the pacer were accidentally cremated after the patient's death. As an oxide, the fuel is in its most stable form, whereas, as a metal, if it were released during a fire, it could oxidize and powder, possibly dispersing into the atmosphere. The Nuclear Regulatory Commission, the agency responsible for biosphere safety, now requires that *all* nuclear batteries use the oxide form.

Plutonium-238 is formed in a nuclear reactor by the bombardment of ${}_{93}Np^{237}$ with a neutron. The neptunium nucleus captures the neutron and emits a beta particle, or electron, thus gaining a proton and becoming ${}_{94}Pu^{238}$. The neptunium is produced by separation from reactor fission products. Both isotopes are completely man-made and do not exist in nature. As part of the neptunium irradiation, other isotopes of plutonium result and act as impurities in the fuel.

A typical fuel composition used for nuclear batteries is shown in Table 2

TABLE 2 Typical Fuel Composition for Nuclear Batteries

Plutonium-238 content, wt % powder	78.88
Plutonium-238 isotopic content, at. % of Pu	90.19
Total impurities excluding oxygen, wt % of Pu	0.70
Fission product activity, curies g Plutonium	9.3×10^{-6}

Plutonium Isotopes	*Content*
Plutonium-236, ppm	0.26
Plutonium-238, wt %	90.14
Plutonium-239, wt %	9.21
Plutonium-240, wt %	0.607
Plutonium-241, wt %	0.029
Plutonium-242, wt %	<0.01

	Impurities		*Impurities*
^{241}Am, wt % powder	< 0.01	Mq, ppm Pu	<10
Np, wt % powder	< 0.14	Mn, ppm Pu	<50
U, wt % powder	< 0.01	Ni, ppm Pu	<50
Th, wt % powder	< 0.28	Pb, ppm Pu	<10
Al, ppm Pu	<10	Si, ppm Pu	<100
B, ppm Pu	<20	Sn, ppm Pu	<50
Be, ppm Pu	<10	Zn, ppm Pu	<200
Ca, ppm Pu	<50	Mo, ppm Pu	<10
Cd, ppm Pu	<10	Na, ppm Pu	<20
Cr, ppm Pu	<10	P, ppm Pu	<500
K, ppm Pu	<200		
Cu, ppm Pu	<20		
Fe, ppm Pu	<100		

and is produced at the Oak Ridge Laboratories of the Energy Research & Development Administration.

The plutonium-236 is the most undesirable contaminant in the fuel, since it contributes appreciably to the gamma ray radiation from the fuel. It is radioactive, with a half-life of 2.8 years and decays to uranium-232, which is an alpha emitter with a 72-year half-life. The uranium-232 produces a chain of decay products or daughters, one of which is thallium-208, which produces high-energy gamma radiation. The Pu-239, which is itself radioactive and decays by alpha emission, has such a long half-life, 24,306 years, that it does not contribute appreciably to the energy of the fuel.

3.2. Types of Radiation

The radiation from the fuel emanates from six sources:

1. $_{94}Pu^{238}$ and $_{94}Pu^{240}$ decay by both alpha emission and by spontaneous fission. In this process, found in the very heavy elements, the atom disintegrates into two other elements and simultaneously gives off neutrons as part of the fissioning. This occurs spontaneously, with a half-life of 5×10^{10} years for Pu^{238} and creating 2.33 neutrons on the average for each atom of Pu^{238} that fissions. Although the rate is very low compared with the alpha disintegration, the neutrons created contribute to the radiation level from the fuel.

2. When the Pu^{238} decays by emitting an alpha particle, the alpha particle can collide with the nucleus of light elements, such as oxygen, lithium, beryllium, carbon, boron, fluorine, sodium, magnesium, aluminum, and silicon. When this occurs, a neutron is given off as a result of the collision. These neutrons add to the radiation emanating from the fuel. Since the fuel is an oxide, there are a large number of oxygen atoms that can participate in the (α,n) reaction. Naturally occurring oxygen is composed of isotopes that have eight protons and eight, nine, or 10 neutrons; oxygen-16, oxygen-17, and oxygen-18. Fortunately, the oxygen-16 nucleus has a low cross section for capture of an alpha particle, and by exchanging the oxygen-17 and oxygen-18 for oxygen-16, the (α,n) reaction can be reduced by a factor of 4.4, or almost to the level of the pure plutonium metal. The Nuclear Regulatory Commission, for this reason, requires that only $_8O^{16}$ be used in the PuO_2 fuel form.

3. When the (α,n) reaction occurs, the neutrons emitted not only add to the dose rate, but can cause additional fission of the plutonium isotopes, which fission in turn produces more neutrons. Thus, a self-multiplication of the neutron rate occurs.

4. As the $_{94}Pu^{238}$ decays, photons, or gamma rays, are emitted. Gamma rays are electromagnetic rays, as are X-rays, but are higher in frequency. A photon is a particle of energy, whose energy is:

$$E = h\nu$$

where h = Planck's constant and ν is the frequency of the gamma photon.

5. Gamma photons are released as the daughter products of the Pu^{236} and Pu^{241} isotopic decay.

6. Gamma photons also are created from alpha particle collision reactions whereby neutrons or protons are released as a result of the collision.

The effects of the various radiation particles are discussed under Section 10, "Radiation Effects."

The oxide fuel is pressed into a pellet. The pellet is fired at 1400°C to sinter it into a hard mass. The pellet appears as a hard, glossy ceramic. This produces a pellet of approximately 68% of the theoretical density of PuO_2 of 11.46 g per cm^3. The pellet is then contained within a fuel capsule.

TABLE 3 Helium and Thermal Power Generation Rates for a "Typical" Pu-238 Fuel

Fuel Composition: 0.0012% Pu-236, 80.5% Pu-238, 14.9% Pu-239, 2.9% Pu-240, 0.8% Pu-241, 0.1% Pu-242, 0.5% Np-237, 0.29988% Others

Years Elapsed	Cumulative Moles of Helium Generated per Gram of Original Pu-238 in Charge
0	0.00000
0.5	0.16231-4[a]
1	0.32474-4
2	0.64529-4
3	0.96437-4
4	0.12818-3
5	0.15959-3
6	0.19072-3
7	0.22175-3
8	0.25246-3
9	0.28306-3
10	0.31320-3
11	0.34316-3
12	0.37302-3
13	0.40253-3
14	0.43213-3
15	0.46110-3
16	0.48996-3
17	0.51864-3
18	0.54724-3
19	0.57533-3
20	0.60334-3
21	0.63132-3
22	0.65891-3
23	0.68630-3
24	0.71347-3
25	0.74044-3
50	0.13513-2
75	0.18557-2
100	0.22716-2
200	0.33253-2
300	0.38160-2
400	0.40468-2
600	0.42122-2
800	0.42582-2
1000	0.42778-2
1500	0.43067-2
2000	0.43334-2

[a] Denotes 0.16231×10^{-4}

3.3. Helium Release

The plutonium nucleus disintegrates by release of an alpha particle or helium atom nucleus, and these nuclei form helium atoms with the addition of two electrons from the environment. In addition, the alpha decay of the other plutonium isotopes and their daughters contribute slightly to the helium buildup in pressure.

This helium release as a function of time for a representative plutonium fuel mixture has been calculated by Dr. Harold J. Garber on the federally funded Radioisotope Powered Cardiac Pacemaker Project. (4) The cumulative moles of helium per gram of fuel and per gram of Pu^{238} in the fuel are given in Table 3. This helium generation by the Pu^{238} and other impurities and daughters is important in analyzing the effects of time and temperature of the fuel capsule. Although the impurity and daughter contribution to the helium decay have been included in the analysis, their contribution to the total buildup is minor. The amount of helium released as a function of time is shown in Figure 1 and illustrates that after approximately 10 half-lives of the Pu^{238}, or 878 years, the helium release becomes negligible, because of the extremely long half-lives of the remaining material.

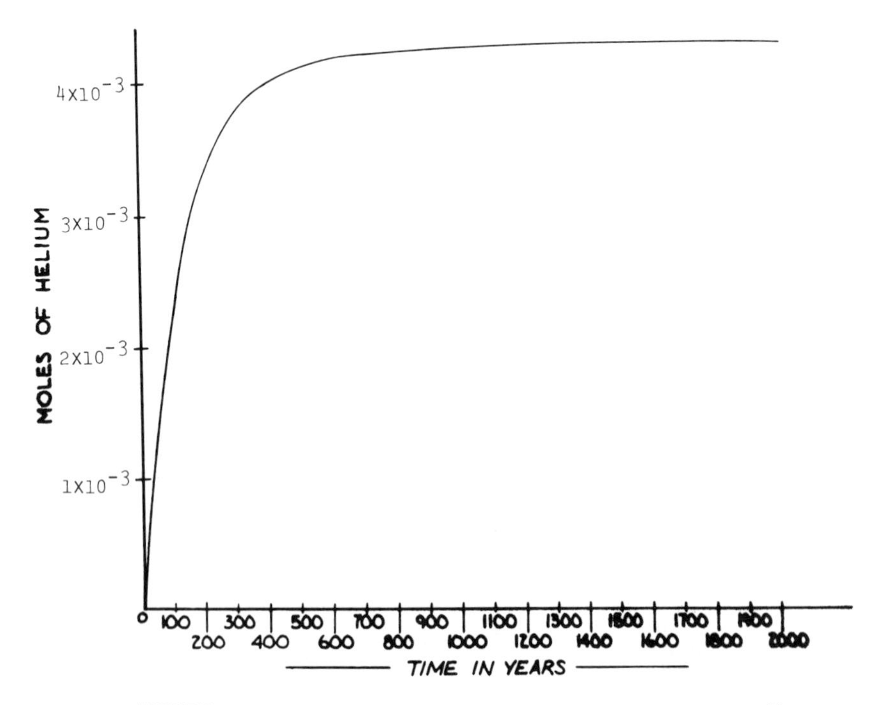

FIGURE 1. Helium Released from Plutonium-238 as a Function of Time.[4]

4. THERMOELECTRIC GENERATOR SYSTEMS

4.1. Nuclear Battery Subsystems

A representative plutonium-powered thermoelectric nuclear battery is shown in Figure 2. The plutonium dioxide fuel pellet is shown within a double-walled spherical fuel capsule. This capsule prevents the helium given off by alpha decay from escaping and protects the fuel from any accidental release into the biosphere. The fuel capsule is heated by the energy released in the radioactive decay process, and this heat increases the temperature of the heat susceptor, which in turn heats the end of a thermopile. The opposite end of the thermopile is cooled by a heat sink, whose temperature is close to body temperature of 37°C. Thus, the thermopile operates with a hot junction heated by the fuel capsule and a cool junction cooled by the body. Insulation surrounds the thermopile and fuel capsule, preventing the heat from the isotope from being lost to the cool environment surrounding the battery. The design of each of the three major subsystems—namely, the fuel capsule, thermopile, and insulation—varies somewhat from manufacturer to manufacturer, but the subsystems are essentially the same, as can be seen in Figure 3, the Hittman Associates nuclear battery, the Atomcell, and Figure 4, the Alcatel GIPSIE nuclear battery. The design of each of the subsystems if heavily dependent on the design constraints and the power output required by the implantable device.

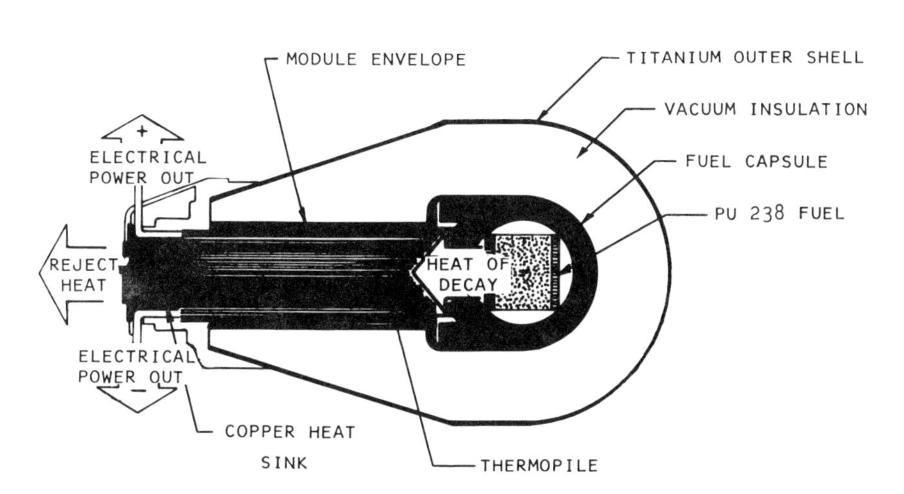

FIGURE 2. Plutonium Fueled Isotopic Battery Schematic—Coratomic, Inc. C-101 Battery.

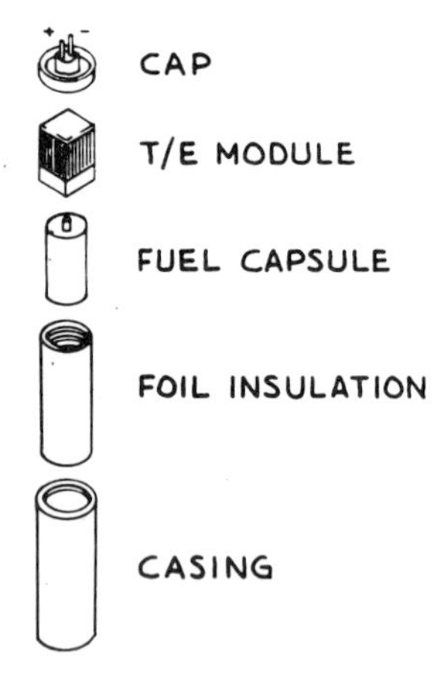

FIGURE 3. Plutonium Fueled Isotopic Battery Components—Hittman Corporation Atomcell.[20]

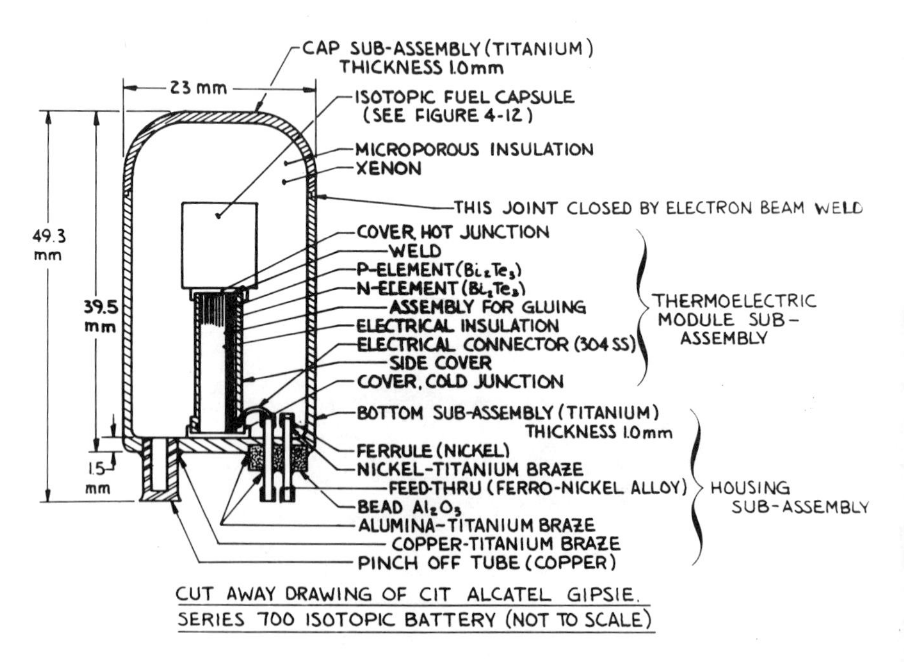

FIGURE 4. Plutonium Fueled Isotopic Battery Schematic—Alcatel GIPSIE Series 700 Battery.[46]

4.2. Biosphere Protection

Since the nuclear pacer has as its thermal source either the isotopes of Pu^{238} or of Pm^{147}, both of which would be hazardous if ingested into the lung in metallic, vapor, or fine powder form, the Atomic Energy Commission, and subsequently the Nuclear Regulatory Commission, undertook to determine the types of accidents that could lead to the possible release of the fuel into the environment. They then postulated a series of qualification tests that would simulate these accidents to allow manufacturers to qualify their nuclear batteries against the possible release of the fuel. The effort to develop a standard was worldwide, the international body in charge of nuclear safety being the Nuclear Energy Agency (NEA), an agency of the International Organization for Economic Cooperation and Development, with headquarters in Brussels, Belgium. The major work in the United States on determining these credible accidents was accomplished by a government developmental program, The Radioisotope Powered Cardiac Pacemaker Program, funded by the AEC, now the Energy Research and Development Agency. The tests were developed for nuclear pacers, but are, of course, applicable to other implantable devices.

4.2.1. Fire Accidents

Since a high-temperature industrial, hospital, or storage fire engulfing a pacer is conceivable, criteria were established to assure that the fuel capsule would not release fuel in the event of such a fire. The test consists of subjecting the pacer to an 800° oxidizing environment for 30 min, followed by a water quench to simulate the use of fire hoses, and then applying a 1000-kg load to simulate the crushing effect of a collapsing building.

The pacer containing the capsule is included in the testing, since, hypothetically, the pacer structure surrounding the nuclear battery could have an effect on the fuel capsule during the test.

4.2.2. Cremation Accidents

The most severe temperature test simulates an accident that could occur if a pacer were accidentally left in a body to be cremated. This test requires that a pacer be placed in a furnace at a temperature greater than 1300°C for 2 h without release of the fuel.

4.2.3. Corrosion Accidents

Another hypothetical accident is the loss of a pacer in an ocean, possibly as a result of a patient drowning, a burial at sea, the accidental crash of an aircraft, or the sinking of a vessel carrying the pacer to foreign countries. It must be certain that the pacer fuel would not be released as a result of corrosion of the fuel capsule by seawater.

4.2.4. Impact Accidents

Another accident that can be hypothesized is the fall of a pacer from an aircraft with subsequent impact on an unyielding surface such as a sidewalk or concrete structure. It can be hypothesized also that in the explosion of an aircraft, the fuel capsule might be blown from the interior of the pacer, although this is very improbable. To guarantee environmental safety in such contingencies, the NRC guideline requires that a fuel capsule be able to withstand impact at a velocity of 50 m/s or greater, assumed to be the terminal velocity of an object falling freely from a high altitude.

4.3. Operating Environment Design Requirements

During the AEC's Radioisotope Powered Cardiac Pacemaker Program, eight nuclear pacemakers were implanted in dogs in 1968. Of these eight, four failed because of structural defects in the nuclear battery. The failures were attributed to the mechanical shock experienced by the pacers, and a program was undertaken to determine the shock levels a nuclear battery could experience during a 20-year lifetime in a human, as well as during a test period in animals. As a result of this study, a series of qualification tests simulating the environment the pacer could expect to see during its lifetime was developed.

4.5.1. Torsional Vibration

During the dog test program, accelerometers were placed on the dogs, and linear acceleration and torsional or angular accelerations were measured. To simulate the measured values, a Miller Model G paint shaker was modified. A pacemaker containing the nuclear battery was first taped to the top of a 1-gal empty paint can using masking tape and then the can was placed in the canholder of the paint shaker. The edge of the can was against the edge of the canholder so that the packer moved in as large an arc as possible. The pacer was then tested for a total of 432,000 cycles, or for 4 h. The can was then rotated so that the X axis of the pacemaker was horizontal at the start. After one-half hour the can was rotated 90° so that the Y axis was horizontal, to provide an equal amount of shaking time in the X and Y axis. The Z axis of the pacer was parallel to the cylindrical axis of the paint can. The peak acceleration in the circumferential axis was 45 g and in the radial axis, 10 g. The shaking frequency is approximately 16 Hz. The output of the pacer, or battery, was monitored on an oscilloscope during the entire 4-h shaking time to detect any transients in the output.

4.5.2. Low-Level Shock

During the normal testing of the pacer in the dog, the pacer was found to be repeatedly impacted by the walls of the cage and protuberances in the cage. Although the pacers were implanted, these shocks were suprisingly high, at a level of 50 *g* with about a 14-ms duration. A miniature accelerometer was used to sense the shock pulse. This test was repeated in each of the three pacer axes for a total of 3000 50-*g* shocks.

4.5.3. Implant Impact Shock

The shock that simulated a 6-ft-high male fainting and falling forward impacting on a ceramic tile floor with a pectoral area implant under the skin was a 545-*g*, 1.3-ms shock. The 545-*g* shock test was done on the same fixture as the low-level shock test using a drop height and shock pad that produced a shock pulse of 545-*g* amplitude and 1.3-ms duration. Thirty-two shocks were done in each axis for a total of 96 545-*g* shocks.

4.5.4. Drop Shock

Although it was considered extremely unlikely, if a pacer or implant were dropped on a ceramic tile floor of a hospital, the battery should not fail. To simulate this potential accident, the pacer containing the battery was dropped 5-ft to a bare concrete floor. This was an estimated shock of 4500 *g*, and the point of impact was random. The battery operated after the drop.

4.5.5. Vibration

During the use of the nuclear battery while in the implant environment or while being shipped, the battery may be subjected to vibrations found in automobiles, trucks, or airplanes. The shipping tests for this vibration were accomplished by testing the battery on a vibratometer and vibrated at a 4-*g* level with a frequency sweep of 20–500–20 Hz and a sweep time of 10 min. The vibration was sinusoidal and was done in all three pacer axes. The battery output was monitored continuously on an oscilloscope and the *g* level was monitored on a vibration meter.

4.5.6. Temperature Cycling

In shipping products in interstate commerce, the product frequently experiences ranges of temperature. The standard shipping test was to place the battery in a 60° oven for 2 h, cool it for 1 h to room temperature, and then pace it in a cold chamber of -40°C for 2 h. The most severe problem in this test was that of thermal expansion and hot junction overtemperature. As the battery heats to 60°C, a temperature transient at the hot junction will be experienced. This might cause, depending on the reaction time of the various thermal components, the hot junction temperature to increase beyond safe limits and to fail.

5. THERMOPILE DESIGN

5.1. Seebeck Effect

The Seebeck or thermocouple effect occurs where two dissimilar electrical conductors, either metals or semiconductors, are connected at separate junctions, and these junctions are held at different temperatures. When this occurs, a Seebeck effect voltage is produced, proportional to the temperature difference between the two junctions, and represented as the open-circuit voltage, or:

$$v_{oc} = (\alpha_n + \alpha_p)\,\Delta T \tag{6}$$

where
v_{oc} = open-circuit voltage of a couple, V
α_n = Seebeck coefficient for the n material, V/°C
α_p = Seebeck coefficient for the p material, V/°C
ΔT = temperature gradient, or $T_h - T_c$, °K
T_h = hot junction temperature, °K
T_c = cold junction temperature, °K

The n material refers to a metal or semiconductor with an excess of electrons, or negative. The p material refers to a material with the absence electrons, or an excess of holes or positive. An n material is joined to a p material to maximize the open-circuit voltage produced.

5.2. Thermal and Electrical Performance

The open circuit voltage V_{oc} will produce a current flow in the circuit, and this current flow will be determined by the electrical resistance of the thermocouple and the resistance of a load that can be placed in series with the thermocouple. The couple appears electrically as if it were a battery with open-circuit voltage and an internal resistance. The current, from Ohm's law, is:

$$i = \frac{V_{oc}}{r_L + r_C} \tag{7}$$

where i = the current in the couple,
r_L = the load resistance, Ω
r_C = the couple's resistance, Ω

and

$$r_C = \rho_n \frac{l_n}{A_n} + \rho_p \frac{l_p}{A_p} \tag{8}$$

where ρ_n, ρ_p are the electrical resistance of the n and p legs of the thermocouple and l_p is the length of the n and p thermocouples, and A_p is the area of the n and p thermocouple.

The electrical power that such a couple produces is:

$$p_c = r_L i^2 \tag{9}$$

where p_c = the couple's electrical power, W
i = the couple's current flow, A

The heat difference between the junctions causes electrical power to be produced and thus thermal power is converted to electrical power.

The thermal power that is used by the couple is a function of the current and voltage flowing, as well as its thermal conductivity, k. The higher its thermal conductivity, the higher the thermal power consumed.

The thermal power (Watts) lost through the thermocouple is given by:

$$q = T_h (\alpha_n + \alpha_p)i + \left(\frac{k_n A_n}{l_n} + \frac{k_p A_p}{l_p}\right) \Delta T - \frac{i^2}{2}\left(\frac{\rho_n l_n}{A_n} + \frac{\rho_p l_p}{A_p}\right) \tag{10}$$

where k_n, k_p are the thermal-conductivities of the n and p material, W/cm-°C; ρ_n, ρ_p are the thermocouple resistivities of the n and p material, Ω-cm; l_n,l_p = n and p thermocouple length, cm; and A_n,A_p = n and p thermocouple area, cm^2.

The first term is the thermoelectric cooling term due to the current flow through the Seebeck voltage; the second term is the thermal loss due to the thermal conductance of the material; and the third term is the effect of joule, or resistive heating of the couple due to its internal resistance.

The ratio of the electrical power output to the thermal power input is the efficiency of the system and is, therefore, from Eqs. (9) and (10):

$$n = \frac{i^2 r_L}{\mathrm{T}_h(\alpha_n + \alpha_p)i + \left(\frac{k_n A_n}{l_n} + \frac{k_p A_p}{l_p}\right) \Delta T - \frac{i^2}{2}\left(\frac{\rho_n l_n}{A_n} + \frac{\rho_p l_p}{A_p}\right)} \tag{11}$$

The efficiency is thus a function of the current, hot and cold junction temperatures, areas and lengths of the couples, and the inherent properties of the couples, α, ρ, and k.

Given T_h and T_c constant, as the load is varied the current varies, as can be seen from Eq. (7), and the power output and efficiency vary, as can be seen from Eqs. (9) and (11). As with all electrical source systems, the power output reaches a maximum when the internal impedance matches the load impedance. It can be shown by differentiating Eq. (11) and setting the change of efficiency as a function of area to zero that the efficiency is a maximum when:

$$\frac{A_p}{A_n} = \sqrt{\frac{\rho_p k_n}{\rho_n k_p}} \tag{12}$$

If the conditions of Eq. (12) are met, the thermocouple pair will deliver its highest efficiency, but not exactly where the load is matched. In practice, the manufacturing problems and the close approximation of properties lead the designer to make the areas of the thermocouple elements the same.

The maximum efficiency of the thermocouple pair, derived from Eq. (11), can be shown to be:

$$n = \frac{2\ \Delta T(100)}{3T_h + T_c + 8Z_s} \tag{13}$$

where

$$Z_s = \frac{(\alpha_n + \alpha_p)^2}{KR} \tag{14}$$

and

$$K = \frac{A_n}{l_n} k_n + \frac{A_p}{l_p} k_p \tag{15}$$

$$R = \rho_n \frac{l_n}{A_n} + \rho_p \frac{l_p}{A_p} \tag{16}$$

and A_n and A_p are optimized according to Eq. (12).

From this equation it is evident that Z_s, called the "figure of merit" of the system, is purely a function of the materials used. The higher the Z_s, the higher the efficiency.

5.3. Material Characteristics

To characterize the value of different materials, it may be assumed that the couple is composed of a material where both the n and p legs have the same characteristics. Thus, from Eqs. (14), (15), and (16):

$$Z_m = \frac{(\alpha_n + \alpha_n)^2}{\left(\frac{A_n}{l_n} k_n + \frac{A_n}{l_n} k_n\right)\left(\rho_n \frac{l_n}{A_n} + \rho_n \frac{l_n}{A_n}\right)}$$

or

$$Z_m = \frac{4\alpha_n{}^2}{(2k_n)(2\rho_n)}$$

$$Z_m = \frac{\alpha_n{}^2}{k_n \rho_n}$$

This is called the material figure of merit, Z_m, with a high figure of merit desirable. The figure of merit of eight materials that have been used in thermoelectric generators is shown in Figure 5, with the Seebeck coefficient, thermal conductivity, and resistivity shown in Figures 6–8, respectively.

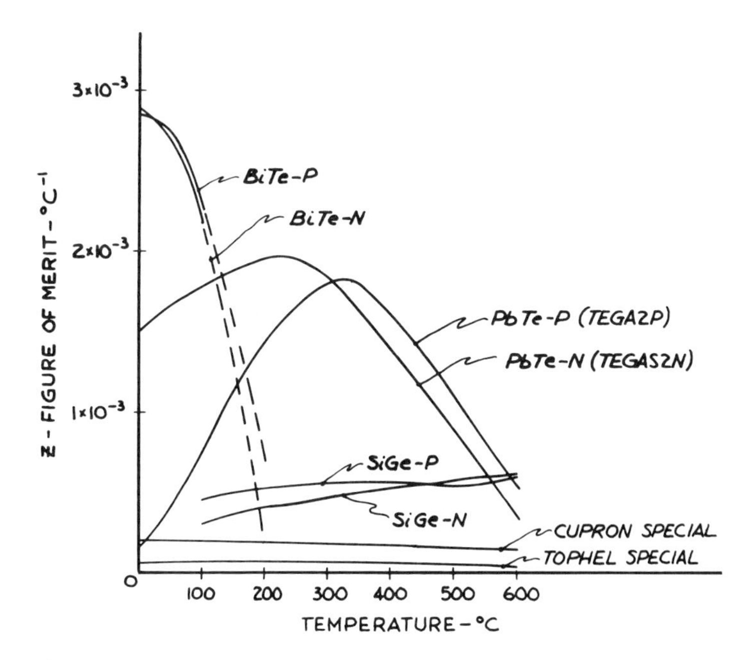

FIGURE 5. Thermoelectric Materials Figure of Merit as a Function of Temperature.

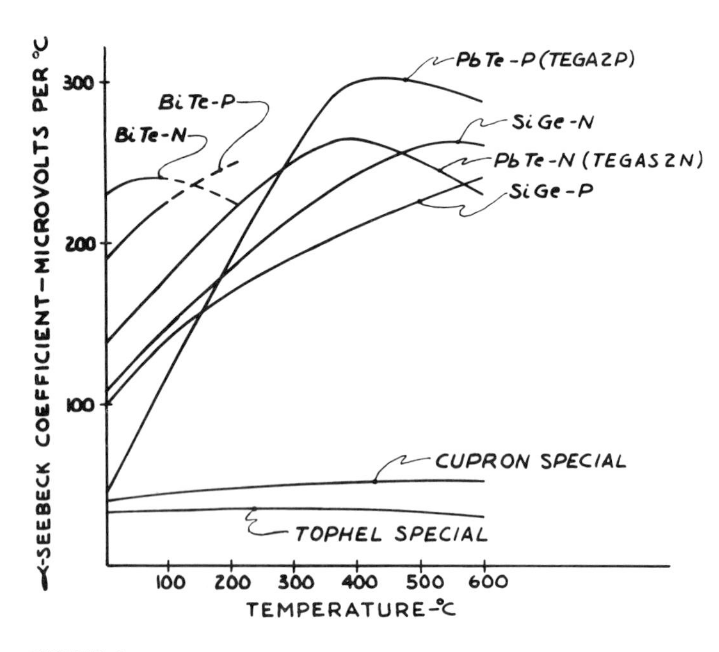

FIGURE 6. Thermoelectric Materials Seebeck Coefficient as a Function of Temperature.

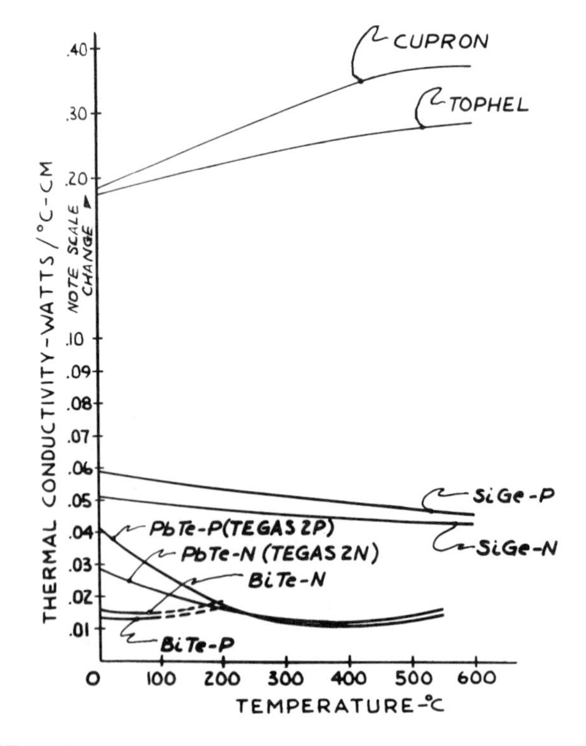

FIGURE 7. Thermoelectric Materials Thermal Conductivity as a Function of Temperature.

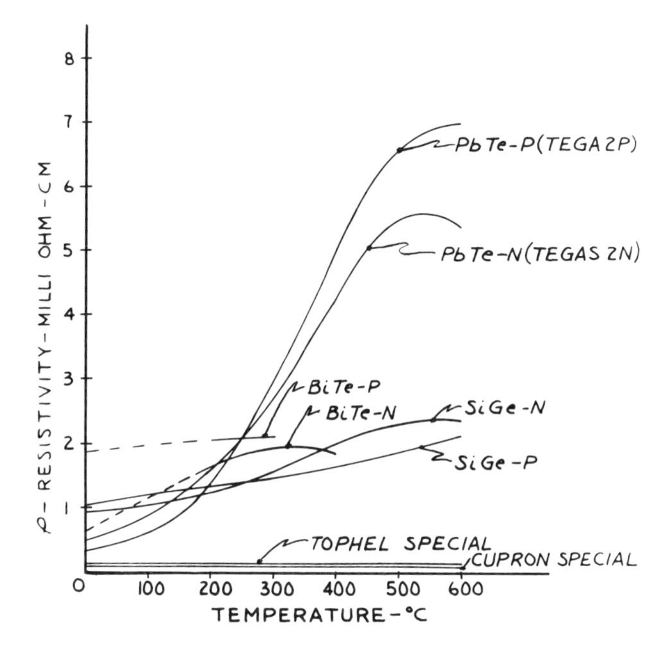

FIGURE 8. Thermoelectric Materials Electrical Resistivity as a Function of Temperature.

5.4. Design Optimization

From this graph it can be seen that bismuth telluride has the highest figure of merit and that the metallic couples cupron special and tophel special have the lowest. The efficiency as a function of temperature for the four different materials is shown in Figure 9. A cold junction of 38° is assumed, 1° above body temperature. It is evident that the bismuth telluride material is the best material up to 250°C, lead telluride is best between 250° and 575°C, and silicon germanium is superior above 575°C. Before the best material can be selected, the impact of fuel capsule support structure and insulation performance on the battery performance must be evaluated.

Although a nuclear battery could be manufactured with one thermocouple, the output voltage would be so low that it would be difficult to electronically step this voltage up to the voltage required for the electronic subsystems of implantable devices, which range from 2 to 9 V. For this reason the thermocouples are connected in series, the interconnection of thermocouples being called a thermopile. Thermopiles for implantable systems have consisted of 40–672 thermocouples. The output voltage of a single thermocouple, for example, with a hot junction of 100°C and a cold junction of 38°C, and at the peak power point, is 0.002 V for Tophel-Cupron, 0.012 V for bismuth telluride, and 0.004

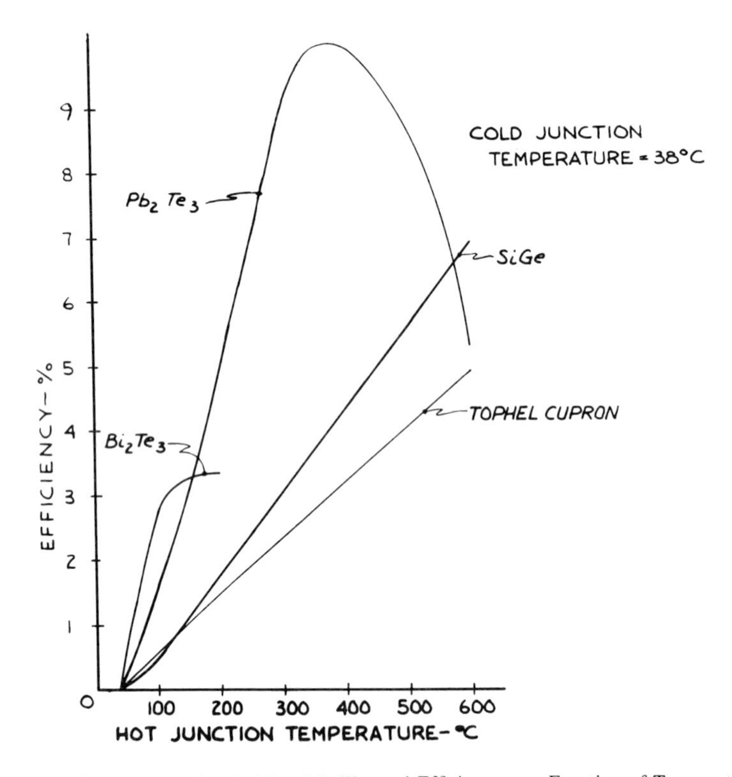

FIGURE 9. Thermoelectric Materials Thermal Efficiency as a Function of Temperature.

V for lead telluride and silicon germanium. Thus, the need to step up the voltage by connecting the thermocouples in series is obvious.

6. INSULATION DESIGN AND SELECTION

The fuel capsule and thermoelectric module must be insulated to prevent the thermal energy created by the decay of the isotope from bypassing the thermoelectric module and being wasted. If improperly insulated, the thermal drop, or ΔT, across the thermopile will be small, and insufficient power will be generated by the thermopile. Thus, the selection of insulation is an important factor in nuclear battery design.

The heat loss through a thermal conductor is given by:

$$q = k\frac{A}{l}\,\Delta T, \tag{17}$$

where q = heat loss, W

k = thermal conductivity of the insulation, W°C cm

A = area of the insulation perpendicular to heat flow, cm^2

l = path length in the direction of the heat flow, cm

ΔT = thermal drop through the insulation, °C, or T_h-T_c

The thermal conductivity of an insulator is due to three components of heat transfer in the insulation: conduction, convection, and radiation. Conduction is the loss through solid material; convection the loss due to mass transport of a gas, such as air, argon, or helium; and radiation the loss due to the transfer by radiant energy from one surface to another.

Shown in Figure 10 are thermal conductivities of insulation used for thermoelectric generators. These materials are generally porous or fibrous materials to minimize the loss due to conduction and to eliminate convection, since the gas

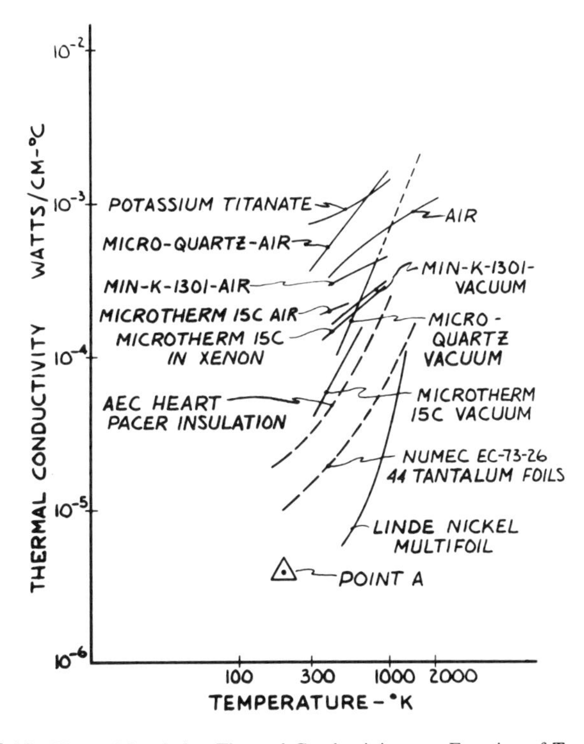

FIGURE 10. Thermal Insulation Thermal Conductivity as a Function of Temperature.

cannot move freely through them. Radiation loss is reduced by providing a sufficiently large number of thermal reflectors, or shields, so that the radiant energy is thereby reduced. By placing the insulation in a vacuum, the conduction loss through the insulation due to the gas is eliminated and the conductivity of the insulation reduced. As can be seen from Figure 10, the conductivity of Microquartz, Min-K 1301, and Microtherm 15C is substantially reduced by operating in vacuum by up to a factor of 5.

By placing a series of foils surrounding the hot zone, the equivalent thermal conductivity can be greatly reduced, since the foils repeatedly reflect the thermal energy. The heat loss through a series of parallel foils is given by:

$$q_{R/A} = \frac{5.67}{(n - 1)(2/\varepsilon - 1)} \left[\left(\frac{T_h}{1000} \right)^4 - \left(\frac{T_c}{1000} \right)^4 \right] \tag{18}$$

where q_R = heat loss in watts
A = foil area, cm^2
n = number of foils
T_h = hot foil temperature, K
T_c = cold foil temperature, K
ε = foil emissivity, dimensionless

As an example, using this equation, the thermal loss through 20 foils placed 0.010 in. apart, with an emissivity of 0.3 and a hot junction temperature of 100°C, a cold junction temperature of 38°C would be:

$$q_{R/A} = \frac{5.67}{(20 - 1)(2/.3 - 1)} \left[\left(\frac{373}{1000}\right.^{4} - \left(\frac{311}{1000} \right)^4 \right]$$

$$= 526.7 \times 10^{-6}\ \mathrm{W/cm^2}$$

These foils would have a thickness of 0.2 in. (20×0.010), or 0.508 cm^2. The equivalent conductivity for this case can be found by solving for k from Eq. (17), or

$$q = \frac{kA}{l} \Delta T \qquad \text{or} \qquad 526.7 \times 10^{-4} = k \frac{1}{0.508} 62$$

$$k = 4.316 \times 10^{-6}$$

at an average temperature of 69°C (342 K). From Figure 10, the equivalent conductivity labeled point A is close to the conductivity of the Linde nickel multifoil insulation. This insulation is used mainly in cryogenic vacuum applications and operates on the principle of repeated reflections from a series of foils. In a practical application in a nuclear battery, conductivities substantially higher than the Linde multifoil insulation are experienced for manufacturing, assembly, and geometry reasons.

From the curve the superiority of a vacuum insulation is clear, as in that of the multifoil insulations. The decision to use a vacuum insulation with its attendant practical problems depends on the thermal analysis of the complete battery, as will be discussed in Section 8.

7. FUEL CAPSULE DESIGN

7.1. General Description

The fuel capsule of the nuclear battery must be designed to withstand the increase in pressure of the helium over the useful life of the device, and at temperatures anticipated during a fire or cremation accident. In order to meet these conditions, double encapsulation of the fuel is necessary, since the only alloys that can withstand the 1300°C temperature are alloys of tungsten and tantalum, which would oxidize severly at the 1300°C temperature in air. Thus, the inner capsule is designed as the pressure vessel; the outer capsule, as an oxidation barrier. The platinum/rhodium alloys are selected for this purpose. In order to prevent diffusion between the inner and outer capsules, an oxide diffusion barrier is used such as aluminum oxide or thorium oxide.

The inner capsule is designed to be as small as possible, consistent with being able to withstand the helium pressure at temperature and with a reasonable margin of safety. This reduces the thermal insulation loss.

7.2. Helium Pressure

The helium quantity is a function of the amount of Pu^{238} atoms in the fuel and the time since encapsulation, as discussed in Section 3.3. The pressure of the helium within the capsule can then be calculated if the volume of the helium space and the temperature are known.

Although the ideal gas law can be utilized, at high pressure the helium pressure deviates from the ideal. This increase in helium pressure was carefully studied under the AEC Radioisotope Powered Cardiac Pacemaker Program, and the pressure/volume/temperature relationship for helium was thoroughly studied for pacer fuel by Dr. Harold J. Garber. (4) In reference 4 an empirical equation has been derived and is given below:

$$P = 14.7\left[\frac{82.0567T}{v - \Lambda} - \frac{3285.3}{(v + 6.84)^2}\right] - 11.673/v \qquad (19)$$

where $\Lambda = 12.732e$

P = pressure inside capsule, psi

T = temperature, K

v = specific volume, cm^3/g-mol

In order to calculate the pressure from Eq. (19), the specific volume within the capsule must be determined. The temperatures of interest are those for cremation, 1300°C; industrial fire, 800°C; normal hot junction operation; and room temperature, 23°C. Once the helium pressures at the various temperatures are determined, the wall stress can be obtained. This wall stress is a function of capsule geometry and capsule material.

7.3. Capsule Material

Several materials have been investigated for use in plutonium-fueled nuclear batteries. They include the nickel base super alloys, including Hastelloy C and Inconel 718, and the refractory alloys, including Ta-10W, T-111, and T-222, both tantalum/tungsten alloys. In the early 1970s the super alloys were selected, since at that time cremation temperatures were thought not to exceed 1100°C, but as studies of temperature extremes in various furnaces in crematoriums around the United States and Europe progressed on the AEC program, it was discovered that the temperature the fuel capsule must withstand during cremation was 1300°C, thus ruling out the use of these oxidation-resistant alloys. The yield strength of the refractory alloys is shown in Figure 11. From this graph it is evident that the refractory alloys are most suitable at the cremation temperature of 1300°C, while the super alloys have no strength, since they melt in the range between 1270°C and 1427°C. The Ta-10 tungston alloy melts at 3047°C. As can be seen from Figure 11, the yield strength at 1300°C ranges from 15,000 to 30,000 psi. As a conservative design rule, the design stress at the operating temperature should be 75% of this value.

7.4. Capsule Geometry

Both spherical, cylindrical with flat ends and cylindrical with round-end geometries can be used as fuel capsules for nuclear batteries. A cylindrical geometry fuel capsule from the GIPSIE–Alcatel battery is shown in Figure 12, the cylindrical fuel capsule for the AEC pacer is shown in Figure 13, and the

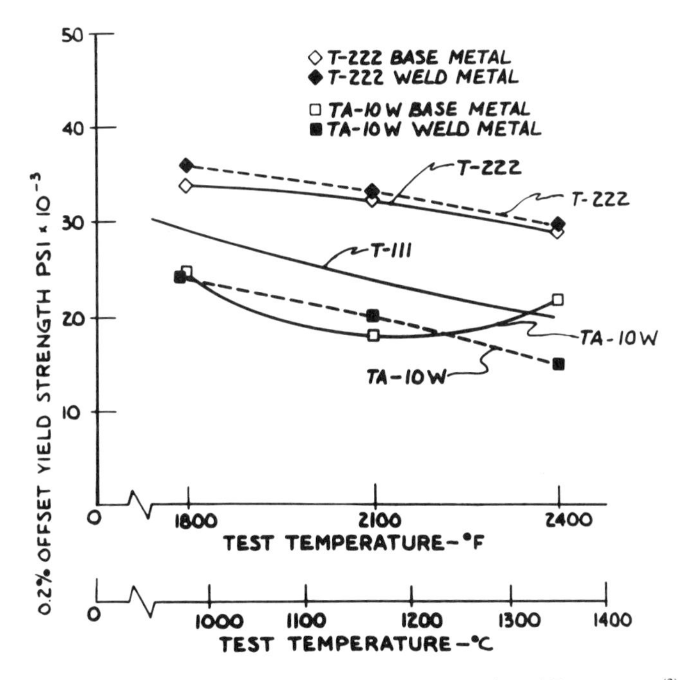

FIGURE 11. Tantalum Alloys Yield Strength as a Function of Temperature.[3]

spherical fuel capsule for the Coratomic nuclear battery is shown in Figure 14. Figure 15 shows the Coratomic capsule components and a completed capsule. In the design of the fuel capsule, it is desired to keep the capsule as small as possible to minimize the hot zone area, thereby keeping the thermal losses as low as possible; to keep the fuel loading and hence radiation as low as possible, minimizing weight; and to simplify the structural support necessary to support the capsule during extremes of mechanical shock.

7.5. Capsule Stress Analysis

To design the minimum-sized capsule, a tradeoff exists between the wall of the capsule and the Helium void volume. The capsule wall thickness is determined by the equations of stress for the geometry selected. For a sphere, the thick wall pressure vessel equation for wall stress is:

$$S_s = \left[\frac{b^3 + 2a^3}{2(b^3 - a^3)} \right] p \qquad (20)$$

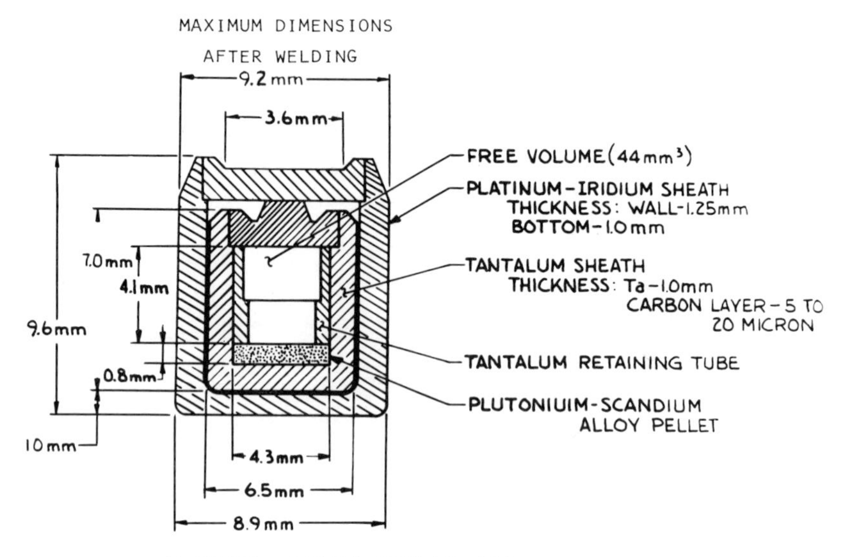

FIGURE 12. Plutonium 238 Fuel Capsule—Alcatel GIPSIE Series 700.[46]

where S_s = maximum stress in capsule wall, on the inside surface, psi

b = outside radius, in.

a = inside radius, in.

p = internal pressure, psi

Since S_s is a function of the material selected and cremation temperature, Eq. (7) can be solved for the outside radius b, giving

$$b = 3\sqrt{\frac{2a^3(p + s)}{2s - p}} \tag{21}$$

since the thickness of the capsule, $t = b - a$, Eq. (21) becomes

$$t = 3\sqrt{\frac{2a^2(p + s)}{2s - p}} - a \tag{22}$$

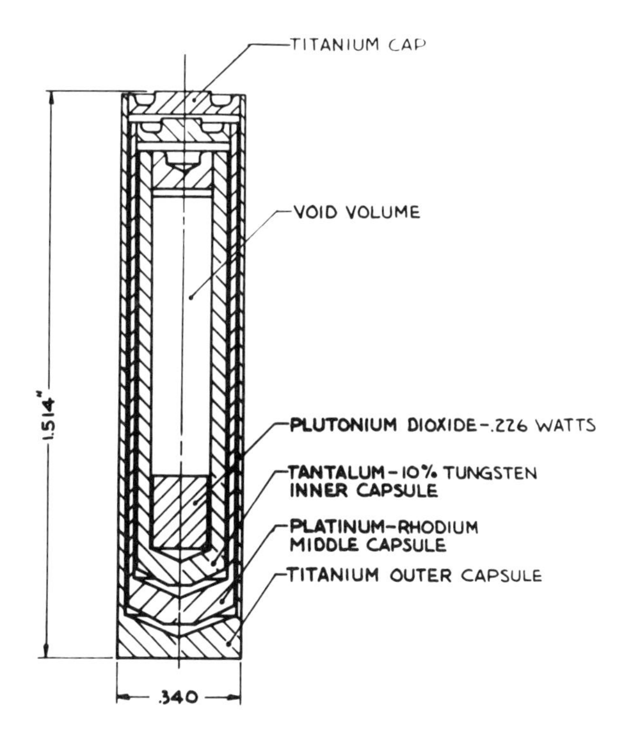

FIGURE 13. Plutonium-238 Fuel Capsule—U.S. Atomic Energy Commission.[1]

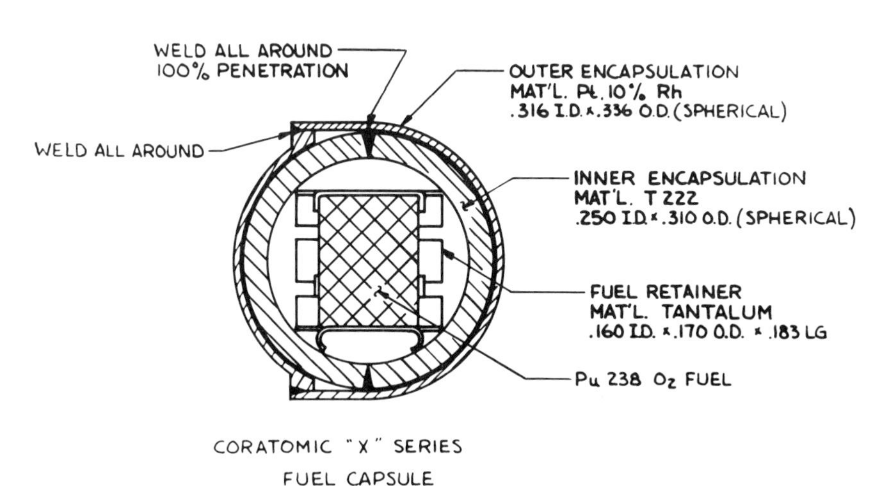

FIGURE 14. Plutonium 238 Spherical Fuel Capsule—Coratomic X Series.

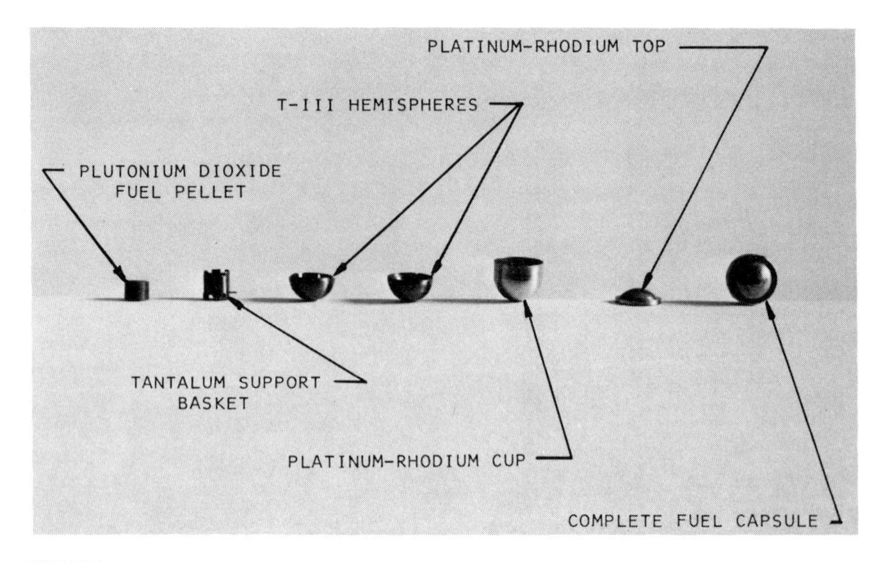

FIGURE 15. Photograph of Plutonium-238 Spherical Fuel Capsule and Components—Coratomic X Series.

With the inside radius given, the inside volume of the sphere is known, and the optimum-sized capsule can be found for different levels of thermal power. In order to determine the minimum radius capsule, a thermal power level must be assumed. Knowing the fuel power density characteristics and assuming a certain capsule diameter, the void available within the capsule for helium to occupy can be determined. Any capsule support structure, such as retaining plates, rings, or baskets, for the fuel subtracts from the available void space. With the void volume known, the amount of Pu^{238} known, and the helium buildup time assumed, the gram-moles of helium can be determined from Table 3, and the specific volume, or cm^3/gram-mole of helium, can be calculated. With the temperature known, the internal pressure can be calculated from Eq. (19). With the internal pressure known, the capsule wall thickness can be computed from Eq. (22).

A plot of wall thickness versus capsule internal radius is shown in Figure 16 for the following assumed values:

Capsule temperature = 1300°C
Helium buildup time = 100 years
Wall stress = 17,700 psi
Fuel power density = 0.37684 W/g of PuO_2
Power level = 50, 150, 250 mW at 20 years

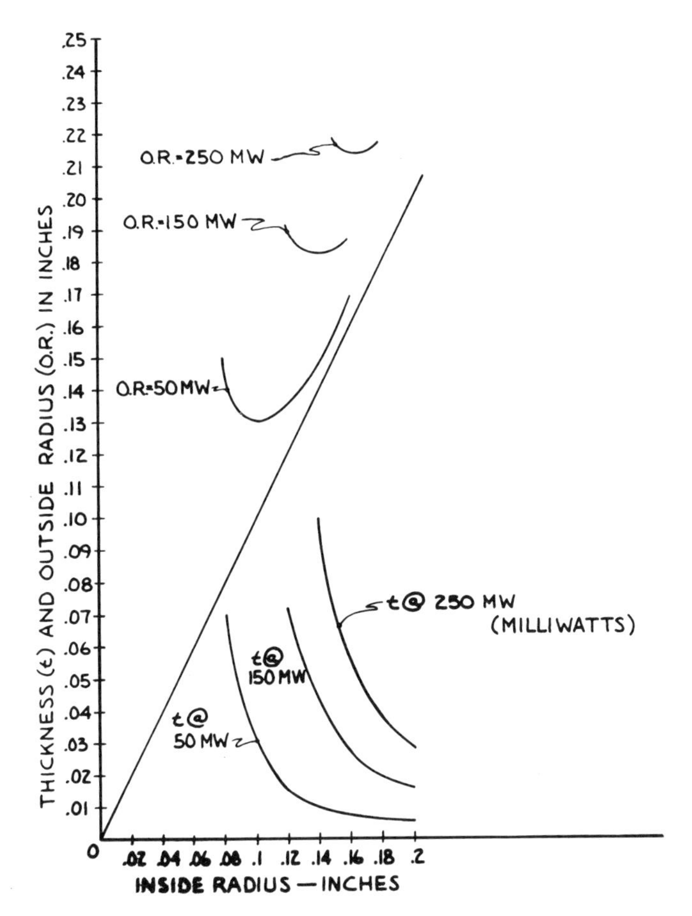

FIGURE 16. Plutonium Spherical Fuel Capsule Pressure Vessel Thickness and Outside Radius as a Function of Inside Radius.

A disk 0.002 in. thick and of the capsule diameter is assumed to support the fuel capsule

From Figure 16, for each power level, the wall thickness can be seen to decrease as the internal radius increases, since the helium pressure decreases with larger capsule void volume. The sum of the wall thickness and internal radius, or the external radius, can be seen to reach a minimum, or optimum point, for different power levels as a function of inside radius. Diameters are shown as a function of power level in Figure 17. The corresponding pressure as a function of inside radius is shown in Figure 18 at various power levels, and at the minimum volume point.

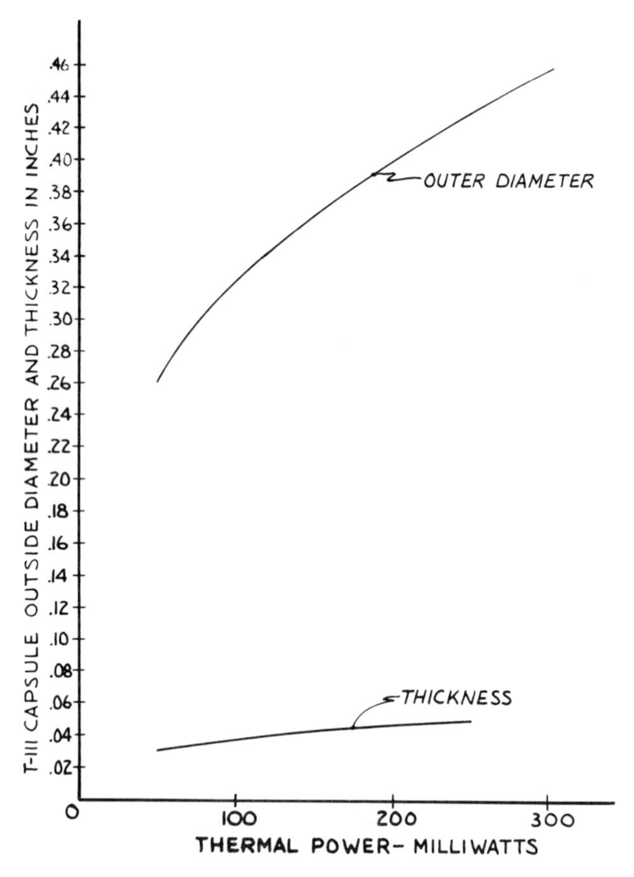

FIGURE 17. Plutonium Spherical Fuel Capsule Pressure Vessel Minimum Diameter and Thickness as a Function of Fuel Thermal Power.

For a cylindrical geometry, the thick wall cylinder stress is given, from reference 5, as

$$S_c = p \left[\frac{b^2 + a^2}{b^2 - a^2} \right] \tag{23}$$

where S_c = maximum wall stress, on inside surface, psi

b = outside radius, in.

a = inside radius, in.

p = internal pressure, psi

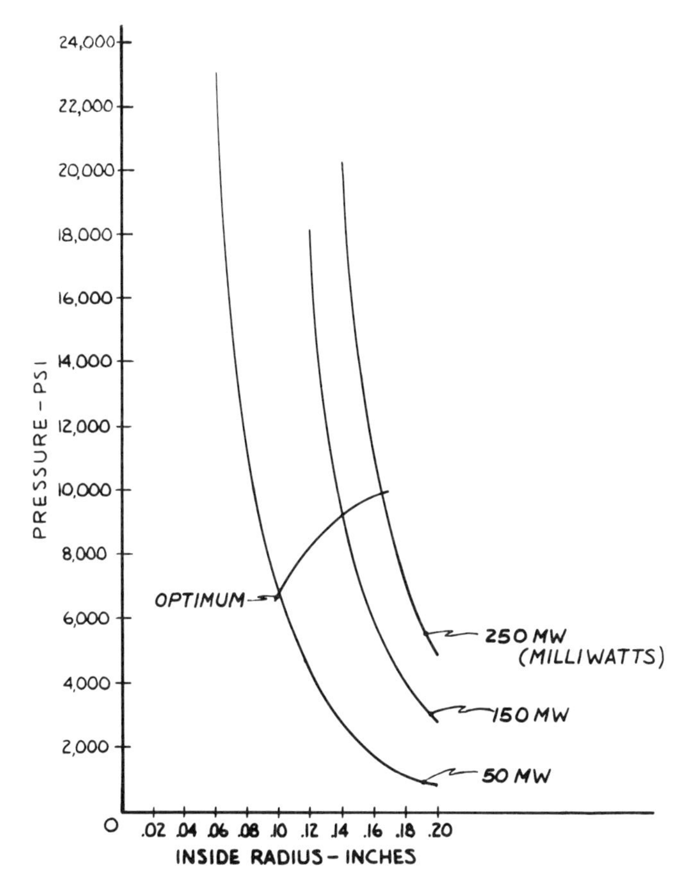

FIGURE 18. Plutonium Spherical Fuel Capsule Pressure Vessel Internal Pressure as a Function of Inside Radius.

From this equation the wall thickness can be found to be:

$$t = a\sqrt{\frac{s/p + 1}{s/p - 1}} - a \tag{24}$$

If the cylinder has a flat end, or plate, the maximum plate stress is:

$$s = \frac{3pa^2}{4t^2} \tag{25}$$

where a = the plate radius, in.

p = the internal pressure, psi

t = the thickness, psi

From this equation, t is found to be:

$$t = a\sqrt{\frac{3p}{4s}} \tag{26}$$

In comparing a cylinder with a sphere, at 150 mW for example, the optimum cylinder outside diameter is found to be 0.222 in., versus the 0.182-in. diameter of the sphere, with a pressure of 1838 lb versus the 9200-lb pressure of the sphere. Thus, the cylinder diameter is 22% greater than the sphere diameter, as might be expected, since a sphere is the optimum-strength pressure vessel. If the cylinder has flat ends, the thickness of the cylinder ends from Eq. (26) is found to be 0.056 in., giving a length of 0.312 in. versus the 0.182-in. diameter of the sphere, or a length 75% longer than the diameter of a sphere.

Obviously, the spherical shape provides the smallest fuel capsule dimensions for containment of the released helium.

The effects of temperature and time on a typical fuel capsule are shown in Figures 19 and 20. This specific capsule has an internal radius of 0.127 in., a fuel volume of 0.027 cm^3, and a void volume for helium buildup of 0.0986 cm^3. The amount of argon required to seal the capsule has been included, adding 0.77 molar percent at 100 years of helium buildup. Figure 19 shows the buildup in pressure within the capsule as a function of time. Figure 20 shows the resulting wall stress as a function of time.

From Figure 20 it can be seen that the yield stress is achieved at the 1300°C cremation temperature at 115 years, which is 5.7 times the useful life of the device. At 20 years the wall stress is only 7000 psi, or one fourth of the yield stress. This excessive safety factor of approximately 4 compared with the normal 1.33 safety factor used in nuclear submarine design, Naval or Aerospace Radioisotope Thermoelectric Generators, illustrates the conservatism used in the design of implanted devices.

The yield stress is reached at 390 years at the fire temperature, which is a factor of 19.5 over the useful life of the device with a stress safety factor of 7.3. At room temperature the wall stress never exceeds 10,000 psi, providing an incredible margin of safety in the ultimate environment the device will experience.

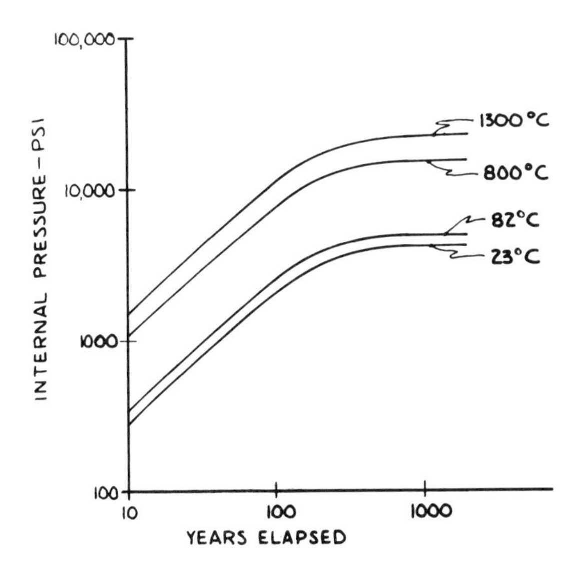

FIGURE 19. Plutonium Spherical Fuel Capsule Pressure Vessel Internal Pressure as a Function of Time and Temperature—Coratomic C-101 Fuel Capsule.

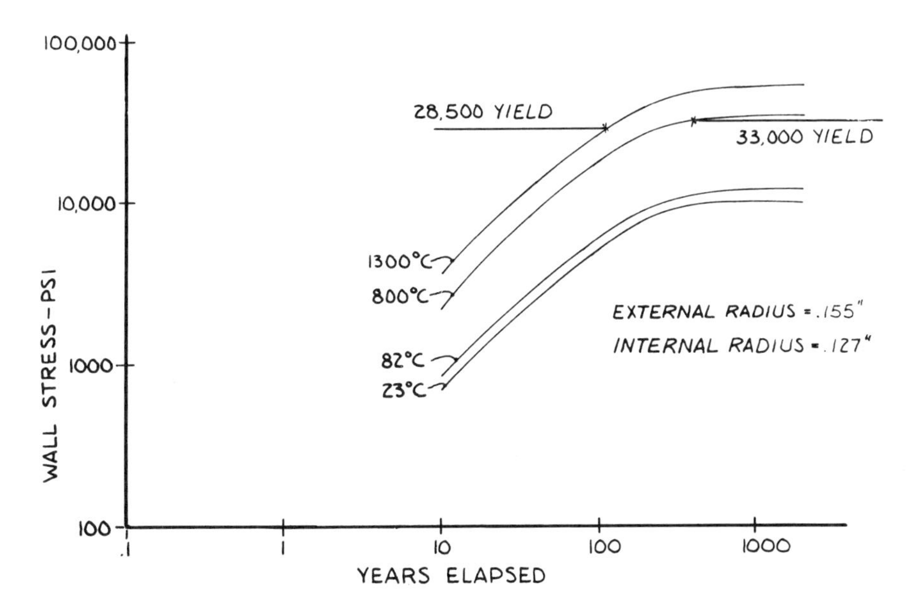

FIGURE 20. Plutonium Spherical Fuel Capsule Pressure Vessel Wall Stress as a Function of Time and Temperature—Coratomic C-101 Fuel Capsule.

7.6. Credible Accident Testing

As discussed in Section 4.2., the fuel capsule, operating within the implantable device it is powering, such as a pacemaker, must survive a series of credible accidents to assure that the fuel is not released to the environment. The most severe test is that of cremation, and nuclear batteries now produced must survive a cremation test at 1300°C for 1.5 h. In practice, this is accomplished by either an induction-heated furnace or a wire-wound furnace. Shown in Figure 21 is a photograph of a Coratomic pacer after the cremation test, and in Figure 22 are shown the pacer and nuclear battery disassembled showing the unaffected platinum/rhodium sheathed capsule at the center of the oxidized remains of the pacer and battery assembly.

The fire and crush test is also a severe test, and the capsule materials have to be sufficiently ductile to survive the extreme pressure. Shown in Figure 23 is a photograph of a fuel capsule that has been heated in air for 0.5 h at 850°C and placed between the jaws of a press. Figure 24 shows the fuel capsule after it has been crushed in the press, and Figure 25 shows it after it has been removed from the press. The ductile deformation and flattening are evident. A helium

FIGURE 21. Nuclear-Powered Cardiac Pacemaker after Cremation Test—Coratomic C-101 Pacer.

FIGURE 22. Nuclear-Powered Cardiac Pacemaker after Cremation Test Showing Intact Plutonium-238 Fuel Capsule—Coratomic C-101 Pacer.

FIGURE 23. Plutonium-238 Fuel Capsule in Compression Test Fixture Prior to Crush Test—Coratomic C-101 Fuel Capsule.

FIGURE 24. Plutonium 238 Fuel Capsule in Compression Test Fixture After Crush Test—Coratomic C-101 Fuel Capsule.

leak check, applied to the crushed capsule, indicated continued hermeticity on both external and internal capsules.

Spherical capsules, impacted at greater than 50 m/s, are shown in Figure 26. Impacts were tested at different altitudes with regard to the weld region to assure that weld position was not a factor. Hermeticity is measured for both outer and inner capsules, at a helium leak rate of less than 10^{-8} torr per atmosphere-second, indicating no breach of the encapsulant.

Corrosion testing in aerobic (oxygenated) and anerobic (unoxygenated) water has been conducted with platinum/rhodium capsules immersed for two years. As part the requirement for safety against fuel release, the Nuclear Regulatory Commission requires for pacemakers that the capsules be tested in not less than 0.1 liter/cm^2 of specimen surface area of seawater. Both oxygenated (aerobic) and unoxygenated (anerobic) water must be used. Galvanic corrosion between the fuel capsule material and other metals in the implantable device system must also be tested. No visible corrosion could be detected in the surface or due to weight change of the capsules after two years in seawater, including scanning

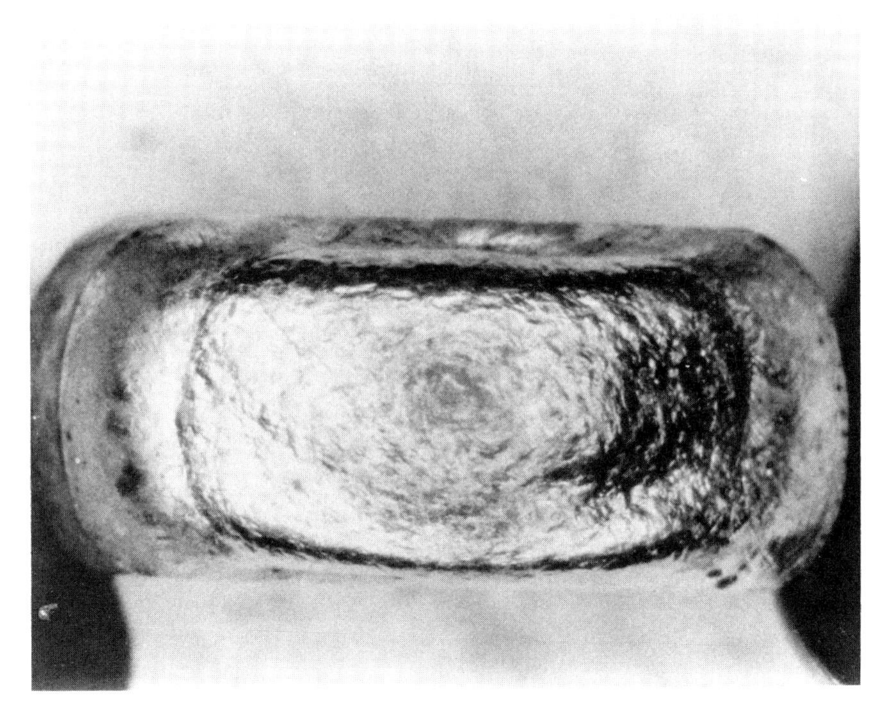

FIGURE 25. Plutonium-238 Fuel Capsule After Crush Test—Coratomic C-101 Fuel Capsule.

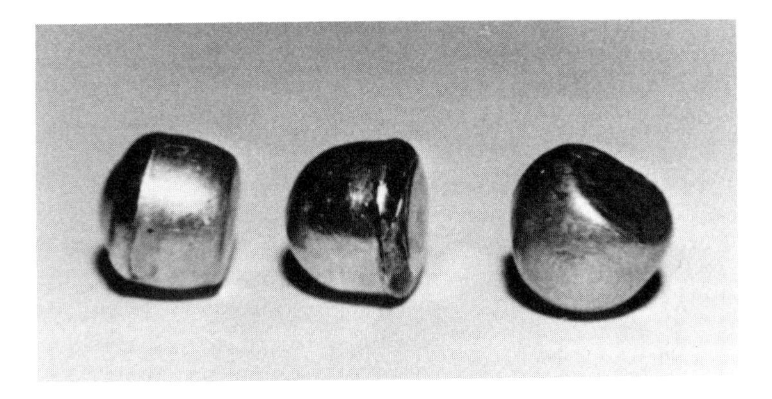

FIGURE 26. Plutonium-238 Fuel Capsules After Impact Test—Coratomic C-101 Fuel Capsule.

electron microscope examination of the surface. A worst-case analysis was undertaken in the analysis of the results by examining scanning electron microscope photographs of the test specimens. From this analysis it was established that within the limits of experimental error, the corrosion penetration into the specimens was between 0 and 0.003 in. in 10 half-lives, or 878 years. The capsule wall thickness is .010 in., or the factor of safety for corrosion release using Pt-10% Rh is 333%. The weight loss test results are given in Table 4.

Similar results were obtained in galvanic specimen tests, indicating no corrosion. The materials tested in combination were Ti versus T-222, Pt-10% Rh versus Ti. The test results are shown in Table 5.

Although not a part of the Nuclear Regulatory Commission or International Nuclear Energy Association test requirements for nuclear licensing, the AEC Radioisotope Powered Cardiac Pacemaker Program studied the possible release of fuel from firearm impact. The capsule was tested under high-velocity impact with the following results:

44 magnum impact—no breach of the capsule
12 gauge shotgun—no capsule breach
22 long rifle—no breach inner capsule
High-velocity 222 Remington—no breach inner capsule

As a result of the rigorous testing against a variety of hypothetical accidents, nuclear batteries have been proved to be safe for the environment and the release of fuel into the biosphere virtually impossible.

TABLE 4 Corrosion Tests of Fuel Capsules in Seawater Environment—Coratomic C-101 Fuel Capsules[a] (740-day exposure to ocean water)

Capsule Number	Test Condition	Weight (G)			Leak Check atm cm^2/s	
		Before Test	After Test	Change	Before Test	After Test
X25	Aerobic	3.7321	3.7323	+0.0002	$<3 \times 10^{-9}$	$<3 \times 10^{-9}$
X28	Aerobic	3.9483	3.9484	+0.0001	$<3 \times 10^{-9}$	$<3 \times 10^{-9}$
X37	Aerobic	3.8875	3.8874	−0.0001	$<3 \times 10^{-9}$	$<3 \times 10^{-9}$
X46	Aerobic	3.7957	3.7956	−0.0001	$<3 \times 10^{-9}$	$<3 \times 10^{-9}$
X51	Aerobic	3.7773	3.7773	0.0000	$<3 \times 10^{-9}$	$<3 \times 10^{-9}$
X26	Anaerobic	3.7689	3.7689	0.0000	$<3 \times 10^{-9}$	$<3 \times 10^{-9}$
X27	Anaerobic	3.7225	3.7226	+0.0001	$<3 \times 10^{-9}$	$<3 \times 10^{-9}$
X63	Anaerobic	3.8150	3.8148	−0.0002	$<3 \times 10^{-9}$	$<3 \times 10^{-9}$
X91	Anaerobic	3.7709	3.7706	−0.0003	$<3 \times 10^{-9}$	$<3 \times 10^{-9}$
X109	Anaerobic	3.8268	3.8266	−0.0002	$<3 \times 10^{-9}$	$<3 \times 10^{-9}$

[a] Surface area of each capsule is 2.6 cm^2.

TABLE 5 Corrosion Results on Test Specimens—Coratomic C-101 Fuel Capsules (740-day exposure to 300 ml of ocean water)

Test No.	Test Condition (Room Temp.)	Materials in Test	Type of Corrosion	Test Specimen Dimensions (in.)	Test Specimen Thickness (in.)		Test Specimen Weight (g)		Material Certification No.	Comments
					Before Test	After Test	Before Test	After Test		
5	Aerobic	Pt-10Rh	General	0.39 × 0.39 × 0.010	0.0100	0.0101	0.5166	0.5165	726	
6	Anaerobic	Pt-10Rh	General	0.39 × 0.39 × 0.010	0.0099	0.0098	0.5121	0.5123	726	
7	Aerobic	T-222	General	0.315dia × 0.171	0.1709	0.1712	3.9818	3.9818	472	Slightly tarnished
8	Anaerobic	T-222	General	0.315dia. × 0.147	0.1465	0.1465	3.4624	3.4624	472	Slightly tarnished
9	Aerobic	Ti	General	0.39 × 0.39 × 0.016	0.0156	0.0154	0.1851	0.1852	13	Slightly tarnished
10	Anaerobic	Ti	General	0.39 × 0.39 × 0.016	0.0157	0.0157	0.1918	0.1920	13	Slightly tarnished
							Weight of Welded Assy.			
11	Aerobic	Ti	Galvanic	0.24 × 0.24 × 0.016	0.0157	0.0156			13	Slightly tarnished
		T-222		0.24 × 0.24 × 0.030	0.0325	0.0326	0.6237	0.6236	906-1	
12	Anaerobic	Ti	Galvanic	0.24 × 0.24 × 0.016	0.0158	0.0157			13	Slightly tarnished
		T-222		0.24 × 0.24 × 0.030	0.0325	0.0325	0.6217	0.6217	906-1	
13	Aerobic	Pt-10Rh	Galvanic	0.24 × 0.24 × 0.010	0.0100	0.0100			726	Slightly tarnished
		T-222		0.24 × 0.24 × 0.030	0.0323	0.0322	0.8262	0.8262	906-1	
14	Anaerobic	Pt-10Rh	Galvanic	0.24 × 0.24 × 0.010	0.0100	0.0100			726	Slightly tarnished
		T-222		0.24 × 0.24 × 0.030	0.0326	0.0325	0.7639	0.7640	906-1	
15	Aerobic	Pt-10Rh	Galvanic	0.24 × 0.24 × 0.010	0.0098	0.0098			726	Slightly tarnished
		Ti		0.24 × 0.24 × 0.016	0.0157	0.0156	0.8273	0.8274	13	
		T-222		0.24 × 0.24 × 0.030	0.0325	0.0325			906-1	
16	Anaerobic	Pt-10Rh	Galvanic	0.24 × 0.24 × 0.010	0.0100	0.0100			726	Slightly tarnished
		Ti		0.24 × 0.24 × 0.016	0.0157	0.0158	0.8225	0.8226	13	
		T-222		0.24 × 0.24 × 0.030	0.0326	0.0326			906-1	

8. THERMAL ANALYSIS

To design a nuclear battery for implantable applications, the power output required to operate the device must first be established. Once this is known, the battery may be designed with the objective of utilizing the least possible fuel for the power output. Since the nuclear battery delivers a slowly declining power over a long period of time, the longevity of the device must be known to determine the initial fuel loading, which must be higher than that required at the end of the device's life to account for the fuel decay.

To illustrate the design procedure, the design of a battery delivering 1 mW of power at the beginning of its life, or 450 μW after 40 years will be described. This power level is typical of that used for pacemaker power supplies.

Since the output voltage of a nuclear battery is low and must be stepped up by dc-to-dc conversion, the maximum number of thermoelectric elements in series should be used to provide the highest voltage possible. The electrical power per thermocouple is given by:

$$p_c = \frac{1}{4} ZK\Delta^2 \tag{27}$$

where Z, K, and Δ were previously defined.

K is proportional to the area and inversely proportional to the length. To provide as many couples as possible to increase the voltage, the electrical power per couple should be reduced as much as possible by increasing the individual couple's length and reducing its area. The couple length is determined by the battery geometry, and the cross-sectional area by the strength limitation of the material. The tellurides are made by power metallurgy hot pressing or casting and then cut to size. It the materials are reduced in thickness much below 0.013 in., or a square element 0.013 × 0.013 in., the grain size of the materials are approached, and the elements become very brittle and could fail in the high-stress environments that implantable devices encounter. The length of the couples is limited by the size of the battery, and if they are manufactured in too great a length, the breaking stress of the element increases, which also would cause failure in a mechanical shock environment. For this reason thermocouple lengths of 0.5–1 in. are generally used. This does not apply to the metallic couple systems, which utilize ductile wires of great length and small diameter.

If Eq. (27) is used to determine the power per couple, the total number of couples as a function of temperature can be calculated. With this information the thermoelectric module performance can be computed.

The results of this analysis for a 1-mW power output are shown in Figure 27, which shows the optimum number of couples as a function of temperature. Figure 28 shows the output voltage at the peak power point as a function of temperature. The module output voltage is given as:

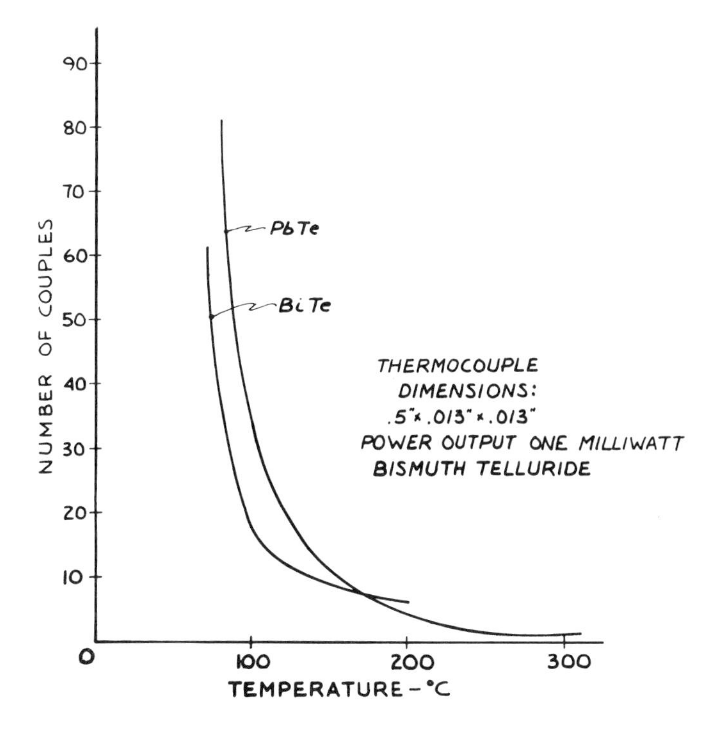

FIGURE 27. Lead and Bismuth Telluride Number of Thermocouples as a Function of Temperature.

$$V_{pp} = \frac{(\alpha_n + \alpha_p)}{2} N \Delta T$$

where V_{pp} = module voltage at peak power, V

N = number of thermocouples in the thermopile

α_n = Seebeck coefficient for n material, V/°C

α_p = Seebeck coefficient for p material, V/°C

ΔT = Temperature drop across the module, °C

From the curve of voltage versus temperature, it can be seen that the voltage becomes very low at the higher temperatures, making dc-to-dc conversion increasingly difficult. Although a separate subject, a dc-to-dc converter cannot, with today's technology, operate below an input of 0.1 V. Thus, the lead telluride systems are impractical above 250°C.

The thermal power required to produce the electrical power output is strictly

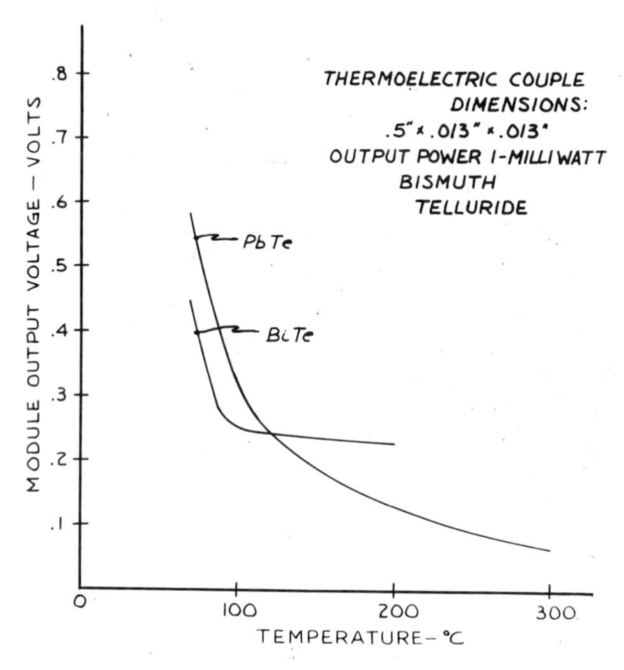

FIGURE 28. Lead and Bismuth Telluride Thermoelectric Module Output Voltage as a Function of Temperature.

a function of the efficiency of the thermoelectric material, as previously described in Eq. (13), and is shown as a function of temperature in Figure 9. Thus, to provide 1 mW of power, the thermal power required for the thermoelectric material is given by Eq. (28):

$$n = \frac{P_e}{Q_t}(100) \tag{28}$$

where n = thermoelectric material efficiency, %

P_e = electrical power output, W

Q_t = thermal power input, W

To this loss must be added the thermal loss of the insulators electrically isolating the elements from each other. Many insulation materials are possible, depending on the manufacturing techniques and hot junction temperature selected, ranging from ceramics to elastomers. Ceramics are used for high-temperature systems; epoxies and silastics, for lower temperature systems. These electrical insulators, necessary to insulate the elements of the thermopile to allow them to be connected in series to produce voltages as high as possible, constitute a parasitic power loss.

In addition to the electrical insulation thermal loss, the thermal losses of the thermopile support structure must also be included.

To protect the thermopile from mechanical stress, a stainless steel or titanium envelope is provided surrounding the array of thermoelectric elements and was shown previously in Figure 2. This envelope also contributes to the mechanical support of the fuel capsule and prevents it from injuring the thermoelectric elements under high vibration or acceleration or shock loads. A silastic or epoxy filler occupies the space between the elements and the cylindrical envelope. In a typical thermopile, the electrical insulator between elements will be approximately 0.002 in. thick, the clearance between the edge of the thermocouple array and the envelope 0.010 in., and the envelope 0.002, in. thick. A silastic, epoxy, or polyurethane can typically be used between the thermopile and the envelope.

The loss through each of these structures is calculated by the standard equation for heat transfer:

$$g = k\frac{A}{l}\,\Delta T \tag{29}$$

where k = thermal conductivity of the insulators, W/°C-cm

A = area of the insulator perpendicular to heat flow, cm^2

l = length of the insulator parallel to heat flow, cm

ΔT = thermal drop across the thermopile, °C

Since the number of thermocouples varies for each hot junction temperature, as shown in Figure 27, the insulation and envelope areas are also different for each hot junction. Assuming a thermal conductivity of 0.0052 W/cm^2-°C for the support material inside the envelope and between the thermocouples of an epoxy, and 0.0726 W/cm^2-°C for the titanium envelope, these losses for a lead telluride nuclear battery are plotted in Figures 29 and 30 and for a bismuth telluride system in Figure 31 and 32. The thermocouple and electrical insulation loss decreases as a function of hot junction temperature, since the efficiency of the couples increases, requiring less power for a given electrical output, and since the electrical insulation area decreases, since the number of couples decreases. The envelope and filler loss first decreases as the envelope size rapidly decreases, but then increases slightly as the thermal drop becomes more dominant than the decreasing envelope size.

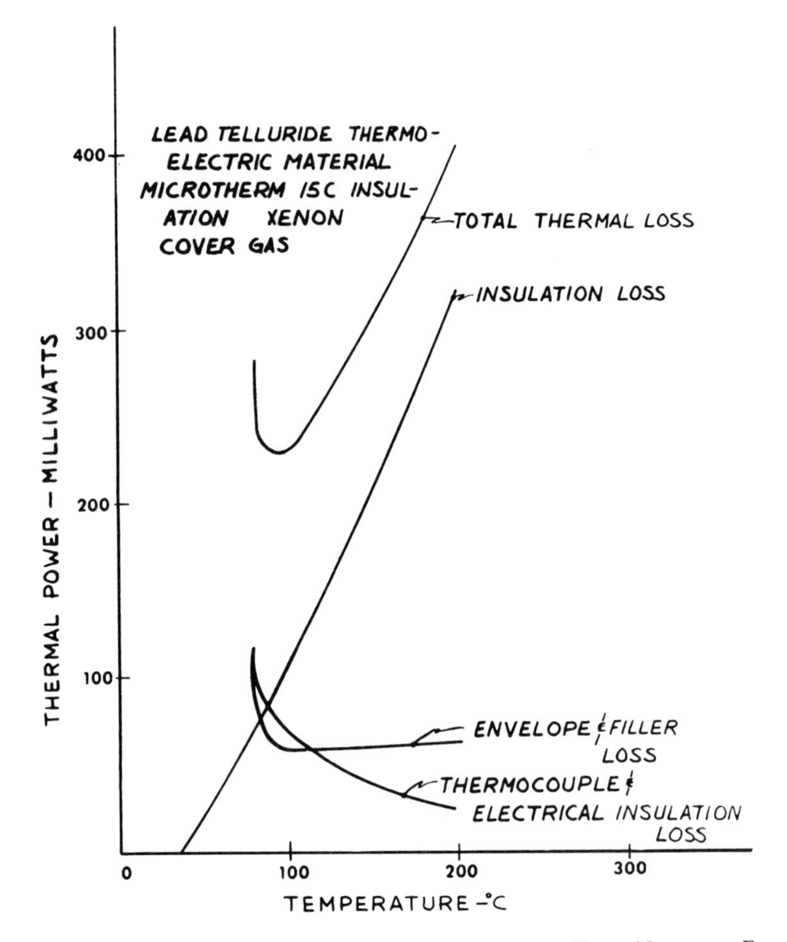

FIGURE 29. Lead Telluride Plutonium-238 Powered Nuclear Battery Thermal Losses as a Function of Temperature—with Gas-Filled Insulation.

The insulation loss must be added to the other losses to compute the total thermal loss. The insulation loss is also given by Eq. (29), but for spherical, cylindrical, or conical shapes l would be the average path length, and the area would be given by:

$$A_{ms} = \sqrt{A_i A_O}$$

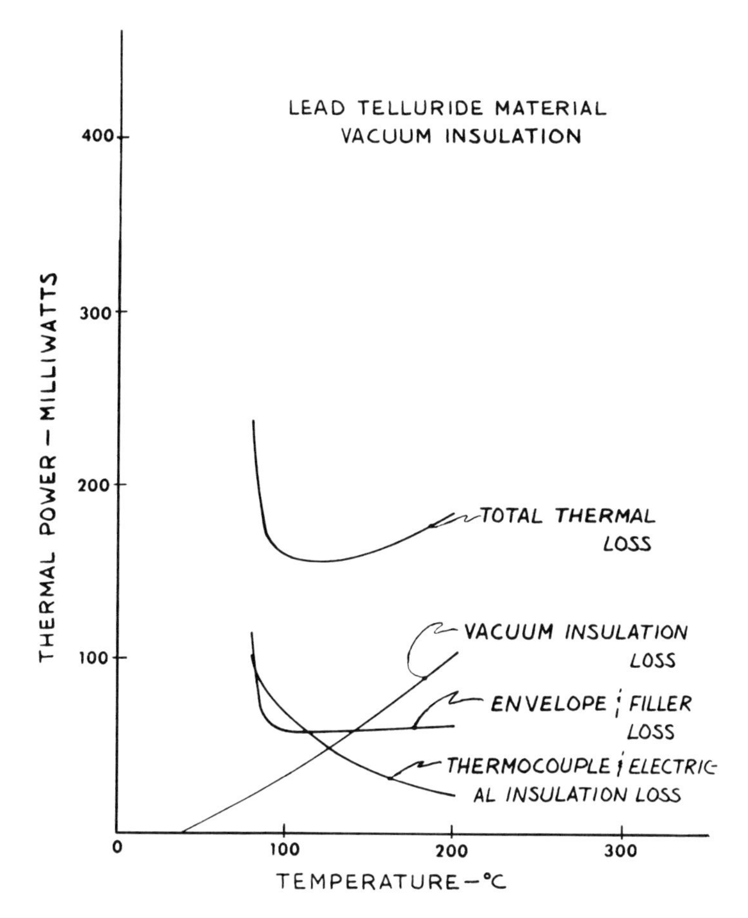

FIGURE 30. Lead Telluride Plutonium-238 Powered Nuclear Battery Thermal Losses as a Function of Temperature—with Vacuum Insulation.

where A_{ms} = the effective area for spherical or conical sections
A_i = the inside area perpendicular to heat flow
A_o = the outside area perpendicular to heat flow.

For cylindrical shapes:

$$A_{mc} = \frac{A_i - A_o}{ln\ A_i/A_o}$$

where A_i and A_o are inside and outside areas of the cylindrical insulation.

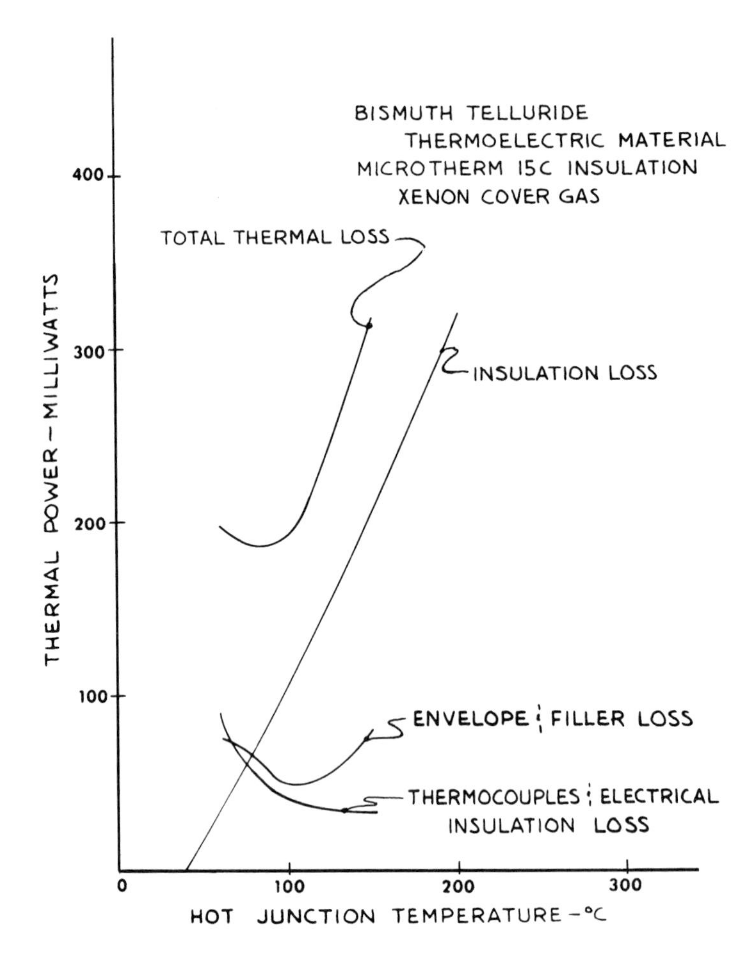

FIGURE 31. Bismuth Telluride Plutonium-238 Powered Nuclear Battery Thermal Losses as a Function of Temperature—with Gas Filled Insulation.

Using the thermal conductivity as a function of temperature for the insulations as given in Figure 10, the loss for a typical nuclear battery, the Coratomic C-100 battery shown in Figure 2, is also shown in Figures 29–32. From these figures the total loss for vacuum and nonvacuum bismuth telluride and lead telluride thermocouple nuclear batteries as a function of temperature is shown. From these curves it can be seen that the vacuum-insulated systems are superior in requiring less thermal energy, and hence less isotope, and the bismuth telluride systems are superior to the lead telluride systems. It can also be seen that an optimum hot junction temperature exists for both systems, as a result of the trade-off between reduced losses due to thermocouple efficiency as the hot junc-

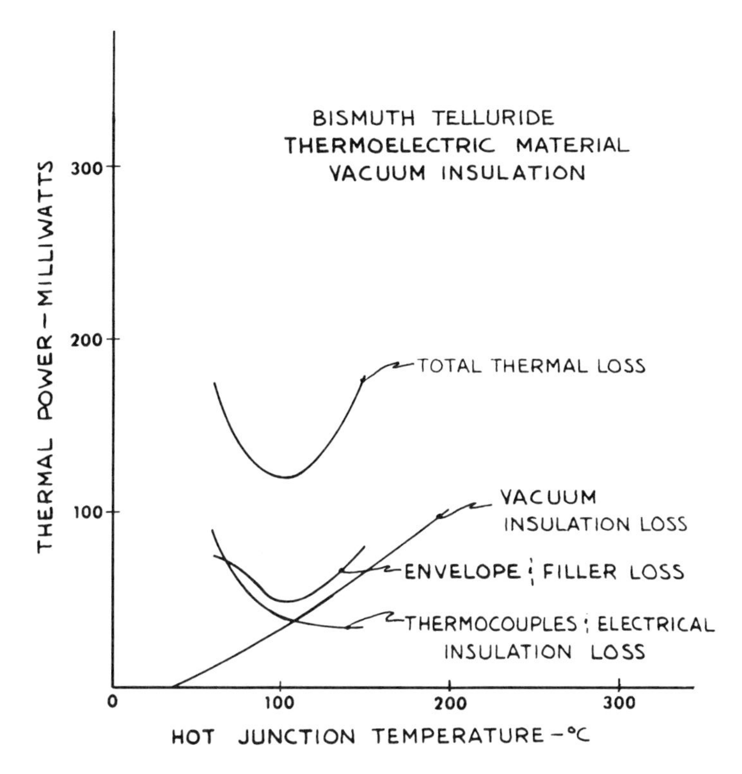

FIGURE 32. Bismuth Telluride Plutonium-238 Powered Nuclear Battery Thermal Losses as a Function of Temperature—with Vacuum Insulation.

tion temperature increases, counterbalanced by increased losses due to insulation as the hot junction temperature increases.

With regard to the bismuth telluride system, the optimum hot junction temperature is 100°C from Figure 32. At this temperature the output voltage is 0.25 V, possibly too low for the dc-to-dc converter. Although requiring slightly more thermal energy, a lower hot junction temperature will provide higher voltage. This trade-off of voltage versus isotope required must be made by examination of the voltage required for a specific application.

In Table 6 the operating parameters at the minimum thermal power point are given.

It is interesting to note that the thermocouples consume only 30.3% of the total power, the other parasitic losses consuming the greater percentage of the power. The overall system efficiency is 1 mW electrical ÷ 120 mW thermal, or only 0.83%.

In spite of this, the specific power (or watt-hours per gram) of the nuclear battery, compared with other stored energy systems, is remarkable. The C-100

TABLE 6 Parameters at Optimum Hot Junction Temperature

Electrical power output	1 mW at BOL
Hot junction temperature	100°C
Cold junction temperature	38°C
Number of thermocouples	18
Output voltae at peak power	0.25 V
Open-circuit voltage	0.5 V
Current at peak power	4 mA
Thermal losses	
Thermocouples	36.3 mW (30.3%)
Thermocouple electrical insulation	3.8 mW (3.2%)
Metal envelope	21.6 mW (18.0%)
Thermopile support filler	27.6 mW (23.0%)
Vacuum insulation	30.7 mW (25.5%)
Total	120 mW (100%)
Pu-238 O_2 Required	0.21 g

battery, for example, which weighs 16 g and delivers 790 μW at BOL, has a specific power of 9.2 Wh of energy per gram over a period of 40 years. A typical lithium battery used for pacemaker applications, weighing 20 g and delivering 3.9 Wh, stores 0.2 Wh of energy per gram.

9. ELECTRICAL CHARACTERISTICS

The nuclear battery provides a current voltage curve representative of a voltage in series with an internal resistance. The open-circuit voltage is directly proportional to the temperature drop across the thermopile, and this is directly proportional to the amount of thermal power of the isotope. As the isotope decays, this power decays exponentially, as discussed in Section 2.2., and so does the output voltage. This voltage decay is shown in Figure 33. The resistance of the thermopile remains approximately constant as the temperature of the hot junction slowly drops; thus, the short-circuit current that is proportional to the voltage if the internal resistance is constant also drops at the same rate as the voltage. The peak power at one half of the open-circuit voltage and short-circuit current point thus drops as a function of the square of the isotope decay. This percentage of power decline is also shown in Figure 33.

The actual current voltage curves of a typical Coratomic battery at the beginning of life and at 20 and 40 years are shown in Figure 34. The power at these times is also plotted. It can be seen that the peak power declines from a beginning of life (BOL) value of 0.79 mW to 0.59 at 20 years and 0.42 at 40

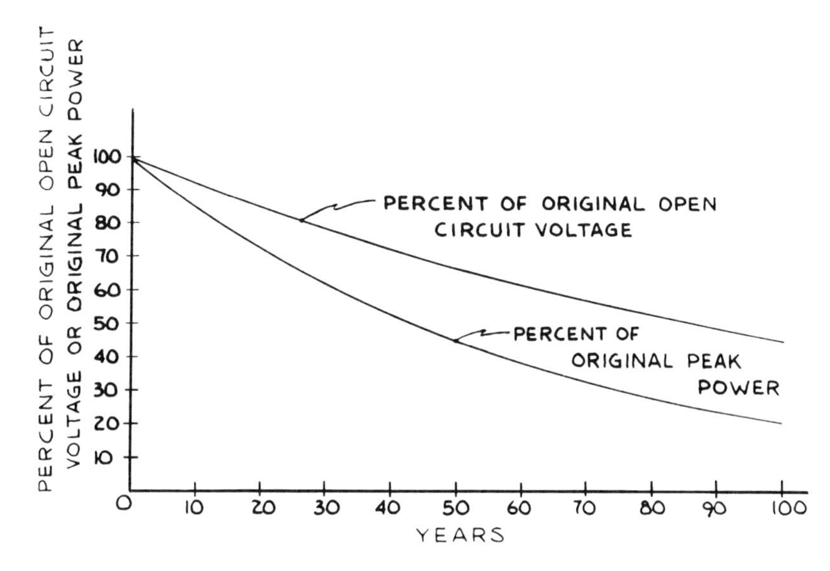

FIGURE 33. Plutonium-238 Powered Nuclear Battery Open-Circuit Voltage and Peak Power Decay as a Function of Time.

years, while the voltage at peak power declines from 0.43 V at BOL to 0.35 V at 30 years and 0.3 V at 40 years. The circuit designer must select which end of life power and voltage is needed for proper dc-to-dc and circuit performance at the end of life, and the battery must be designed accordingly. It should be noted that the power curve remains reasonably flat around the peak power point, and this factor gives the circuit designer a reasonable degree of latitude in selecting the required voltage. At the 40-year time period, for example, a 33% increase in voltage from 0.3 to 0.4 V only results in a 10% decrease in power, form 0.42 mW to 0.38 mW. The shorter the desired battery life, of course, the less the performance penalty of the nuclear battery.

10. RADIATION EFFECTS

10.1. Somatic Effects

As discussed in Section 3.2., plutonium fuel emits neutrons, gamma particles, and alpha particles. The alpha particles become helium nuclei and are stopped within the confines of the fuel capsule. Neutrons and gamma protons emanate from this source. Since the source is very small, as these particles

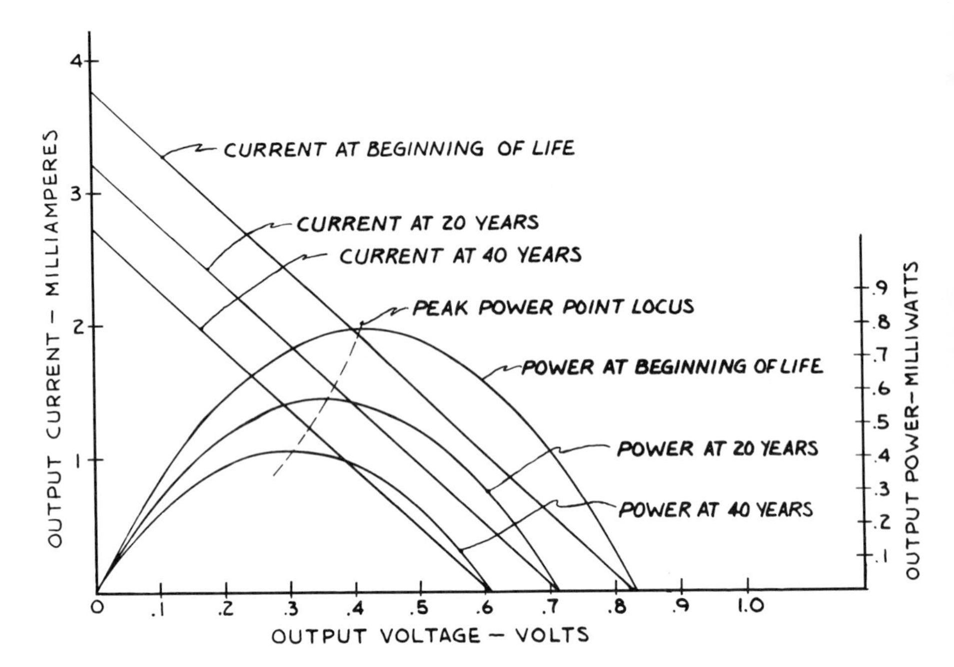

FIGURE 34. Plutonium-238 Powered Nuclear Battery Current and Power as a Function of Voltage and Time—Coratomic C-101 Nuclear Battery.

emanate from it they spread out in an increasing spherical pattern in the same manner as radiant energy emanates from the sun into the solar system. Since the neutrons and gamma particles interact with matter as they pass through it, the number of protons and neutrons per square centimeter as a function of distance do not follow exactly the inverse square law that governs solar energy emanating from the sun. Therefore, measurements must be made of the neutron and gamma dosage as a function of distance from the source in the body. This has been done with a nuclear source within a Medtronic Model 9000 pulse generator at Battelle Pacific Northwest Laboratories. Measurements were taken in phantoms simulating the body tissue and with dosimeters and detectors used to measure the number of protons and neutrons per unit time. The rate of proton and neutron emission is converted to a unit known as a roentgėn, which is an amount of ionizing radiation. Factors must be applied to the roentgen to determine its effect on human living tissue. This unit is called the Roentgen Equivalent Man (REM), and it is the unit that is used to compare radiation effects from a variety of sources in the nuclear industry as well as to determine health effects on the human populace.

A curve showing the actual measured dose as a function of distance from the center of the source is given in Figure 35, and the result of the measurements

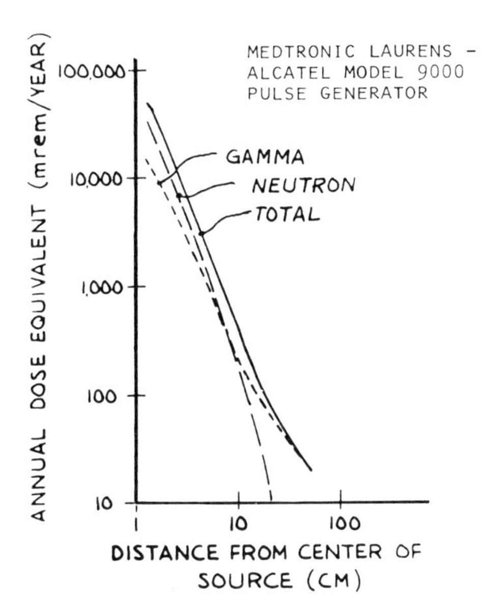

FIGURE 35. Nuclear Radiation from a Plutonium-238 Powered Nuclear Battery as a Function of Distance from the Fuel Source.[12]

made at various body locations and organ locations is given in Table 7. In Table 7 the actual dose has been corrected for the buildup of daughter products, and the resultant radiation emanating from them over a period of years, and is for a 250-mg source of plutonium oxide. In the same table the doses stipulated by the National Council for Radiation Protection and the International Council of Radiation Protection–Radiation Workers is also shown. This is considered a limit for radiation workers. As can be seen from the table, the dose rates to critical organs and to the whole body is extremely low compared with the allowable radiation dose to radiation workers.

It should also be pointed out that the radiation workers receive uniform doses over their entire body surface, whereas in the case of the pacemaker, the farther the organ or tissue is from the pacer, the less the amount of radiation. It should also be mentioned that certain tissues are more resistant to radiation than others, such as muscles and connective tissue, which might be in intimate contact with the implanted device and the heart and sternum.

10.2. Genetic Effects

A genetic risk can occur because of radiation of the gonads. A recent study issued by the Council on Radiation Protection based on examination of survivors

TABLE 7 Dose Equivalents to Organs for 5-, 10-, 15-, and 20-Year Periods from a Pulse Generator that Contains 250 mg of PuO_2 and Is Located Above the Left Pectoral Muscle

Location	Dose Equivalent, rems (neutron and gamma) after 20 years	NCRP, ICRP Limit for Radiation Workers After 20 Years
Thyroid	17.3	300
Left axillary lymph nodes	15.9	300
Right axillary lymph nodes	2.5	300
Sternum	14.4	300
Left pectoral muscle (base of breast)	5.2	300
Right pectoral muscle (base of breast)	2.5	300
Heart	4.6	300
Liver	2.0	300
Spleen	3.5	300
Stomach	2.3	300
Left kidney	1.7	300
Right kidney	1.6	300
Left ovary	0.87	300
Right ovary	0.84	300
Uterus	0.78	300
Testes	0.62	300
Spine (average)	4.8	300
Torso (average)	4.8	300
Whole body (average)	2.6	300

of the Nagasaki and Hiroshima atomic bomb explosions found that the estimated genetic risk if substantially lower than originally anticipated, possibly by a factor of 4.6. Thus, with a device implanted in the pectoral region with the gonads far from the source, the hazards are miniscule.

Another enlightening comparison on the possible hazard of the isotopic battery can be seen by comparing the dose of a battery with doses normally assumed routinely by a populace in the United States. Table 8 shows the annual dose equivalents for various radiation sources compared with annual dose equivalents from the pacemaker. It can be noted that an average jet air crew body dose equivalent is substantially greater than that delivered by a pacemaker, as is the natural background radiation in Colorado. Thus, implantation of a pacemaker is less significant than joining an air crew or moving to Denver, Colorado. For these reasons the radiation impact of a nuclear battery is negligible to the user.

TABLE 8 Annual Dose Equivalents from Various Radiation Sources Compared with Annual Dose Equivalents from A Pulse Generator.[a]

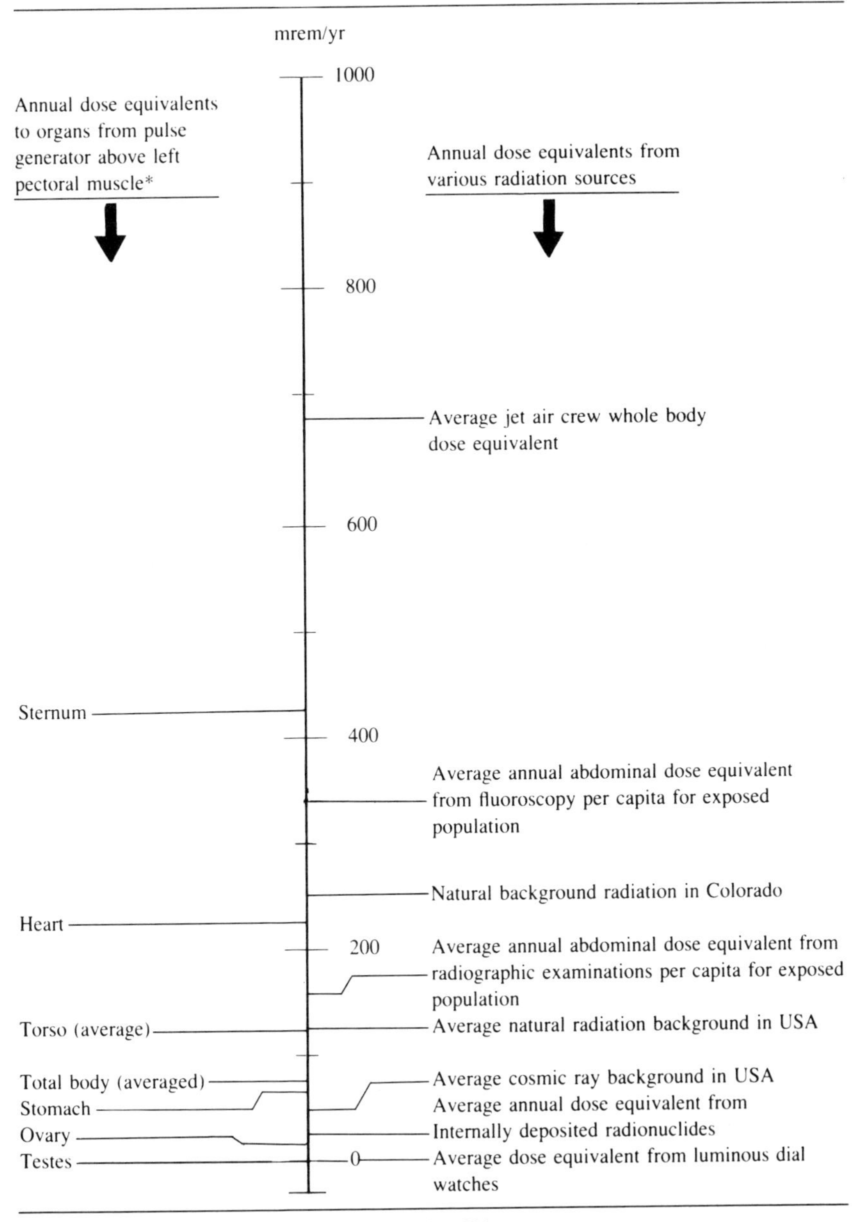

* From a 2-year-old source containing 0.26 ppm plutonium-236.

[a] *Source:* L. W. Brackenbush, G. W. R. Endres, and B. I. Griffin, *Radiation Doses from the Medtronic Laurens-Alcatel Model 9000 Pulse Generator*, Report 2311201653, Amendment 2, Pacific Northwest Laboratories, October 1973.

10.3. Public Exposure

As part of the Nuclear Regulatory Commission's environmental impact assessment, (12) the Nuclear Regulatory Commission also studied the effect of a nuclear battery on household members, work associates, nonwork associates, physicians, transportation workers, bus riders, and the total population assuming 10,000 pacemakers implanted. In all cases the effects were negligible.

It can thus be concluded that the radiation effect of a nuclear battery should not be a factor in determining its use.

11. LICENSING REQUIREMENTS

Nuclear batteries in implantable devices are regulated by two agencies in the United States: the Food and Drug Administration, which is responsible for the safety and efficacy of the device, and the Nuclear Regulatory Commission, which is responsible for the protection of the biosphere from accidental release of the fuel or for radiation exposure to the populace. In foreign countries individual nuclear regulatory commissions exist in each country, and the overall use of the devices in the world is controlled by the Nuclear Energy Agency, which is part of the Organization for Economic Cooperation and Development. This body developed the standards that were discussed for safety under credible accident conditions earlier. At the present time, nuclear batteries have been approved for use by the United States Nuclear Regulatory Commission and have been proved to be safe and effective under a series of controlled clinical investigations that have been under way since 1970.

More controls are placed upon devices using nuclear batteries than on those using conventional chemical batteries.

The manufacturer of the device must report the patient and status of the patient every six months to the Nuclear Regulatory Commission. In this manner the Nuclear Regulatory Commission is assured that the isotope has not been lost. The NRC's objective is to maintain 100% accountability of the radioactive plutonium material.

Responsibility for maintenance of the patient's medical records and recovery of the isotope lies with the hospital, which must be licensed by the Nuclear Regulatory Commission for the receipt of the isotopic material. Part of this licensing procedure requires that the hospital have a radiologist in attendance and health physics personnel who understand the use of radioactive materials. Discussion is continuing at this time with regard to recovery and transferability of records, should the patient move. The possibility that the manufacturer assumes the follow-up responsibility for the patient exists, or the Nuclear Regulatory Commission may set up a special group to oversee the location and return of the radioactive material.

On death of the patient, the implanted device must be returned to the manufacturer in order to maintain accountability. The common method of implementing this approach is to rely on the methods established under the Anatomical Gift Act, in which organs can be removed and transferred to a proper medical authority on death of a patient. An Anatomical Gift Act card is normally kept in the patient's wallet for this purpose; in the case of a minor, the parents must agree to the return of the device on the death of the patient, since minors are not legally responsible for their own actions. The patient is also required to wear a necklace or a bracelet delineating the fact that he is the wearer of a device using a nuclear battery. The medical alert jewelry presently used for a variety of diseases has been appropriately used in the case of pacemaker wearers for this purpose.

The patient receiving a device containing a nuclear battery must be a sufficiently responsible member of society to report every six months and to maintain continuous contact with his physician in case a problem results. During clinical studies of the pacemakers, several irresponsible patients received pacemakers and were lost into the general populace for periods of time because of drug use or other mental idiosyncrasies. This, of course, creates a problem in locating and disposing of the device and in maintaining adequate accountability for its location.

12. APPLICATIONS OF NUCLEAR BATTERIES

The only implantable application of the nuclear battery to data has been as a power source for a cardiac pacemaker. Six different pacemakers have utilized six different nuclear batteries in the time period since the first human implant in 1970 to the fall of 1982. Many of these batteries are now obsolete, but for purposes of completeness they are discussed in order of their chronological use.

The Biotronik pacemaker shown in Figure 36 utilizes the premethium-147 isotope, which is a beta emitter. As the beta particles emitted from the isotope collide with the atoms of a semiconductor, electrons are released on collector plates and used as a source of electrical current.

A clinical evaluation program was started in Germany and The Netherlands in November of 1972 with Biotronik pacemakers powered by the Model 400 Betacel batteries, shown in Figure 37. Because of the short half-life of the premethium-147, the anticipated life of this unit was approximately nine years, and its weight was heavy compared with the plutonium devices introduced in the same time period, because of the requirement for high-density shielding to reduce the high level of gamma energy emanating from premethium-147. It is presently obsolete and is no longer being implanted. Two hundred forty-six implantations of the Betacel have been made.

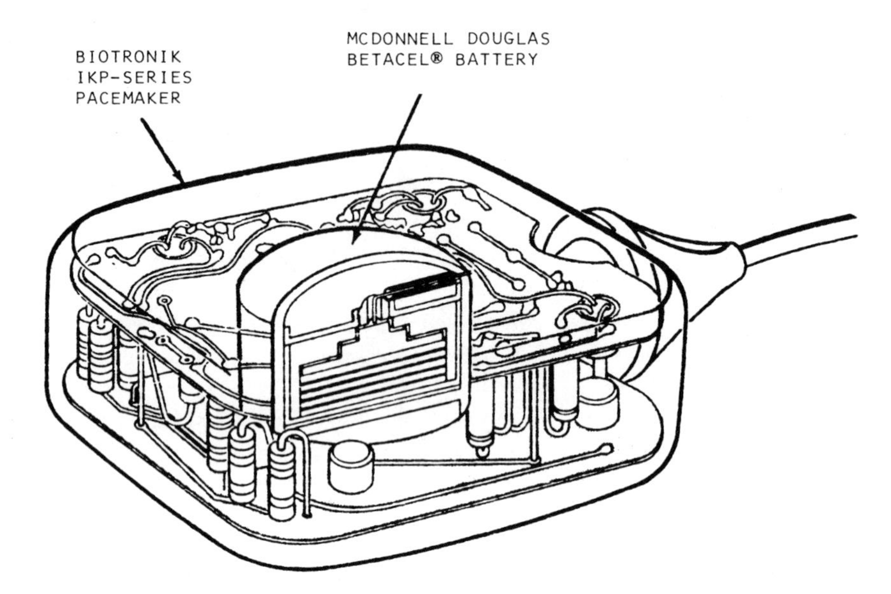

FIGURE 36. Promethium-147 Powered Cardiac Pacemaker—Biotronik Pacemaker with McDonnell Douglas Betacel Battery.[19]

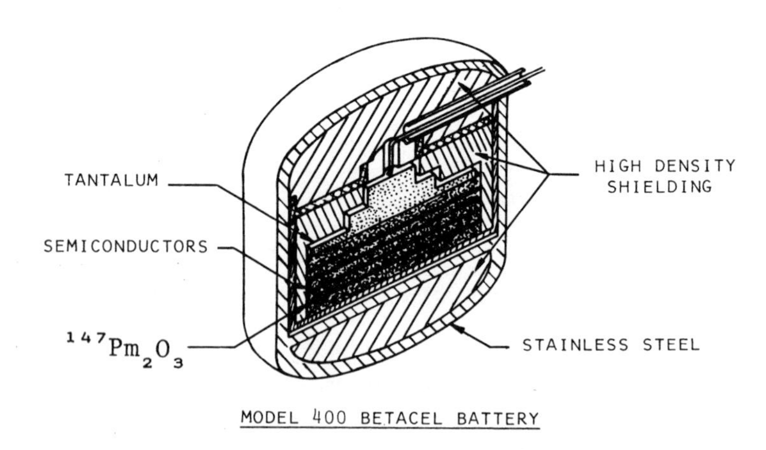

FIGURE 37. Promethium-147 Powered Nuclear Battery—McDonnell Douglas Model 400 Betacel Battery.[18]

The AEC pacer, which was designed in 1966, was the first pacer implanted in animals in 1968. It was first implanted in humans by Dr. Victor Parsonnet in 1973 in the United States. The AEC pacer initially used plutonium metal, which was later changed to plutonium oxide because of the cremation hazard of the metal. The pacer is shown in Figure 38. It was the first pacer to utilize vacuum insulation and used tophel cupron thermocouples in electrical series as the thermopile to obtain an output voltage of approximately 6 V. The thermopile and insulation were integrated into a spiral wrap surrounding a cylindrical fuel capsule. This device utilized 500 mg of plutonium oxide because of the relatively low efficiency of the tophel cupron, which is a metallic thermocouple material, compared with the semiconductor materials. This nuclear battery is no longer being implanted. A total of 126 such units were implanted beginning in April of 1973.

The first pacemaker using a nuclear battery implanted in humans was the Metronic Model 9000, implanted at Broussais Hospital in Paris in April of 1970 by Drs. Piwnica and Laurens, and it contained the G.I.P.S.I.E. 1 nuclear battery. Figure 39 shows this pacemaker and the nuclear battery. One thousand eight hundred forty-seven of the Model 9000 pacemakers have been implanted. This

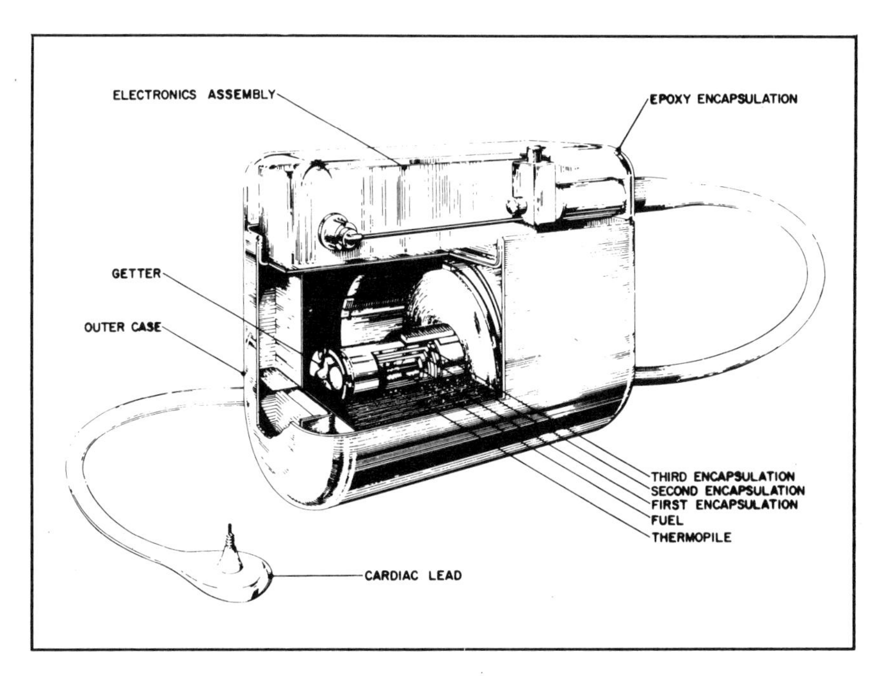

FIGURE 38. Plutonium-238 Powered Cardiac Pacemaker, Atomic Energy Commission.[36]

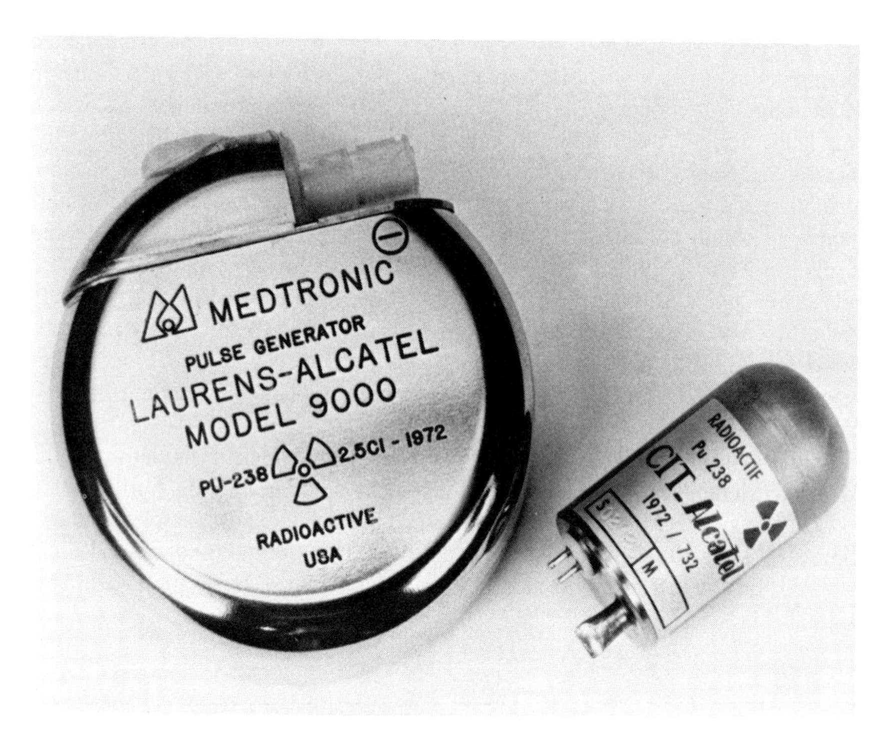

FIGURE 39. Plutonium-238 Powered Cardiac Pacemaker and Battery, Medtronic Model 9000 Pacemaker and Alcatel GIPSIE Series 700 Battery.[16]

model has been superseded by the Model 9090, which has been implanted in approximately 180 patients in France.

The worldwide total of the Medtronic/Laurens/Alcatel systems is approximately 2169 implants.

The Cordis 184A and B pacers utilized the Hittman Atomcell nuclear battery, and a photograph of the pacer and the battery is shown in Figure 40. Approximately 230 of these units have been implanted. The Hittman battery is no longer in production.

The Harwell Devices nuclear pacemaker was implanted in England in small numbers and was quickly withdrawn from the market.

The Coratomic C-101 pacer was first implanted in 1974 and uses the Coratomic nuclear battery. A photograph of the pacer is shown in Figure 41. Four hundred and fifty-five of the Coratomic nuclear-powered models have been implanted. The Coratomic programmable C-101-P was first implanted in June of 1981 under a clinical protocol; 29 of these units have been implanted to date. A photograph of the C-101-P is shown in Figure 42.

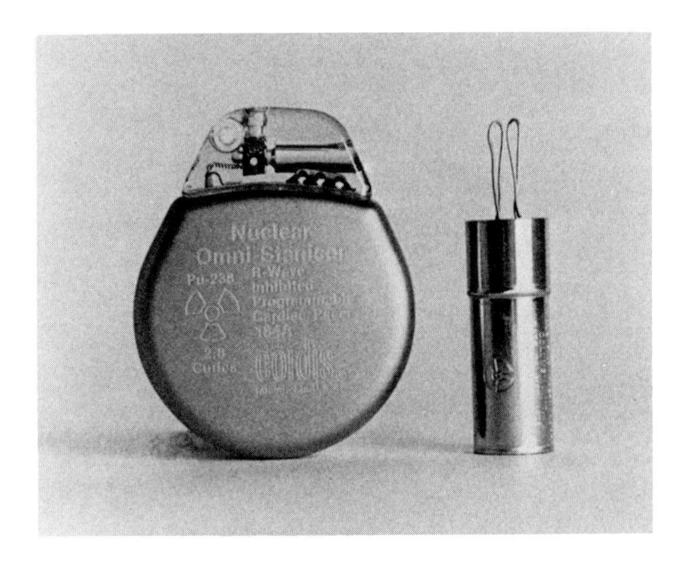

FIGURE 40. Plutonium-238 Powered Cardiac Pacemaker and Battery, Cordis Programmable Nuclear Pacemaker Model 184A and Hittman Atomcell Battery.[21]

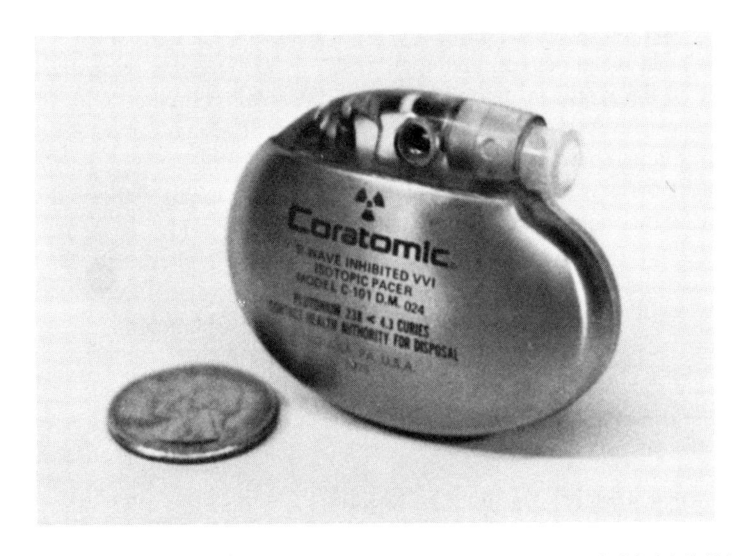

FIGURE 41. Plutonium-238 Powered Cardiac Pacemaker—Coratomic Model C-101.

FIGURE 42. Plutonium-238 Powered Cardiac Pacemaker—Coratomic Nuclear Multiprogrammable Pacemaker Model C-101-P.

Table 9 summarizes the nuclear battery characteristics of the various batteries discussed, and Table 10 describes the pacemaker characteristics. Shown in Table 11 are the estimated number of implants made through January of 1982. From this table it can be seen that approximately 3000 nuclear pacers have been implanted in the world to date. This number is small compared with the large number of pacemakers presently being implanted. The nuclear pacemaker has not taken a large share of the market, mainly because of the high performance of chemical battery systems, especially the lithium iodide battery discussed elsewhere in this book.

TABLE 9 Nuclear Battery and Pacemaker Characteristics

	Gipsie 1	Gipsie 2	Atomcell	AEC	Coratomic
Battery diameter, mm	23	15.8	17.0	32	18.1
Battery length, mm	44	36.7	45.7	60	29
Battery volume, cm^3	16.2	6.5	10.3	60	5.3
Plutonium dioxide weight, g	0.155	0.160	0.14	0.5	0.25
Battery output voltage, V	0.64	0.36	0.3	5	0.36
Battery output power, mW	303	305	600	200	570
Battery weight, g	30	20	35	100	16

TABLE 10 Nuclear Pacemaker Characteristics

	Medtronic 9000	Medtronic 9090	Cordis 184A	AEC Pacer	Coratomic C-101	Coratomic C-101-P
Pacemaker thickness, mm	26	18	23	32	19	19
Pacemaker height, mm	—	—	50	50	47	51
Pacemaker diameter, mm	70	56	—	—	—	—
Pacemaker length, mm	—	—	68	60	59	64
Pacemaker weight, gm	170	80	125	120	48	61
Pacemaker volume, cm^3	90	43	52	86	33	39

Future applications of nuclear pacers may be found in the new DDD pacers presently being introduced and possibly for use in implantable drug dispensers, such as the artificial pancreas. The typical VVI pacer of the type implanted to date using nuclear batteries consumes approximately 120 μW. The new dual-chamber pacers being introduced consume approximately 240 μW, since they are essentially two electronic circuits operating in separate chambers of the heart. As an example, the Coratomic battery would power this type of a pacemaker for 40 years, whereas lithium-battery-powered systems operate from two to five years if both chambers are pulsing continuously. Therefore, there may be increased application in the near future for the nuclear source utilized with a dual-chambered pacemaker.

The implantable insulin-dispensing devices presently being tested utilize lithium batteries that last for approximately two years. The same system utilizing a nuclear battery would run for approximately 20 years; therefore, at some point in the future nuclear batteries may find application in these devices.

13. NUCLEAR BATTERY RELIABILITY

Nuclear battery reliability, which has been measured in the clinical studies utilizing more than 3000 pacemakers, has been exemplary, if not incredible. Only one failure has been reported in more than 92,000 device months with the

TABLE 11 Worldwide Implants of PU^{238} Pacemakers Through January 1986

AEC	126
Coratomic	490
Cordis	230+
Harwell/Devices Ltd.	76
Medtronic (Laurens Alcatel)	2,160+
	3,082+

GIPSIE 1 battery, which is an average failure rate of 0.001% of 0.004% at 90% confidence.

Based on a sample of 57 Cordis 184A pacemakers reported in PACE Magazine (Jan.–Feb., 1983), the calculated failure rate of the pacer/Hittman battery system is 0.046% per month, with a failure rate at 90% confidence of 0.089% per month. The failure rate of the Hittman battery may be considerably less than this, since the actual number of failures due to the battery has not been reported.

No failures have occurred with the Coratomic nuclear battery in over 27,000 unit months, which results in a calculated battery failure rate of 0 with a failure rate at 90% confidence of 0.009% per month.

When compared with the failure rate in the same time period of mercury and lithium batteries, the nuclear battery has obviously performed in an exemplary manner.

REFERENCES

1. Reference Study and Test Plan for Phase II, August 15, 1969, Radioisotope-Powered Cardiac Pacemaker Report Number 3731-17, Contract AT(30-1)-3731.
2. S. A. Kolenik, Statistical Reliability Analysis Summary, Radioisotope-Powered Cardiac Pacemaker, Contract At(11-1)-3057, submitted to the Atomic Energy Commission Materials Branch, Directorate of Licensing (November 1972).
3. Determination of the Weldability and Elevated Temperature Stability of Refractory Metal Alloys, NASA-CR-1607.
4. Harold J. Garber, Pressure-Volume-Temperature Relationships for Helium, Technical Memorandum, Radioisotope-Powered Cardiac Pacemaker Project, NUMEC 3731-4.5-3.
5. Raymond J. Roark, *Formulas for Stress and Strain,* McGraw-Hill, 4th Ed. (1965).
6. Platinum Clad Isotope Fuel Capsule for Space Applications, U.S. Atomic Energy Commission Report NAA-SR-12578.
7. H. J. Garber, Detailed Study of the Radioactive Decay Kinetics of Plutonium-238 and Related Parent and Daughter Nuclides, Radioisotope-Powered Cardiac Pacemaker Project, NUMEC 3731-4.5.7.
8. J. E. Selle, J. J. English, P. E. Teaney, and J. R. McDougal, The Compatability of 238 PuO_2 with Various Refractory Metals and Alloys, Interim Report MLM-1706, October 23, 1970.
9. Quarterly Progress Report for Radioisotope Powered Cardiac Pacemaker Program, August 1, 1973–October 31, 1973, ARCO-3057-15.
10. A Review of Determinations of Radiation Dose to the Active Bone Marrow from Diagnostic X-ray Examinations, DHEW Publication (FDA)-74-8007.
11. Interim Radiation Protection Standards for the Design, Construction, Testing and Control of Radioisotope Cardiac Pacemakers, Organization for Economic Co-operation and Development (OECD), Nuclear Energy Agency (NEA) Paris, 21st May, 1974, C(74)101.
12. Final Generic Environmental Statement on the Routine Use of Plutonium-Powered Cardiac Pacemaker, July 1976, U.S. Nuclear Regulatory Commission, Office of Nuclear Material Safety and Safeguards, NUREG-0060.
13. David L. Purdy, Design of Isotopically Powered Thermoelectric Generators, *American Nuclear Society* (July 1968).
14. David L. Purdy and Zalman M. Shapiro, Design of Isotopic Generators, 8th Japan Conference on Radioisotopes, November 16, 1967, Paper Number C/E-6.

15. Robert F. Hicks and Roland W. Ure, *Science and Engineering of Thermoelectricity*, Interscience, New York (1961).
16. Michael Alais, Rene Berger, Rene Boucher, Kenneth A. Gasper, Paul Laurens, A Plutonium-238 Fueled Cardiac Pacemaker *Nucl. Technol.*, Vol. 26 (1975).
17. T. H. Smith, J. Greenborg, W. E. Matheson, Ing. M. Schaldach, A Benefit/Risk Analysis of the Betacel Battery Nuclear-Powered Cardiac Pacemaker, McDonnell Douglas Astronautics Company Paper WD2110, (May 1973).
18. M. L. Smith, T. A. Golding, W. E. Matheson, Model 400 Betacel Battery Qualification Tests, Donald W. Douglas Laboratories Report No. DWDC-727-078 (July 1972).
19. T. H. Smith, J. Greenborg, W. E. Matheson, Benefit/risk analysis of cardiac pacemakers powered by Betacel promethium-147 batteries, *Nucl. Technol.*, *26* (May 1975).
20. Atomcell Brochure, Hittman Corporation (1976).
21. Hittman Corporation Annual Report (1976).
22. M. S. Hixson and Paul Laurens, Design Criteria and Two Year Clinical Results of Pu^{-238} Fueled Demand Pacemaker, Medtronic, Inc. Report (1972).
23. Francoise Marchand, personal communication, Hopital Broussais, 96, rue Didot, Paris, France, September 2, 1982.
24. Paul Laurens, Nuclear-powered pacemakers: An eight year clinical experience, *Pace 2* (1979).
25. Victor Parsonnet, personal communication, Newark Beth Israel Medical Center, Newark, NJ, September 28, 1982.
26. P. Laurens, P. Gavelle, A. Piwnica, C. Farge, Ch. Dubost, and P. Maurcie, Nuclear Pacemaker Clnical experience: Nine Year Follow-up of 300 Patients, Proceedings of the VIth World Symposium on Cardiac Pacing, Montreal (October 1979).
27. J. Hixson, P. Laurens, P. Gavelle, and J. Van Leusen, 10 Years Technical and Clinical Results of Pu^{-238} Fueled Demand Pacemaker, AAMI Annual Meeting, San Francisco (13–17 April 1980).
28. Paul Laurens, Pierre Thomas, and Gerard Koehly, Les progres dans les piles isotopiques pacer stimulateurs cardiaques, *Cardiostem* (February 1978).
29. D. L. Purdy, C. A. Bodenschatz, and H. J. Garber, Comparison of vented and unvented ^{238}Pu heat sources, *Am. Nucl. Soc. Trans. 14*(2), (1971).
30. Thomas S. Bustard, A nuclear battery for the cardiac pacemaker, *Am. Nucl. Soc. Trans. 14*(2), (1971).
31. R. L. Schimmel, Industrial status of ^{238}Pu sources, *Am. Nucl. Soc. Trans. 14*(2), (1971).
32. C. A. Bodenschatz, M. G. Blair, G. W. Maurer, and J. F. Hursen, Safety aspects of fuel capsules for medical purposes, *Am. Nucl. Soc. Trans. 14*(2), (1971).
33. G. B. Pleat and W. J. Lindsey, Future production of ^{238}Pu, *Am. Nucl. Soc. Trans. 14*(2), (1971).
34. Nicholas P. D. Smyth, George J. Magovern, William J. Cushing, John M. Keshishian, Leo C. Kelly, and Martin Dixson, Preliminary clinical experience with a new radioisotope powered cardiac pacemaker, *J. Thorac. Cardiovasc. Surg. 71*(2), (1976).
35. Nicholas P. D. Smyth, George J. Magovern, William J. Cushing, John M. Keshishian, Preliminary Experience With a New Radioisotopic Powered Cardiac Pacemaker, Proceedings of the Vth International Symposium on Cardiac Pacing, Tokyo (March 1976).
36. David L. Purdy, The development of an isotopic cardiac pacer, *Engineering: Cornell Quart. 9*(4), (1975).
37. David L. Purdy, George J. Magovern, and Nicholas P. D. Smith, A new radioisotope powered cardiac pacer, *J. Thorac. Cardiovasc. Surg. 69*(1), (1975).
38. Nicholas P. D. Smyth, David L. Purdy, Diane Sage, and John M. Keshishian, A new multiprogrammable isotopic powered cardiac pacemaker, *Pace 5* (1982).
39. H. C. Carney, Comparison of Strontium-90 and Plutonium-238 Milliwatt Thermoelectric Generators, Power from Radioisotopes, Proceedings of OECD Nuclear Energy Agency and the Junta de Energia Nuclear of Spain, Madrid (June 1972).

40. M. Alais, B. Etieve, A Stahl, P. Thomas, The GIPSIE Radioisotopic Generator for Use in Cardiac Pacemakers, Power from Radioisotopes, Proceedings of OECD Nuclear Energy Agency and the Junta de Energia Nuclear of Spain, Madrid (June 1972).
41. R. Bomal, A. Manin, K. Steinschaden, Prospects for Radiovoltaic Energy Conversion, Power from Radioisotopes, Proceedings of OECD Nuclear Energy Agency and the Junta de Energia Nuclear of Spain, Madrid (June 1972).
42. D. Schalch and A. Scharmann, Practical Limits of Radiophotovoltaic Energy Conversion, Power from Radioisotopes, Proceedings of OECD Nuclear Energy Agency and the Junta de Energia Nuclear of Spain, Madrid (June 1972).
43. Frank G. Gatt, A Tritium Nuclear Cardiac Pacemaker, Power from Radioisotopes, Proceedings of OECD Nuclear Energy Agency and the Junta de Energia Nuclear of Spain, Madrid (June 1972).
44. F. N. Huffman, F. A. Molophia, J. C. Norman, Thermal and Radiation Effects of ^{238}Pu Fuel Capsules on Dogs and Primates, *Trans. Am. Nucl. Soc.* 1971 Winter Meeting, *14*(2), (1971).
45. W. J. Schull, M. Otake, and J. V. Neel, Genetic effects of the atomic bombs: A reappraisal, *Science 213,* 1220 (1981).
46. Medtronic, Inc., Technical Report on the Medtronic™ Model 9000 Isotopic Pulse Generator, submitted to the United States Atomic Energy Commission, June 1, 1974.

Index

Accelerated test, 85, 167
  electrical discharge, 93, 94
  feasibility diagram sample size, 92
Acceleration factor, 86, 89
Acceleration methods
  Arrhenius, 88
  empirical, 87
  physicochemical, 88
  statistical, 87
Alkali metals, 223
Alkaline cell, sealed, 266
Alkaline earth metals, 223
Anode, 54
  chemical, 54
  coating, 146, 218, 228, 229, 234, 246, 250–252, 256
  current collector, 54
  electrochemical, 54
  film, SEI (solid electrolyte interphase) 218, 250–251
Antitachyarrythmia, 48
Asystole, 9
Atrioventricular node, 8, 9
Atrium, 7
Autoclavable devices, 48
Automatic implantable defibrillator, 21–22
AV synchrony, 11

Balanced cell, 39, 40, 266
Battery
  burn-in, 85
  can-negative, 59
  can-positive, 59
  case-neutral, 59
  components, 53–60
  equivalent circuit, 116
  introduction, 51–60
  mechanism of operation, 53–60
Battery (*cont.*)
  modeling
    classification of, 117
    lithium–iodine cell, 120
    overview, 113
    role of, 117
  non-electrochemical, 13
  primary, 51
  secondary, 51
Biogalvanic cells, 48
Bladder stimulator, 5
Blood flow, 8
Blood pressure regulator, 5
Bone growth and repair, 22–26
Bone healing, 22–26
  devices, 25–26
  early studies, 23
  implantable stimulator, 24
  mechanism, 23–24
  physiology of, 23–24
Bradycardia, 9
Brain stimulator, 5
Bromine, 48

Cadmium electrode, 277
Calcium, 248
Calcium–lithium alloy, 249, 255
Calorimetric estimate of efficiency, 239–246
Calorimetry, 68–72, 86, 98, 216, 234–246
Capacity density, 221–233
Cardiac conduction system, 8, 9
Cardiac muscle fiber, 8
Cardiac output, 11
Cardiac pacemaker, 7, 215, 343
  first implant, 9
    of lithium battery, 44
  lead/tissue/electrode impedance, 19
  pacing impedance, 18–20

Cardiac pacemaker (*cont.*)
  power requirements, 12
  threshold energy, 18–21
    power consumption, 18–21
    safety factor, 18–21
Cardiac pacing leads, 14
  bipolar, 14
  dislodgement, 14
  fixation mechanism, 14, 15
  unipolar, 14
Cardiac physiology, 7
Cardiac stimulator, 5
Cathode, 55, 182, 223
  chemical, 55
  current collector, 55
  depolarizer mix, 55
  electrochemical, 55
Cell loading, 62
Cell, 51
Ceramic insulator, 59
Certified cell, 267
Constant load discharge, 95
Corrosion, 7
Creativity, 38, 40, 42, 45
Crimped seal, 58
Cumulative survival analysis, 103, 104
Current capability, 61, 62
Cut-off-voltage, 57

Data analysis, 101
Defibrillator, 21–22
Degradation rate, 89
Deliverable capacity, 61, 66, 67
Dendrites, 73
Depolarization, 8
Devices
  battery manufacturers, 32
  biotelemetry, 28–30
  bone growth and repair, 22–26
  business aspects, 31–32
  CNS stimulator, 26
  defibrillator, 21–22
  drug delivery, 26–28, 48
    cardiovascular disease, 28
    chemotherapy, 28
    chronic pain, 28
    diabetes, 26–28, 349
  future directions, 32–33, 349
  list, 2–5
  pacemaker, 7–20
  pacemaker manufacturers, 31–32
  pain control, 30

Devices (*cont.*)
  power requirements, 6, 7, 208, 216
  sensors, 30
Discharge curve, 62
Discharge product, total, 122
Discharge, stages of, 121
Disposal of Li batteries, 256
Dose from nuclear battery, 337

Efficiency, 238–245
  cumulative, 238–245
  electrochemical, 230–248
  instantaneous, 238–245
  packaging, 228–230
  volumetric, 228–230
Electrochemical anode, 54
Electrochemical cathode, 55
Electrochemical energy conversion, 53, 228–248
Electrode potentials, 264
Electrolyte, 55
Electrolyte concentration, 233
Electronic circuit, 5, 6
Electrophysiology, 1
End-of-life indicator, 57, 68, 78, 248–250
Energy conversion, 53, 228–248
Energy density, 47
  lithium–liquid oxidant cells, 221–233
  lithium–solid cathode cells, 182
Energy losses, 228–248
Enthalpy, 56, 60, 235
Entropy, 56, 60, 235–239, 244
Epoxy impregnation, 268
Epoxy resin, 269
Equivalent circuit, DC, 116
Ethylene oxide sterilization, 48
Evaluation methods, 83
Eyring model, 90

Factorial experiment, 100
Feedthrough, 58, 101, 97
  degradation of glass, 98
Ferrule, 59
Fibrillation, 14
First pacemaker implant,
  lithium battery, 44
  mercury–zinc battery, 261, 280
  nickel oxide–cadmium battery, 9, 276
  nuclear battery, 345
Free energy, 56, 60, 133, 135, 235–239
Fuel capsule, 311
Fuel cell, 48

Future cells, 42, 175–179, 211–212, 282–283
Future devices, 32, 48, 349

Galvani, 1
Gamma radiation, 7
General Electric, 261
Gland stimulator, 5
Glass insulator, 59
Glass seal degradation, liquid electrolyte, 98
Greatbatch cell, 44
Greatbatch, Wilson, 9, 44, 104

Heart
  electrical stimulation, 9
  pacemaker, 5
  pacer, 7
Hemoglobin, 8
Hot spot, 255
Hydrogen overvoltage, 263
  zinc amalgam electrodes, 265

Implant instruments
  closed-loop control signals, 2–5
  remote-controlled manipulation, 2–5
  stimulation, 2–5
  telemetry, 2–5
Implant manipulator, 5
Implantable blood pressure monitor, 46
Implantable device power requirements, 216
Implantable devices, 1–36
  list, 2–5
  power needs, 208
  use of cupric sulfide–Li cells, 210
  use of lead iodide, lead sulfide–Li cells, 211
  use of manganese dioxide–Li cells, 211
  use of silver chromate–Li cells, 209
  use of vanadium pentoxide–Li cells, 210
  see also Devices
Implantable glucose sensor, 46
Implantable pulse generator, 7
Injury potential, 1
Innovation: *see* Creativity
Internal energy, 235–239
Internal resistance, 61, 116
Invention, 41, 44
Iodine cathode, 141
  pellet, 144
  thermal, 143
Iodine complex, 43
Iron, 251
Isotope half life, 288
Isotopes, 290
Isotopic battery, 47, 297

Key events in pacemaker cell development, 37

Lapicque Equation, 17
Least squares regression, 124
LeClanche cell, 40, 58, 261, 272, 273
Left ventricle, 8
Life expectancy of patient, 14
Life of device, 6
Lifetime, pacemaker, statistical, 130
Lithium cells, 12, 13
Lithium–BCX cell, 219, 223–225
Lithium–bromine cell, 76, 77, 136–138
Lithium–copper sulfide cell, 79, 196–200, 210, 223–225
  cell chemistry and cathode preparation, 196
  cell construction and performance, 197
  electrolytes for, 196
Lithium–CSC cell, chlorine as soluble cathode, 217, 223
Lithium–halogen cell, 133
  kinetic characteristics, 135
  thermodynamics, 134
Lithium–iodine battery, 47
  longevity projections, 102, 125
  reliability data, 109
Lithium–iodine cell, 59–75, 94, 96, 133, 139–175, 217–220, 223–225, 231
  case grounded, 156
  central anode, 153
  central cathode, 154
  current drain limitations, 164
  design, 151
  discharge behavior, 157
  historical, 44
  materials testing, 97
  model 702C, 45
  modeling, 120–125, 167–171
  pacemaker statistics, 107
  performance domain, 170
  reliability, 171
  self discharge, 165
  variability, 174
Lithium–lead iodide, lead sulfide cell, 77
Lithium–liquid oxidant cells, Wilson Greatbatch Ltd., 219, 223, 256
Lithium–liquid oxidant systems, 217–219
Lithium–manganese dioxide cells, 80, 203–204, 211, 223–225
Lithium–phosphoryl chloride cells, 223–225
Lithium–silver bismuth chromate cells, 195

Lithium–silver chromate cells, 79, 187–194, 209–210
  cell chemistry, 187
  cell construction and types, 190
  cell performance, 193
  discharge mechanism, 187
Lithium–solid cathode cell
  physicochemical properties of solvents, 184
  specific conductivities of electrolytes, 185
Lithium–solid electrolyte cells
  lithium iodide based electrolytes, 204
  metal salt cathodes, 206
Lithium–sulfur dioxide cells, 217, 223–225
Lithium–sulfuryl chloride cells, 217, 223–225
Lithium–thionyl chloride cells, 63, 77, 96, 215–259
  abusive conditions, 252
    casual storage, 255
    charging, 255
    overdischarge, 253
    short circuit, 252
  acetylene black, 220
  alloy anodes, lithium–calcium, 248, 249
  anode, 218–222, 228–229
  anode coating, polymeric, 218, 228, 229, 234, 250–252, 256
  anode film, SEI, 218, 250–251
  anode limited cell, 252, 254, 255
  balanced cells, 217, 221–233
  binders, 220
  bromine, 219
  bromine–chloride, 217, 219, 255
  carbon blacks, 220, 225–226
  carbon cathode element, 219, 224, 226
  carbon limited cell, 225, 254
  cathode materials, 220, 222–223
  cathodes, 218–226, 230, 252, 254
  conductivity, 221, 224, 233, 253, 254
  corrosion, 250–252
  discharge mechanism, 217–220, 248
  electrolyte concentration effects, 227–233
  end-of-life indication, 248–250
  energy density, 221–233
  principles of operation, 219–220
  safety, 252–256
  specific capacity, 222–228
  thermodynamic properties, 234–248
  voltage delay, 250–252
  Wilson Greatbatch Ltd., 219, 223, 256
Lithium–vanadium pentoxide cells, 80, 199–203, 210
  cell discharge reactions, 201–202
  construction and performance, 202
Lithium–vanadium pentoxide cells (*cont.*)
  electrolytes for, 200
Lithium tetrachloroaluminate, 219, 227
Load voltage, 61
Longevity, 268
Longevity model, 126

Mallory, P.R., 261
Maximum available capacity, 67, 170, 233
Mechanical stress, 95
Mercuric oxide electrode, 263
Mercuric oxide–zinc primary cell: *see* Mercury–zinc battery
Mercuric oxide–zinc rechargeable cell, 275
  accelerated testing, 281
  calendar life, 282
  clinical prototype, 280
  cycle life, 282
  design, 280
  hermetic encapsulation, 281
  history, 279
  hydrogen generation, 281
  overcharge, 281
  performance, 281
  recharge time, 282
  self discharge, 281
Mercury immobilization, 266
Mercury–zinc battery, 12, 40, 69, 73, 261
  energy density, 12
  gassing, 12
  self discharge, 12
Mercury–zinc pacemaker cell: *see* Mercury–zinc battery
Mercury–zinc pacemaker, 47
Microcalorimetry: *see* Calorimetry
Microelectronics, 7
Model of self discharge, 118
Muscle stimulator, 5

Negative terminal, 54
Nernst equation, 265
Nickel oxide electrode, 277
Nickel oxide–cadmium cells, 74, 275
  construction, 277
  elevated temperature design, 278
  gas evolution, 277
  implant experience, 276
  life, 278
  overcharge, 277
  pacemaker, 278
  reactions, 277
  recharge, 278
  recharge interval, 279

Nickel oxide–cadmium cells (*cont.*)
  self-discharge, 279
  specifications, 279
Nuclear battery, 45, 285–352
  mechanism, 285–287
  types, 287

Open-circuit voltage, 60
Osteogenesis, 22–26
Oxidation, 54

Pacemaker, 5, 261, 267
  asynchronous, 11
  demand, 11
  dual-chamber, 11
  indications for therapy, 11
  malfunction, 7
  technology, 9, 10
    circuitry, 10
    power source, 10
Pacemaker lead, 15
  transcutaneous, 9
  transvenous endocardial, 9
Pacemaker longevity, 268
Pacemaker longevity model, 126–130
Pacemaker manufacturers, 31–32
Pacemaker power source, rechargeable, 276
Pacemaker survival, 105
Pacemaker syndrome, 11
Pacing physiology, 9
Pain suppressor, 5
Parasitic drain, 269
Peak power, 336
Perfusion, 8
Peripheral organs, 8
Peripheral venous circulation, 7
Physiology, 1, 9
Plutonium, 292–296
Polarization, 65
  modeling of, 119
Poly-2-vinylpyridine, 59, 71, 75, 76, 133
Polymeric anode coating: *see* Anode coating
Pourbaix diagram, 263
Power requirements, 6, 7, 208, 216
Power source, 7
  external, 9
Premature failures, 7
Primary cells, 51
Prostheses, 1–36
Pulmonary veins, 8

Qualification protocol, 99
Quality assurance, 85

Radiation, 294
Radioactive decay, 285
Random failure rate, 106, 108
Rate capability, 61, 62
Rated capacity, 67
Real time testing, 96
Rechargeable battery, 47
Rechargeable cells, 51, 74, 275–284
Redox reactions, 54
Reduction, 54
Refractory, 8
Reliability, 84, 101–109, 349
  lithium–silver chromate cell, 209
  nuclear battery, 349–350
  standards, 84
Replacement interval, 58
Resistance, internal, 116
Right atrium, 7
Right heart, 7
Right ventricle, 7
Ruben cell, 45
Ruben laboratories, 39
Ruben, Samuel, 38, 261
Ruetschi, P., 270, 271

Safety, 80
  nuclear batteries, 299–300, 311, 322–327, 337–342
  lithium–thionylchloride cells, 252–256
Secondary cells, 51, 74
Seebeck effect, 302
Self-discharge, 65, 69, 78, 115, 118
  equation, 124
  modeling of, 118
Sensory stimulator, 5
Separator, 56, 73
Sick sinus syndrome, 11
Sinoatrial node, 8
Solid electrolyte, 56, 134
Solid electrolyte interphase, 218, 250, 255
State-of-discharge indication, 248–250
Statistical variations,
  battery, 127
  circuit, 127
  medical management, 129
  patient/lead interface, 128
Statistics, 105, 107
  data analysis, 103
  role of, 117
Stimulation threshold, 14, 16
Stoichiometric capacity, 66, 170, 224, 225
Stoichiometric capacity density, 60
Strength–duration curve, 17, 18

Sudden cardiac death, 22
Sulfur dioxide, 218, 236, 251–252
Sulfur, 236, 255–256
Sulfuryl chloride, 48, 217, 223–225
Superior venacava, 8
Survival probability, 106

Tachycardia, 14
Telemetry, 5, 48
Terminal voltage, lithium–iodine cell, 125
Terminology, 104
Theoretical energy density, 60
Thermal batteries, 46
Thermal insulation, 308
Thermal power, 328
Thermodynamic properties, 60, 134–135, 182, 234–248
Thermodynamic voltage, 265
Thermoelectric, 302
Thermopile, 70, 302
Threshold voltage, 16
Time, implicit variable, 123

Valence, 53
Ventricle, 7
Ventricular fibrillation, 21–22
Volta, 1
Voltage delay, 250–252

Weibull distribution, 91

Zinc electrode, 263
Zinc–mercuric dioxysulfate cell, 42, 272
Zinc–mercuric oxide cell: *see* Mercury–zinc battery